Middle Iron Age Warfare of the Hillfort Dominated Zone *c*. 400 BC to *c*. 150 BC

Jon Bryant Finney

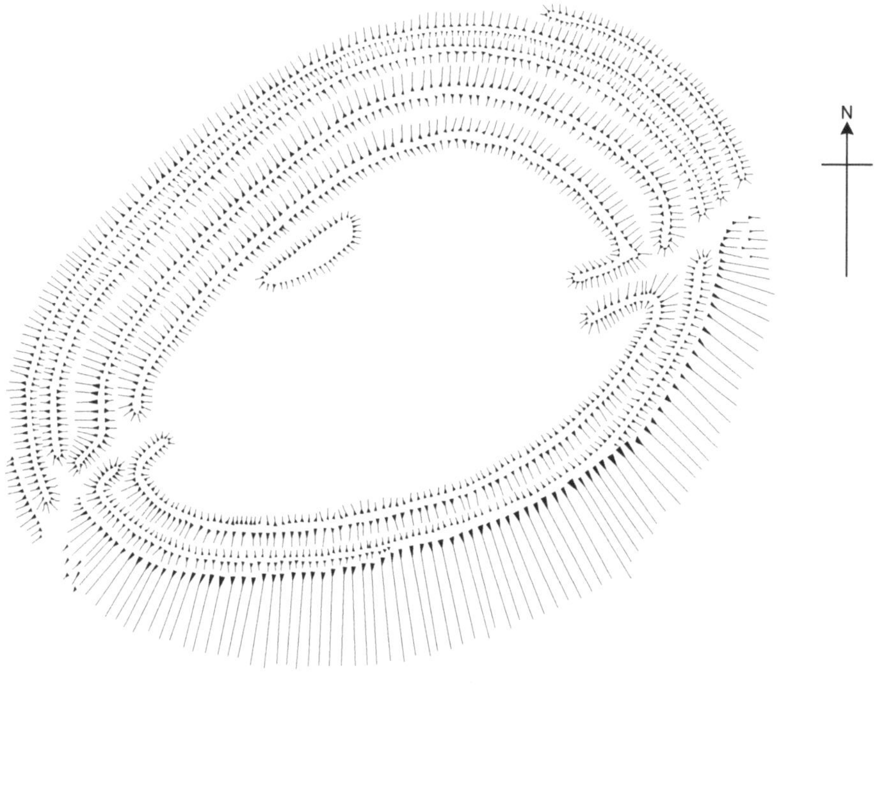

BAR British Series 423
2006

Published in 2016 by
BAR Publishing, Oxford

BAR British Series 423

Middle Iron Age Warfare of the Hillfort Dominated Zone c. *400 BC to* c. *150 BC*

ISBN 978 1 84171 987 0

BAR Publishing is the trading name of British Archaeological Reports (Oxford) Ltd. British Archaeological Reports was first incorporated in 1974 to publish the BAR Series, International and British. In 1992 Hadrian Books Ltd became part of the BAR group. This volume was originally published by Archaeopress in conjunction with British Archaeological Reports (Oxford) Ltd / Hadrian Books Ltd, the Series principal publisher, in 2006. This present volume is published by BAR Publishing, 2016.

Printed in England

BAR
PUBLISHING

BAR titles are available from:

BAR Publishing
122 Banbury Rd, Oxford, OX2 7BP, UK
EMAIL info@barpublishing.com
PHONE +44 (0)1865 310431
FAX +44 (0)1865 316916
www.barpublishing.com

Contents

List of Tables and Illustrations

1. List of Illustrations

List of Tables

List of Drawings

Acknowledgements

According to Chairman Mao the longest journey starts with the first step, then this publication must start with the first acknowledgement.

Firstly my supervisor John Gale, for asking "what if?...", or "perhaps it could be"... and for putting an initial 'spanner in the works' that led me to develop this work as it stands, and not follow a different path. To Mark Maltby my internal examiner, who in a moment of foolhardiness agreed to undertake the minor corrections to my thesis that formed the basis of this publication.

To the following for their kind permission to survey the hillforts in their ownership/control: English Heritage, the Forestry Commission, Shropshire Wildlife Trust, Hampshire County Council, English Nature, Gloucestershire County Council, Overbury Estate, Wiltshire County Council, The Ministry of Defence, and Mrs. S.L Harris.

To the staff of the Haddon Library University of Cambridge and Cambridge City Council for the use of AutoCAD.

Dedication

Despite being dyslexic I actually enjoy the process of writing. However, what I perceive as an elegant piece of prose worthy of any archaeological writer, is more likely to be comparable with a letter from Wol (Milne 2000 [1928], 74-5).

Therefore three people deserve to have this work dedicated to them, as without each of them many hurdles would not have been crossed:

My father Derek Michael Bryant Finney, for reading the second proofs and undertaking the final corrections as provided by,

Nicolas James, a true friend, who without hesitation took on the task of finally polishing the English of this thesis, despite his own heavy workload as a freelance archaeologist.

Finally to Sarah. Without her help, in surveying the hillforts, beating my terrible English into a readable format and, many years ago, buying me my first sling, this work would never had reached fulfilment.

They are my three musketeers 'one for all and all for one' and I am the lucky person who happens to have been that one.

CHAPTER 1

INTRODUCTION

Regionalisation has long been recognised as one of the basic characteristics of the Iron Age of Britain. By the late 1950s, Hawkes' 'A, B, C' of the Iron Age was revised (Hawkes 1959), to include a strong regional element. Further refinement of this concept has led to the widely accepted broad spatial and temporal regionalisation as defined by Cunliffe (1990, Figure 14.38). It is this interpretation that will be used to define the scope, both geographic and temporal, of this study (Figure 1.1). Regionalisation has been successfully applied to agricultural regimes (Grant 1984; Hambleton 1999, Chapter 10; Jones 1996, 37-8; Knight 1984), settlement studies (Collis 1996; Cunliffe 1991, Chapters 11-14; Forde-Johnston 1976), pottery manufacture (Cunliffe 1984b; 1991, Chapter 4, Figure 20.4; Lambrick 1984,169-73; Morris 1996, 43-6), and numismatics (Haselgrove 1996, 74-7; Sellwood 1984). The last two have been used to define potential ethnic groups in southern Britain during the Late Iron Ages.

Yet warfare has essentially remained aloof from this process. There exists a widely accepted overarching interpretation of warfare throughout the Iron Age, from the west coast of Wales to the North Sea coast of East Anglia, from Dorset to the Hebrides, from *c.* 600 BC to *c.* AD 60. This is, in essence, heroic warrior combat, undertaken by a people(s) whose whole culture was orientated towards war. "*The whole race... is war-mad, high-spirited and quick to battle*" (Strabo 4.4.2). This is based upon the ideological construct that in any area of Western Europe that displayed La Tène material there was a correspondingly single Celtic ethnicity, which was nearly identical across space and time. This concept has been the centre of heated and often (regrettably) acrimonious debate in recent years (Collis 1997; Hill 1989; 1995 James 1998; 1999; Megaw and Megaw 1996; 1998).

The purpose of this study is not to enter into this debate but to attempt to define the nature of warfare within the Middle Iron Age of the hillfort dominated zone from the archaeological remains and ethnographic inferences relating to the social significance of weapons, within specific analogies. In the process, this study will attempt to provide a new model for Middle Iron Age warfare in the hillfort dominated zone of lowland Britain rather than using the generalised model derived from continental Classical sources and seventh-eighth century AD Irish vernacular literature. The process of model generation will include comparison of the proposed model with another prehistoric European culture within the same temporal framework. Denmark was also on the fringe of Mediterranean influence during the Iron Age and yet, despite evidence of increasing land pressure, there are no significant fortifications and there is strong evidence of some form of elite. Therefore, Early Iron Age Denmark (400-50 BC) presents a powerful foil for the proposed model for the Middle Iron Age of the hillfort dominated zone.

As the material culture expressed in the hillfort dominated zone was different from other areas of Britain, including the range and type of weaponry recovered, it would not be an unreasonable assumption that the social patterns by which people lived would have also been different. Simply put, it is not reasonable to assume that warfare would be expressed in the same way by people who occupied brochs in western Scotland, or those who built (and occupied) hillforts of southern and western Britain or those who lived in small ditched enclosures common in East Anglia. Each people is likely to have had its own social construct relating to warfare and, as such, each requires a different model to reflect this regionalisation.

Why Study War(fare) at all?

The term warfare is often used within archaeological literature in what appears to be an attempt to reduce the overall impact of armed aggression within a particular culture. This is in many respects an implicit following of Turney-High's ideology, that war in its true sense is only carried out by complex or, preferably, state level societies (Turney-High 1971). However, warfare is in reality the act of undertaking war in, and in relationship to, demographic impact, and can be more devastating, in terms of the percentage of a community involved in the actual fighting, to small non-state societies, than to states themselves (Keeley 1996, 189, Table 2.6).

The basic causes of war represent a major field of research within anthropology. Although today the idea of 'man the primitive savage' instinctively undertaking war has been superceded, there is still a significant debate as to whether biological (e.g. the drive to find a mate and pass on the individual's genes (Chagnon 1990)), materialist (i.e. the competition for basic resources, e.g. land, hunting areas, mineral deposits (Harris 1984; LeBlanc 2003; Vayda 1969)), or social factors (e.g. honour, revenge, appeasement of ancestors/divinities) are the main driving forces behind conflict in pre-state societies. This study adopts the view that, while a certain level of aggression and competition is natural to humans, war is a social construct and not a biological or material necessity (Ferguson 1984, 1-61; Mead 1940, 402-5; Malinowski 1964, 245-68).

It is easy to see war as an aberration, as existing outside the normal behaviour of a given society. This is of course understandable from our early twenty-first century

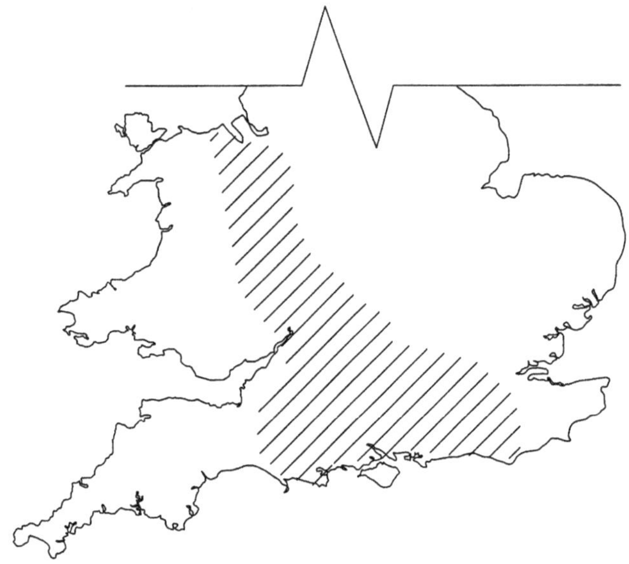

Figure 1.1

The Hillfort Dominated Zone c 400-150BC (after Cunliffe 1990)

perspective. The results of two world wars and more recent conflicts have only confirmed that the horrors of war should only be used as a last resort. This has led to a strange circumstance where many archaeologists do not consider war as part of the social structure of a culture, despite there often being potential evidence for conflict. Mostly this is an implicit statement as conflict either does not appear at all in or does not form a significant part of their interpretations (Burgess 1980; Hodder 1990; Hedges 1984; Thomas 1999; Whittle 1985). As a result, there has been a strong tendency within the archaeological community to pacify the past. For example, it was easier to see the weapons of the Bronze Age as part of a depositional ideology and expressions of political power than to question how that power may have been expressed through violence and warfare (e.g. Barrett and Needham 1988; Bradley 1984).

Recently there has been a reversal of this trend, and more studies relating directly to warfare in prehistory and early historic societies have been published (Carman 1997; Carman and Harding 1999; Hanson 1991; 1998; 2000; Osgood 1998; Osgood *et al.* 2000; Randsborg 1995). These studies have begun to provide a more rounded picture of society in the past, where warfare played a part, sometimes potentially dominant, at other times occasionally expressed. Many of the studies of prehistoric warfare ignore the ethnographic record as part of the archaeological process and expressly disregard the insights that ethnography can give (Osgood *et al.* 2000, 1). These studies often use random (piecemeal) ethnographic examples that have little direct reference to the material remains that are being studied (e.g. Osgood's (1998, 9, 15, 16, 49 etc.) use of Zulu references when dealing with Late Bronze Age warfare). Other studies express the ideology that weapons carry inherent meanings that can be directly accessed without explaining the derivation of these concepts (Harding 1999, 91, 92; Hill 1996,107-8; Randsborg 1995, 38-42).

The available ethnographic record shows that warfare formed a profound part of many cultures, and that peaceful or near peaceful societies are rare (Otterbein 1970; Sipes 1973, 68). By not studying warfare within prehistory there is a strong possibility that a significant part of any culture will be overlooked, which in turn may have implications for other interpretations.

Existing models of Warfare in the Iron Age of Britain

Until recently, almost all interpretations of warfare within the British Iron Age have relied heavily on the Classical authors and the use of vernacular Irish literature of the late first millennium AD (Cunliffe 1991, Chapter 18; Pleiner 1993, 24-33). These descriptions are fraught with difficulties and, as Hill (1989, 18-9) has pointed out, are potentially unlikely to have any direct relevance to those cultures existing in the Middle Iron Age. Although the specifics of weaponry have been studied (Lang 1987; Piggott 1950; Pleiner 1993), remarkably little interpretation of Iron Age warfare, outside of a general acceptance of classical writings as representing a cultural reality (Cunliffe 1986, Chapter 5; 1991, Chapter 18; Harding 1974, Chapter 4; Rankin 1996, Chapter 4; Ritchie and Ritchie 1985), has been attempted. Avery (1986; 1993) gives a detailed account of the ramparts and entrance details of a series of hillforts across Britain. He attempts to interpret the tactical implications of hillforts (and therefore Iron Age warfare) in general but seems to base most of his interpretations on a 'common sense' approach. This inherently assumes that hillforts were central to the practice of warfare in the Iron Age and that their features all derive from the need to improve their defensive nature against ever more sophisticated methods of attack (Avery 1993, 143-4). His work implicitly suggests that Iron Age peoples would have been able to organise their forces relative to the operations required in each successive stage of a siege.

Sharples' (1991a) work on warfare in Wessex fully recognises that different patterns of warfare existed during the period. He concentrates his interpretation within the Middle Iron Age, which is almost totally lacking evidence of warfare, with the exception of hillforts, occasional spearheads and concentrations of pebbles assumed to be sling stones. He does not deal, as Avery has, with the 'practicalities' of how war was undertaken, but focuses on the reasons behind war, the social implications of changes in warfare, and how this may be represented in the archaeological record. Sharples also defines the underlying reason for Iron Age warfare as a desire for land and its agricultural potential. This interpretation seems to be at variance with the available evidence from the Middle Iron Age, with its apparently stable landscape of hillforts, open villages and 'enclosed' settlements, all of which have evidence of long term occupation and limited social differentiation (at least in terms of the range of material goods available if not the quantity). The ethnographic record would also tend to suggest that land appropriation was not one of the driving forces behind pre-state warfare, where there was insufficient surplus population to occupy the newly acquired territory (Drennan 1991, 279-80). There is also, no clear evidence of the emergence of single dominating centres with significantly greater levels of material wealth until the Late Iron Age (Collis 1984a, 154-61; Cunliffe 1984c, 32-9; 1984d), which would perhaps be expected if this mode of warfare predominated within the Middle Iron Age.

Despite the paucity of research relating specifically to the military dimension of hillforts, much recent work has been undertaken to expand the nature of enclosed space within the cultural context of the Iron Age. Particularly influential has been Bowden and McOmish's (1987; 1989) work outlining their ideas of the required barrier, and the resultant reduction of hillforts as primarily defensive monuments by placing them within a broader context of social display and position. This concept has been adopted and expanded by Hill (1996), who has attempted to express not only the alien nature of the Iron Age but who also sees defended enclosures as complex constructions forming part of the social fabric of a non-hierarchical society. The sentiment that hillforts are more than just defended areas is supported by Collis (1996), who also points out the variety of enclosed space and the variability of its use.

Much of the literature devoted to hillforts is typological and the 'military' nature of these monuments is implicit rather than explicit. Apart from Avery's (1986; 1993) and Cunliffe's (1971) work, there has been no attempt to understand hillforts in relationship to warfare. Rather, they are seen as being associated with war but with no clear definition of this association. There also appears to be an underlying concept that in some form the classification of 'types' of hillfort will inevitably lead to an understanding of their use. This is flawed. Although such studies are useful and may indicate relative dating due to the spread of new ideas or even fashion, they fail to appreciate the social nature of hillforts. This study will examine one aspect of these complex monuments, to see if they served any definable military function within a multifaceted social context.

Most models of hillfort use are loose in their terminology. They tend to see hillforts as 'refuges' (Hawkes 1931, 76; Harding 1972, 17), 'cattle corrals' (Hawkes 1931, 67), or 'outposts' (Hawkes 1931, 93). All these interpretations seem to have their basis in, if not state level societies, then those where a significant level of complexity existed and where some form of central control was operating, which had an overview of the strategic nature of fortifications. This does not appear to have been the case in the Middle Iron Age. A new model, attempting to place of hillforts within the social structure of the Middle Iron Age, is overdue. The proposed model will have to encompass the military conundrum that although many hillforts appear to be specifically designed to be truly defensive (rather than merely defendable), their location within the landscape, both physical and social, renders them less than ideal for that purpose. The question therefore is what were they defending?

The use of ethnography

> *The archaeology of warfare cries out for the use of ethnographic analogy* (Arkush & Staish 2005, 16).

Throughout this study there will be an explicit use of ethnography as an interpretative tool. In fact it will form the core for attempting to place weaponry within a social context. As stated above, there is either an implicit use of ethnography within prehistoric warfare studies or an attempt to deny its value (Osgood *et al.* 2000, 1). Whenever a social construct is developed within an archaeological framework, the use of ethnography to generate social data is in essence inescapable, although potentially misleading (Wylie 1985, 81). The use of a piecemeal approach has led some archaeologists to reject analogy altogether (Orme 1981, ix), but, despite its problems, analogy is likely to remain the main source of interpretation for the archaeological record (Murray and Walker 1988).

The use of specific ethnographic analogy is an attempt to reduce the number of potential interpretations within one particular area of focus, thus hopefully increasing the strength of any inferences that can be drawn from the data (Watson 1979). This is in direct contrast with the piecemeal approach, which, while often successful when dealing with larger issues (e.g. the existence or otherwise of ranking within a particular archaeological culture (Wason 1994)), would have the opposite effect when dealing with the ideologies of weaponry and by implication warfare. It is highly probable that the social constructs that surround, for instance, archery, will vary across differing cultural groups. It is unlikely that hunter-gatherers will have the same view of the bow as nomadic herders or settled agriculturalists. Due to their own unique historical trajectory, each will have created a different set of associations for this particular weapon. Therefore, the generation of a general 'law' of the social significance of archery would be impossible, as it is inherently unlikely that any significant similarities could be found among such diverse peoples. By drawing on four historically independent societies that shared a similar economic/social background (i.e. pre-state sedentary agriculturalists), it is hoped that significant inferences relating to the possible social expressions of weapons within the Middle Iron Age of the hillfort dominated zone can be drawn, whilst remaining within the cultural parameters that the archaeological evidence suggests.

CHAPTER TWO

MATERIAL EVIDENCE FOR WARFARE IN THE MIDDLE IRON AGE

Most evidence for the use of weapons in a warfare context from prehistory is ambiguous. What evidence there is for organised violence against another group, or an individual, is open to competing interpretations that are often as parsimonious as that involving warfare. Is the Neolithic body with a leaf-shaped arrowhead within the body cavity from Fengate, Cambridgeshire (Pryor 1984, 19-22, Figure 19, Plates 12-13) a murder victim, a victim of a hunting accident, a ritual/legal killing, or the victim of open aggression in the form of war? More convincing evidence for prehistoric warfare within a British context comes from Tormarton (Knight *et al.* 1972), where a series of wounds including a broken bronze spearhead embedded in the spine of one of the individuals certainly hints at more formal or organised aggression, but it is not proof. Even the use of the sword as a weapon of war, at least in the Bronze Age, has been questioned (Harding 1999, 88-91).

Evidence from the hillfort dominated zone in the Middle Iron Age is equally ambiguous. It is, in reality, often only its context that gives any indication that some of the material remains may have had some form of 'military' function. 'Military' as a term is used here in its loosest sense, indicating anything to do with the organisation or expression of war or warfare. The evidence that is available comes in three broad categories:

1. *Weapons*
Like all human artefacts, these carry social messages, and this will be explored more fully in Chapter Three. The material remains will be discussed in this Chapter.

2. *Settlement*
As landscapes are essentially social expressions, the structure of settlement types may elucidate the type and nature of any warfare that a particular culture may have undertaken.

3. *Morphological*
Evidence of trauma inflicted by particular weapons on individuals is particularly limited within the Middle Iron Age of the hillfort dominated zone, with its apparent lack of widely recognisable formal rituals for disposing of the dead.

The Weapons of War

The evidence for weapons that can be associated with warlike activity within the Middle Iron Age of the hillfort dominated zone is surprisingly limited. Simple agricultural tools, such as hook-shaped cutting tools (Cunliffe 1984b, 346-9, Figures 7.8-7.9), knives (Cunliffe 1984b, 349-51, Figure 7.10), adzes and picks (Cunliffe 1984b, 351-4, Figure 7.12), all of which could technically be used as offensive weapons, but are unlikely to have been, can be excluded. Therefore, the Middle Iron Age only produces three types of artefact that can be associated with war through ethnographic and historical analogy: the spear and, based on large numbers of pebbles, and chalk and clay ovoids, the sling and the sword. The sword stands within this panoply as having only an extremely limited representation within the hillfort dominated zone, when compared with contemporary geographically adjacent cultures. Within the existing models of Iron Age warfare the sword is often referred to, and forms an essential part of the model of 'Celtic heroic' warfare (Cunliffe 1990, 448; Pleiner 1993, 33-5). Its limited representation deserves further exploration. Is this near absence an artefact of archaeological recovery, or a genuine deficit?

The Sword

Swords are a relatively common artefact from the Late Bronze Age, being recovered from watery contexts (Needham and Burgess 1980, Figure 7), within hoards (Coombs 1975) and from settlement sites (Barrett and Bradley 1980, 263). There is a general reduction in the number of swords that are deposited throughout the this period, but the tradition continues well into the seventh century, which can be seen as part of the transition to the earliest Iron Age. Such deposits are found within the geographical area of the hillfort dominated zone. One is from Weymouth Bay, Dorset (Cowen 1967, 450) where a bronze Hallstatt sword was recovered, and two are from Shropshire, one unprovenanced from Brogynton Selattyn, (Cowen 1967, 444) and another from the River Severn at Jackfield (Much Wenlock Museum, Accession No. A.01161 Burns: *pers. comm.*). Both have been dated typologically to the seventh century BC. Stead's work on Iron Age swords (2006) lists twelve possible examples of swords, sword fragments, scabbards and scabbard fragments including chapes etc. from the Hillfort Dominated Zone that potentially could be dated to the Middle Iron Age. Using continental examples as a background Stead (*ibid,* 2-3) has grouped the remains into broad typological categories. Group A (*c.* 400-250 BC) and Group B (*c.* 250-150 BC) falling within the Middle Iron Age. Of these listed in Group A the only sword recovered from a stratigraphic context, was that from the Caburn (Curwen and Curwen 1927, 11, pl. IV), though recorded as being associated with shards (*ibid,* 55) these are not identified. Whereas, the majority of the pottery recovered from the site seems to date from the first two centuries BC and the fist century AD (*ibid.* 45), the sword is typologically dated significantly earlier (Stead's Group A).

Of those placed in Group B three were recovered during modern excavations and the associated pottery again indicates a date that appears to be at variance with the typological description. The sword from Maiden Castle was recovered from a layer dated by Wheeler (1943, 275), to *c*. AD 25. A date later than the proposed typological classification can be provided for the iron scabbard found in the 'massacre level' at Breedon Hill, the only pottery associated with this layer mixed with the crushed skulls (Hencken 1938, 109), was similar to that of the first century BC single vallate enclosure at Salmonsby (Dunning 1931, 490 (line illustration), 491). The Wooley Down chape was recovered from a deposit containing both Iron Age and Romano-British Pottery (Peake and Padel 1934, 35). A number of swords and spears, along with human skeletons were recovered from a large pit excavated at Spettisbury, Dorset, but as these were dug in the middle years of the nineteenth century, their context is poor at best and the inhumations, assuming that they were burials would indicate a Late Iron Age date. The dagger, scabbard and chape fragments from Cadbury Castle (Barrett *et al.* 2000 236-9), serve to highlight the problems associated with typologically dating prestige items. The daggers are un-diagnostic and can be dated as late as the first century AD (*ibid.* 236) and though the chapes are typologically dated to La Tène I and II, one (which is placed by Stead in his group B (2006, 167-8)), was recovered from a context containing Late Iron Age pottery and contains a low level of antimony in the copper alloy, which is a trait of the Later Iron Age (Barrett *et al.* 2000, 273).

It would be easy to invoke the curation of treasured heirlooms, with perhaps an associated history, to account for the potential anonymities in the typological dating, however, this in itself raises other questions. Some of the curation would have to have been for a considerable time (well over one hundred and fifty years, in the case of the Maiden Castle and Wooley Down examples). Assuming that such artefacts carried social meaning, it is also possible that Iron Age smiths continued to manufacture 'out of fashion' swords, scabbards etc. to satisfy a more conservative cliental. Certainly some aspects of insular Iron Age warfare seem anachronistic in comparison to continental trends, e.g. the retention of the chariot, which survived well into the first century AD in Britain (Tacitus *Agricola* bk 35). If such conservatism did extend to the manufacture of swords etc. it is possible that all artefactual evidence in the hillfort dominated zone falls outside the Middle Iron Age.

More secure evidence for sword use comes from morphological trauma. From Danebury, a number of skulls show wounds that in all probability were inflicted with a sword (Craig *et al.* 2005 170-1). All of these date to ceramic phase (cp 7) (260 BC-50 BC (Cunliffe 1995 18)). A single pelvis also shows signs of being damaged with a sword-like weapon but this dates to cp 3 (450-360 BC), which falls outside the Middle Iron Age (Cunliffe and Poole 1991, Table 8.10). One skeleton from Old Down Farm, Andover (Davies 1981, 132-3) displays a series of cuts. It is not clear if these were caused by a single implement (in which case it was likely to be long, sharp, and double-edged and have a thin blade, which could be a sword, dagger or long knife) or a number of weapons, e.g. a knife and spear. The deposit is dated to the 'Middle Iron Age', with no more precise location within this period. All the above would perhaps not be remarkable, except that swords were certainly in use (or at least being deposited) in relatively significant numbers in areas adjacent to, and presumably in contact with, the hillfort dominated zone during the Middle Iron Age.

The Thames Valley has the highest concentration of swords for the whole of southern Britain (Piggott 1950), but even here, there appears to be a break in deposition from the seventh century BC to the fourth century BC (Jope 1961, Figure 8). It is unlikely that this represents an artefact of recovery, as during this period the swords seem to be replaced by Late Hallstatt D Early La Tène iron daggers, which are recovered from similar if not identical contexts (*ibid* 307). Outside this area, swords are recorded throughout the fourth century BC, to the first century AD, in Lincolnshire, Yorkshire and on the western fringes in Somerset (Wait 1985, 22-31, 273-89).

The limited evidence for swords within the hillfort dominated zone is marked within this overall distribution. Given the intensity of archaeological exploration of the Iron Age in the southern hillfort dominated zone (particularly Wessex), the general lack of swords is all the more remarkable. The most parsimonious explanation for the very limited evidence is that they were not widely used during the fourth to late second centuries BC, and may have been circumscribed to very specific areas of social interaction. Almost certainly the swords used in the hillfort dominated zone were not understood in the same social construct as those used outside this area. As will be argued in Chapter Three, the choice of weaponry is entirely social and the mechanical effectiveness of individual weapons is not an important part of that choice.

The Sling

Of the other weapons used by the Middle Iron Age peoples of the hillfort dominated zone, the sling must be the best represented. In fact, given the simply enormous number of pebbles interpreted as sling stones, along with the smaller numbers of baked clay and carved chalk ovoids, the ammunition for slings must be one of the most common archaeological finds other than pottery, from this period. Examples include the 22,600 pebbles recovered from one pit at Maiden Castle (Wheeler 1943, 49), the 11,000 from one pit at Danebury (Cunliffe 1984b, 425) and the 1000 or so from Shipton Hill, Dorset (Farrar 1955,136).

Sling ammunition is represented by four categories of material: natural water-worn pebbles (sea-rounded, riverine and glacially derived); baked clay ovoids; unbaked clay ovoids; and occasionally carved stone ovoids, usually chalk, (see Stead 1968, 166 for a probable exception). Sling ammunition occurs throughout the hillfort dominated zone (Figure 2.1) and, although not found on all sites (where many of the excavations were very limited), it is likely that its use was familiar throughout the entire area. Geographically the utilisation of baked clay slingshot seems to stop north of a line roughly parallel with Bath and London. The Late Iron Age clay shot from West Stow, Suffolk (West 1990, 60-1) and Thetford, Norfolk (Gregory 1991, 148-9) are exceptions. Excluding the carved slingshot from Grimthorpe (Stead 1968, 168), which lies outside the hillfort dominated zone, the occurrence of carved slingshot and baked or unbaked clay shot do appear to have a marked correlation.

The use of carved stone clay shot is recorded within the ethnographic literature (Mockton 1921, 38) as being one method of improving accuracy. This may also account for the relatively standard weights of baked clay shot (see Table 2.1). However, the clay object found at Shearplace Hill, Sydling St. Nicholas, Dorset (Rahtz and ApSimon 1962, 323), offers an intriguing, alternative possibility. The excavators interpreted the object as a possible sling bullet, consisting of reddish clay, composed of two pointed flattened pieces pressed together while wet, made in a double mould or two single moulds. Either way it was not successful (*ibid*, 323). Surface examination of the pellets (SF 2256 and SF 1448 respectively) from West Stow would also suggest this. It is possible to see the edges where a joint may have been smoothed, presumably with a finger (West 1990, 60, Figure 45, 72-3). If the above interpretation has any validity, carved chalk slingshot may have been used as a master for moulds. Unfortunately no such moulds have yet been recovered from an archaeological context but, if they were used, the resulting standardised projectiles would have had the potential to significantly increase a slinger's accuracy.

The various types of ammunition have been interpreted as having differing functions. The lighter baked or unbaked clay and chalk shot were used for hunting or herding (Cunliffe 1987, 165; Perles 2001, 228-31) and the heavier pebbles were used for war (Cunliffe 1984b, 425; Wheeler 1943, 49-50). The use of clay or chalk shot for hunting or herding would correlate well with the presumed requirement for increased accuracy. In the latter case, unbaked shot (sun or hearth dried) would have been particularly effective and the shot would tend to 'explode' upon impact, increasing the overall effect by scattering debris into vegetation or even by striking a wayward animal without doing any particular harm. However, the light weight of such projectiles does not exclude them from use in war. Many of the lead shot recovered from historical sites have masses below 25g (Greep 1987, Figure 5), yet their context and often the inscriptions on their surfaces (Bishop and Coulston 1993, 55) indicate strongly that they were used in conflict. Ethnohistorical records from early Medieval Ireland also suggest that baked clay shot were used in war. The poet Tathlum states that, by mixing the brains of the enemy with the clay of a slingshot, not only was its effectiveness increased but also its accuracy (O'Curry 1873, 252).

The deposition of large numbers of pebbles near to the entrance of Maiden Castle has been used by Avery (1986, 225; 1993, 7) to suggest that they were not sling stones at all but were thrown by hand. Although some of the heavier stones (up to 260g) recovered from Danebury (Cunliffe and Poole 1991, micro-fiche 12:02, Figure 2A), could have been used in this manner, the vast majority would have insufficient mass to have any serious effect against a human opponent. The iconographic evidence from Classical Greece would tend to show that hand-thrown stones were at least 'fist' sized. For example, the Great Melos Amphora (Louvre S1677) shows the gods engaged in combat with giants. An Attic amphora (Louvre G216) shows a psilos with a hand-sized stone, and a third example shows the dispatching of a hare in a hunt scene (British Museum D60). The Greek representations of hand-cast rocks do not appear to be an iconographic metaphor. A figure shown on a pithos from the temple of Artemis Orthia, just outside Sparta, is shown in a posture that indicates he is using the sling (originally this was painted onto the ceramic surface), holding a small well-rounded pebble (Dawkins 1929, Plates XV and XVI).

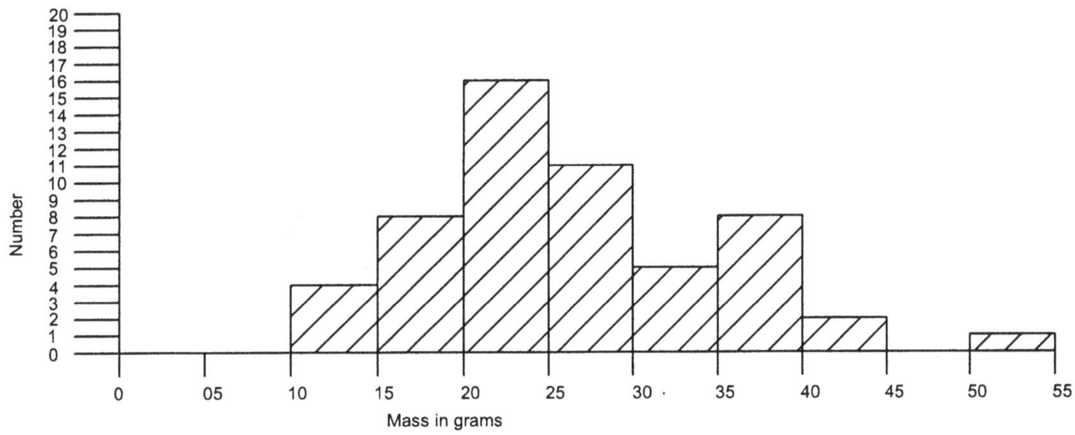

Table 2.1. *Weight Distribution for clay slingshot (all sites). 49% of the sample falls within the 20g-30g. range.*

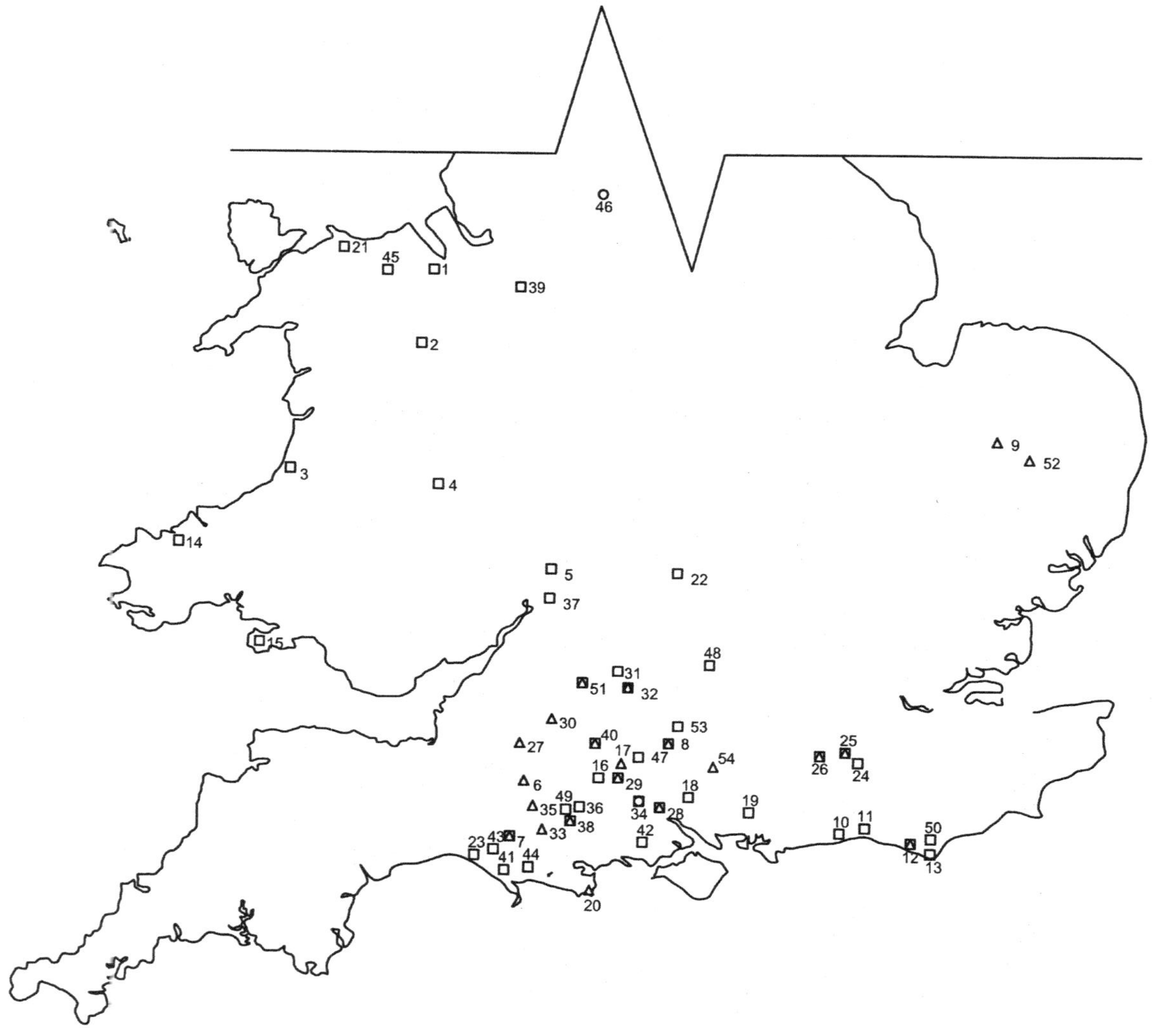

Figure 2.1

Distribution of Slingstones/slingshot in Southern Britain

Key

□ Natural pebble

Δ Clay slingshot

○ Carved stone slingshot

List of Sites.

1. Conway Mountain (Griffiths and Hogg 1956)
2. Pen-Y-Gaer (Hughes and Gardner 1906)
3. Pen Dinas (Forde, Griffiths Hogg and Houlder 1963)
4. Titterstone Clee Hill (St. J. O'Neil 1932)
5. Bredon Hill (Hencken 1934)
6. Cadbury Castle (Alcock 1969)
7. Maiden Castle (Wheeler 1943)
8. Danebury (Cunliffe 1991)
9. West Stow (West 1990)
10. Beltout (Toms 1912)
11. Thundersbarrow (Curwen and Oakley 1933)
12. Caburn (Curwin 1927)
13. Cissbury (Curwen and Williamson 1931)
14. Moel Trigarn (Gould 1900)
15. Harding's Down (Hogg 1973)
16. Yarnbury (Cunningham 1932)
17. Casterley (Cunningham 1914)
18. St Catherines Hill (Hawkes, Myres and Stevens 1929)
19. The Trundle (Curwen 1929)
20. Hengisbury Head (Cunliffe 1987)
21. Pen - Y -Corddyn (Gardner 1910)
22. Lyneham Camp (Bayne 1957)
23. Pilsdon Pen (Gelling 1977)
24. Anstiebury (Thompson 1979)
25. Holmby (Thompson 1979)
26. Hascombe (Thompson 1979)
27. Glastonbury Lake Village (Bulleid and Gray 1910)
28. Boscombe Down East and West (Stone 1936; Richardson 1951)
29. Fisherton (Stevens 1934)
30. Budbury Camp (Wainwright 1970)
31. Fifield Down (Clay 1924)
32. Lidbury Camp (Cunningham 1917)
33. Allards Quary (Williams 1951)
34. Swallowcliffe Down (Clay 1924)
35. Sydling St. Nicolas (Rahtz 1962)
36. Gussage All Saints (Wainwright 1973)
37. Crickley Hill (Dixon 1972)
38. Hod Hill (Richmound 1968)
39. Castle Ditches (Varley 1940)
40 Battlebury Camp (Cunningham 1934)
41 Eggerdon Hill (Gray 1901)
42. Chalbury Camp (Whitley 1943)
43. Shipton Hill (Farrar 1955)
44. Pins Knoll (Bailey 1967)
45. Dinordan (Gardner 1964)
46. Grimthrope (Stead 1968)
47. Stockton (Saunders 1997)
48. Uffington Castle (Lock *et al* 2004)
49. Hambeldon Hill (Cunningham 1895)
50. Bishopstone (Bell 1977)
51. All Cannings Cross (Cunningham 1923)
52. Thetford (Gregory 1991)
53. Balksbury Camp (Wainwright and Davies 1995)
54 Winall Down (Bates and Winham 1985)

Caches of large rocks from the Roman fort at Burnswark (Jobey 1978), worked so that they have flattened sides for ease of stacking, have been interpreted as being stones intended for throwing by hand (*ibid,* 91). Caesar records the slingers of the Balearic Islands as throwing stones weighing up to one mina (500g) with their *brawny arms* (*de Bello Civili* IV, 25). Occasionally this has been interpreted as being the weight of the sling stones cast by these well known mercenaries. However, the use of hand thrown stones is more likely, as the limited ethnographic and historical evidence available[1] would tend to suggest that a weight range of 50-60g was regarded as being an optimum for sling use. This closely correlates with the weights of the stones recovered from both Danebury and Maiden Castle (see Appendix 1.1 for a gazetteer of sites where sling projectiles have been recovered in southern Britain).

Iron Spearheads

Few spearheads have survived from the Middle Iron Age of the hillfort dominated zone. This may be a product of local conditions (Cunliffe 1987, 142). Those that have been recovered tend to be small (less than 200mm long) and simply made. With a few later exceptions (Cunliffe and Poole 1991, 2, 823; Gresham 1939, Plate IV, 3 and 4), where a central strengthening mid-rib is provided, most Middle Iron Age spearheads appear to be a single flat sheet of metal, hammered around a shaft and beaten into a simple leaf shape (this may have been cut out before the socket was formed, making the process even easier). The techniques used are comparable to those in the manufacture of other more obviously agricultural iron tools, recovered from both hillforts and domestic sites (see Chapter Three for a fuller exploration of this difference). Although concentrated within hillforts, spearheads are occasionally recovered from 'domestic sites' (Bulleid and Gray 1917, 381; Williams 1951, 57). This tends to suggest that spears were a fairly widespread weapon and not confined to a particular social group, who may have occupied a particular settlement type. As noted by Cunliffe and Poole (1991, 352), although there is a variety of spearhead shapes from Middle Iron Age contexts, the numbers of each type is so low, often no more than one example for each design, that they defy any typological classification.

The spearheads do, however, tend to share certain characteristics. They are for the most part 'leaf-shaped' (i.e. wide in comparison to their length). This is primarily a subjective observation. The sample available is too small for any meaningful statistical analysis. The spearheads from Spettisbury Ring (Gresham 1939, 122-3) stand out from this classification. As they were excavated in the middle years of the nineteenth century the dating of these artefacts is problematic and their inclusion in the gazetteer is a recognition that some may date to the Middle Iron Age. The reported recovery of at least one of the larger and more complex spearheads with parts of a skeleton (Gresham 1939, 120, Plate IV, 3) does tend to suggest that this artefact at least may have come from a Late Iron Age context, when interments with weapons occur within the archaeological record (Collis 1973).

The broad leaf-shaped nature of the spearheads would suggest that the use of defensive armour was either uncommon or unknown. This shape maximises the blade area for haemorrhaging, which is the primary cause of death associated with slow moving weapons such as spears. A narrow blade, although allowing armour to be penetrated, reduces the cutting area of the blade. This change can been seen at Lagore Crannog (Hencken 1950), where the broad leaf-shaped spearheads of the pre-Norse Irish are replaced with narrow-bladed shouldered types during the eighth to tenth centuries AD (*ibid,* 94-8), presumably partly in response to the use of ring mail by the Vikings. The small size of the spearheads also suggests that they were primarily throwing weapons.

The variability in size would also tend to suggest that spearheads were manufactured to individual requirements, and not from any form of centralised workshop. Their simple manufacture would have been within the capabilities of almost any blacksmith with only a rudimentary knowledge of wrought iron working (Hodges 1989, 86), which seems to have been the case in Middle Iron Age Wessex (Ehrenreich 1986, 182).

Bone Spear Points

It has recently been suggested that some of the so-called bone gouges found on many Middle Iron Age sites across Britain (see Olsen 2003, 102 for a comprehensive list of locations and references), may have been manufactured as spear points (Olsen 2003). This inference is buttressed by the context of their recovery (*ibid,* 109-111) and similar finds in Denmark (Kaul 2003, 148, Figures 4.26, 4.31), also from contexts that strongly infer a military nature. If bone was used for spear points (and Olsen (2003, 93) points out that not all gouges are likely to be spear points), then, given their form (a simple point, rather than a more complex cutting edge), it would be probable that they were used as casting weapons. However, it is not so much the possible functional nature of these artefacts that is of primary concern. If bone was used to manufacture spears, then its control within a social context is likely to have been difficult, if not impossible, and the social significance of the weapon would almost certainly have reflected this.

Settlement Patterns and Types

Agricultural landscapes are social constructs that should in some degree reflect the social ideals of the cultures that created them. The construction of defensive structures within a landscape does not necessarily mean

[1] Unfortunately during the period of research the Museum of Mankind's collections were unavailable for study as they were in storage awaiting re-housing.

that there was a constant threat of war, but would tend to suggest that the people who created them perceived that such a threat existed, or structured their society, in part at least, within a military ideology. The continued building of castles in eastern and southern Britain during the Middle Ages may be an example of this. Fortified sites became more of a status symbol than a defensive requirement (Woolgar 1999, Chapter 4) but the status of being a knight, and all that this military ideal stood for, remained all-important. In opposition to this, a lack of defensive structures does not automatically mean that a culture was mostly peaceful. The city state of Sparta would perhaps be one of the best known examples of this. Despite being an aggressive expansionist polity, whose whole culture seemed to be geared to war (Forrest 1980, 51-54), Sparta had no city walls and there were few defensive structures in the rest of Lakedaimon. The landscape of the hillfort dominated zone during the Middle Iron Age should be no different. The settlement pattern should by its very nature allow certain inferences to be generated relating to the type of war that may have been undertaken.

The evidence can for ease of interpretation be divided into three categories:

1. Hillforts.
2. Enclosed settlements.
3. Open settlements, occasionally called villages.

Each of the above must represent a specific response, by the people who constructed them, to a particular set of social circumstances. However, it is clear that in some cases these circumstances changed through time. For instance, Winnall Down (Winchester) was in its initial phase, during the Early Iron Age, an enclosed settlement. By the Middle Iron Age the original ditch had silted up and an open settlement of possibly up to six houses occupied the site (Fasham 1985, 130).

Hillforts

'Hillfort' is of course a purely descriptive term that has a wide and long currency. However, the use of the term with its overt military connotations, should not be taken as a prejudicing other interpretations of their use.

Hillforts are large impressive monuments that even after over two thousand years of erosion, still dominate the skyline of much of the hillfort dominated zone. There is an enormous literature associated with hillforts, much dealing with typological classifications (Forde-Johnston 1976, Savory 1976), construction types (Cunliffe 1991, Chapter 14; Hawkes 1971), and social constructs (Bradley 1984,138-44, Cunliffe 1971; Hill 1996). It is not the intention of this study to reiterate the detailed nature of this corpus of knowledge (see Avery 1993; Cunliffe 1991, Chapter 14; Harding 1974, Chapter 4; Forde-Johnston 1976 for a comprehensive overview of the development and structural character of hillforts in southern Britain), but to concentrate on the military aspects of the forts, where surprisingly little exploration has been undertaken. The association between hillforts and warfare has often been assumed (Harding 1974, Chapter 4; Hamilton 1968). The sheer size and complexity of their ramparts has been take as proof of the unstable nature of the Iron Age in general, where communities hid behind the hillforts' defences awaiting the next raid from their belligerent neighbours (Avery 1986, 228).

Yet the complex nature of hillforts belies such a simple monocausal explanation. No doubt their role changed through both space and time (this may even have occurred annually within the context of a ritual cycle) and through the perceptions of the individuals involved with them. Much of this social context is unrecoverable but the following sections explore the complexities of hillforts as military installations. This is not to propose that they were primarily or always seen as associated with warfare, but it is to propose that this was one of their facets within the culture of the Middle Iron Age peoples of the hillfort dominated zone.

Ditches and Banks

The massive nature of hillfort earthworks with their internal banks and deep external ditches would infer that they are not merely defendable structures but were purposefully constructed to be defended. By the Middle Iron Age, most hillforts within the hillfort dominated zone were being constructed using a simple dump method which had replaced the more complex 'box rampart', where two rows of posts enclosed rubble fill derived from a fronting ditch. Dump ramparts were commonly revetted at the front either in stone or timber. In the latter case this was often replaced with a true glacis style construction, where the bank and ditch presented a single continuous face (see Cunliffe 1991, Chapter 4; Avery 1993, Chapters 6-10, for summaries). The sheer effort involved in the construction of these ramparts would have been considerable. Calculations of man-hours etc. are fraught with difficulties, but the order of magnitude of many of the hillforts enclosing structures when compared with other settlement types is self-evident (see Table 2.2).

On average, the amount of material excavated from a hillfort ditch per linear metre was four times greater than that required for an enclosed settlement. This would of course require a corresponding increase in labour. However, the massive earthworks of many hillforts are repeated as a series of such ditches and banks.

This apparent 'over-building' is perhaps most sharply drawn at Hod Hill, Dorset. The defences, although strengthened during the Late Iron Age, are in essence of Middle Iron Age construction (Richmond 1968, 11-12). The ditch and bank have a combined height of 8-9m topped with what appears to have been a low breast work constructed of flint nodules (*ibid,* 12-13), though this feature was destroyed by the Romans sometime after AD 43 and used to metal the entrance road into the fort.

Hillforts		Non-hillfort enclosures	
Name	Average cubic excavation of ditch per linear metre	Name	Average cubic excavation of ditch per linear metre
Danebury (Cunliffe 1984a)	24	Fisherwick (Smith 1979)	2.8
Crickley Hill (Dixon 1994)	17.3	Little Woodbury (Bersu 1940)	4.5
Midsummer Hill (Stanford 1974)	8	Owslebury (Collis 1970)	1.7
Croft Ambrey (Stanford 1981)	12	The Collfryn (Britnell 1989)	5.8
St. Catherine's Hill (Hawkes, Myres and Stevens 1930)	16.2	Old Down Farm (Davies 1981)	3.8
Chalbury (Whitley 1943)	20	Winnall Down (Fasham 1985)	1.2
Hod Hill (Richmond 1968)	27	Meon Hill (Liddell 1933)	9.9
Poundbury (Richardson 1940)	11	Pimperne (Harding, Blake and Reynolds 1993)	2.3
Average	16.9		4

Table 2.2 *Volumes of excavated material from hillforts and non-hillforts.*

The flint breastwork might not have constituted part of the Middle Iron Age defences, though a shallow depression on the banks' summit may represent the foundation trench of this feature (*ibid,* 11). No estimate of its original height is given. Following the capture of the hillfort by Roman forces during the initial conquest of southern Britain (AD 43-4), a typical Roman fort was constructed in the north-western corner. The Roman fortifications used the existing defences as part of their circuit on the northern and western sides.

However, completely new earthworks were constructed on the southern and eastern flanks. In comparison to the Iron Age fortifications, these look insignificant. The ditch and bank were built in the 'standard' Roman fashion (Collingwood 1930, 22-9). Including the ditches and ramparts the total width of the Roman defences was approximately 8m, the maximum depth of the ditches was 2.3m (the minimum was 1.5m) and the associated bank was 3m in height (Richmond 1968, 68-9). The fort was occupied for approximately ten years before being abandoned following an accidental fire around AD 51 (*ibid,* 119-21).

The difference in size between the two ramparts is all the more extraordinary considering that the Roman fort was not a mere marching camp, but was occupied for a relatively long period of time. Roman forts were not only constructed to a standard manual (Collingwood 1930, 26), but had evolved over a considerable length of time, and represented a significant and often successful defensive measure against not only 'barbarian' forces but against other (including other Roman) state level polities. These had at their command not only trained, motivated troops, but also sophisticated siege equipment. The defences would therefore presumably have been viewed as sufficient to withstand an assault from such forces, with a reasonably good chance of successfully repelling it. Of course, Roman forts included other measures such as 'lilies' or sharpened stakes set into small pits arrayed to the front of the ditch and bank and the use of sharpened stakes along the bank itself again to break up any assault. However, none of these were beyond the capabilities of Middle Iron Age peoples to produce, and the use of *chevaux de frise* as part of a hillfort's defence has been suggested by Harbison (1971), based on literary sources and archaeological evidence.

Other evidence from the Italian peninsula also highlights the scale of Middle Iron Age hillforts. During the expansion of Rome in the third century BC, they undertook a series of campaigns against the Samnites. It is probable that the Samnites, along with most Italian peoples, were organised at state level. As part of their defensive system they constructed a series of hillforts, using often Cyclopean dry-stone masonry. The walls of these forts still survive in many places and their original height may be estimated. One extreme example found at Treglia (Oakley 1995, 12) had a wall height of ten metres. However, the average seems to have been three to four metres (*ibid,* 12). As with the Roman fort at Hod Hill, these structures were designed to withstand an assault by an organised state level society, even if at this early date they may have lacked siege equipment (Kern 1999, 251-4). Yet their dimensions do not approach those of the majority of hillforts when both the ditch and bank are combined. There is an apparent military conundrum. Many hillforts show levels of multivallation that appear to go well beyond the requirements of simple defence.

Given the pre-state and possibly non-hierarchical (Hill 1995b) nature of Middle Iron Age society, the size of the defences at Hod Hill, and most other hillforts, seems to be out of all proportion to the potential threat from the military forces available to any polity would have been able to pose. Why expend so much energy and resources in creating a defensive system that seems to go so far beyond the requirements of simple military defence? There is undoubtedly an element of expressing status within the construction of larger and more elaborate earthworks and walls (Bowden and McOmish 1987, 80), this could also extend to the erection of stone *chevaux de frise* considering the often enormous amounts of material required. However, there may also be an underlying military purpose. Wheeler suggested that multivallation was the result of the introduction of the sling by 'Iron Age B peoples' (Wheeler 1943, 48-49). The archaeological evidence now confirms that not only was the sling known by the Early Iron Age (Cunliffe 1984b, 425), but also multivallation was an early feature of many hillforts (e.g. Danebury (Cunliffe 1991, 36, Figure 3.25) and Rainsborough (Avery, Sutton and Banks 1967:291)).

Many of the outer works do not follow the line of the inner banks and ditches at the same distance. They seem to relate to the inner works in terms of the slope over which they were constructed. Simply put, the steeper the slope the closer the outer earthworks are to the main ramparts. This makes little sense in terms of the area enclosed by these outer works operating as a stock enclosure. The steeper slopes may not only have presented a danger to stock but would also have made their management more difficult. Also the large steep ditch that fronts the main rampart would have been easily accessible and potentially dangerous, particularly if a large number of cattle was herded together in unfamiliar circumstances, where jostling and gadding would be more common. There is no evidence from Danebury of a fence or palisade between the inner defensive ditch and the outer enclosed area, and the regular deposition of silting material as a counterscarp bank would argue against a hedge being in place. There is evidence of a palisade between the main bank and the surrounding area at Hod Hill (Richmond 1968, 12) but there is no evidence of an outer earthwork that could have operated as a stock enclosure. Perhaps here the palisade was to prevent cattle becoming trapped in the main ditch in a more open landscape.

As Cunliffe (1995, 19) observes the outer bank at Danebury is hardly of defensive proportions. Its dimensions are close to those specified by early Irish law texts as being sufficient for retaining cattle (Kelly 1998, 372, Figure 17). The intention of these outer works may not have been to prevent cattle *leaving* the enclosed area, but to prevent cattle *ingress*. This interpretation is buttressed by the erection of stone *chevaux de frise*, between the outer defensive works (a small ditch and bank) and the inner defensive works, as at Pen-Y-Gaer (Hughes and Gardner 1906). Although Harbison suggests that the *chevaux de frise* were a reaction to the introduction of cavalry (1971, 220), the difficultly of charging even a trained warhorse against bodies of infantry let alone a wall (Machiavelli [1521], 2005, II: 94-5) would strongly argue against this interpretation, and it more likely that *chevaux de frise* are a method of delaying or disordering assailants on foot. The potential defensive nature of the *chevaux de frise* would render this space unsuitable as a holding area for livestock. If this interpretation is correct, the deliberate exclusion of cattle could have allowed the regeneration of scrub growth between the inner and outer banks. This of course would have been pre-empted by the erection of *chevaux de frise.* Ethnographic evidence indicates that the sling was mostly widely used in a semi-vertical fashion, and any obstruction to the front or rear of the slinger would have made the use of the sling ineffective. Initially even herbaceous growth would have been effective, rendering the use of the sling very difficult if not actually impossible within the area defined by the outer earthwork. In the reverse, any obstruction to the front of slingers on the hillfort's ramparts would also have hampered their ability to use the sling. This may go some way to explaining the lack of evidence for breastworks associated with hillfort ramparts. If scrub growth was allowed to regenerate between the inner and outer banks of multivallate hillforts, this would also have offered the defenders protection against any assailants' slingers, as they would have been unable to come within range of the inner bank without entering the area defined by the outer banks. This may have been imperative, if little or no breastwork was feasible due to the use of the sling by the defenders.

Entrances and Gates

Many hillforts have in-turned entrances set back from the main bank. These extend for often considerable distances into the interior of the fort. (See Table 2.3 for tabulated distance for various forts).

As many in-turned entrances narrow as they approach the main gate, they have been interpreted as acting as a funnel, crowding an attacking force which would have lost much of its cohesion and become disorganised (Avery 1986, 222-3). The length of the in-turned walls could also be used by defenders to harry the attacking force as it approached the gate (Cunliffe 1991, 332-3). This makes the assumption that the attackers' primary target would have been the gate and that they would indeed have been able to destroy it, if and when they reached it.

The description by Caesar in *De Bello Gallico* (II, 6) is often quoted in this context and his work is often applied to interpretations of how pre-Roman attacks on a hillfort would have been undertaken for the whole of the Iron Age. However, there are potential flaws with this interpretation. Although normally seen as a truly 'native'

Hillfort	Length of inturn in metres	Reference
Danebury	20	Cunliffe 1984a, 26, Figure 3.14
Midsummer Hill	10	Stanford 1974, Figure 2
Croft Ambrey	16	Stanford 1981, Figure 29
St Catherine's Hill	17	Hawkes, Myres and Stevens 1930, 29, Figure 5
The Trundle	21	Curwen 1929, 119, Plate V
Bury Wood Camp	10	King 1967, Figure 1
Rainsborough	18.5	Avery, Sutton and Banks 1967, 254
The Breiddin	60	Musson 1991, Figure 5

Table 2.3 *Lengths of entrance in-turns*

attack on a hillfort, by forces unaffected by Roman tactics (Avery 1993, 91; Rivet 1971, 189), this proposition is difficult to sustain. Clearly Caesar is describing an internal conflict, but it is entirely possible that the Belgic forces had seen Roman siege methods and have adapted them to their needs, as the use of *testudo* would suggest. The final description is translated to mean that the gates are fired and the fort stormed (Avery 1993, Chapter 17; Rivet 1971, 189). There are inherent difficulties with translocating this description to the Middle Iron Age of the hillfort dominated zone. The passage makes it clear that there is a significant level of military differentiation between troop types within the Belgic forces. It would be no use clearing the ramparts, using (presumably) slingers, only to have these forces then reassemble to attack the gate, as this would give ample opportunity for the defenders to reoccupy the walls or banks. To engage the attackers, 'covering' fire would have to have been constant during an assault. The use of the *testudo* also suggests that the troops assailing the fort had undergone some level of training to maintain the required formation that would make the *testudo* effective. This in turn suggests that the organisation of the Belgic forces was in some respects becoming specialised.

This level of organisation, and the evidence available from both Caesar and the archaeological record, would suggest that, if the Belgae and Gauls had not reached an initial state level society, then they had at least become complex chiefdoms (Burn 1995, 19-21; Wells 1984, Chapter 6). Both of these have greater levels of complexity than any suggested social formation for the Middle Iron Age of the hillfort dominated zone. This level of complexity engenders military organisation and the formation of some form of command structure (Ferguson 1990, 47-51), which may have removed the overall commander from direct involvement in the fighting.

The firing of the gates also poses significant problems within a British context. The approach to the gate was, as outlined above, often protected by an in-turned entrance giving the defenders ample opportunity to attempt to prevent the attackers from reaching the gate without significant risk to themselves. Assuming that the assailants could reach the gate and attempt to fire it, this is not easily undertaken. Given its overall strength and good cleaving structure, it is probable that most gates (whatever form they took – see Stanford 1974, 51) would have been made of oak. Although, like most timber, dry oak burns well, wet oak tends to char forming a barrier to further burning. In any event, the defenders would have had the opportunity to attempt to put out the fire by either dousing it with water (the lack of an internal water supply for most hillforts would be a problem) or by throwing earth or other non-flammable debris over the combustible material. There is clear evidence that some hillfort gates were indeed burnt, but the nature of some of the associated deposits has led to their direct association with assault being questioned and an explanation based in ritual destruction being proposed (Hill 1996, 108).

If the proposition outlined above relating to the relationship between the outer and inner banks has validity, it would potentially prevent any assailing force from being able to attack a fort in the manner described by Caesar, as the attacking slingers would be unable to operate within the area defined by the outer banks. The use of complex out-hornworks, seen in many hillforts, would also prevent this tactic when attacking an entrance. Another feature of many hillfort entrances, at least in the north-western part of the hillfort dominated zone, are the so-called guard rooms. This is a deliberately constructed feature just behind the gate which is normally interpreted as operating either as guard house (Harding 1974, 68) or store room (Stanford 1974, 54). However, as Harding has observed, these structures could not really have had much relevance to the defence of the fort during an all-out attack. This suggests that ingress and exit to or from the fort was being controlled, which in turn suggests a certain level of coercion by sections of the population.

All the above evidence suggests that while the vast majority of hillforts were constructed within the ideology of defence, this was not always the case. As a number of hillforts have been constructed on the side of hills or are overlooked by higher ground, both of which appear to diminish their defensive capabilities. In the latter case the two most widely quoted examples, Chesters, Drem, East Lothian and Kerr's Knowe, Peebleshire (Bowden and McOmish 1987, 78; Collis 1996, 88) clearly lay outside the hillfort dominated zone. Neither hillfort has been excavated: they are typologically dated. If they do date to the Iron Age then the outstanding 'military' question has to be what type of warfare were they built to defend against? There appears to be no archaeological evidence of sling use or archery use in the Pre-Roman Iron Age of Scotland, which would leave the spear or hand-thrown natural stones as the only potential distance weapons. As neither of these has a range in excess of 40m, the 'problem' of being overlooked by higher ground may be a product of our perception of war rather than a practical consideration of the builders. After all, Caernarvon Castle is also overlooked by higher ground some 300m from its Great Court (Payne-Galloway 1981 [1903], 23). It would be difficult to postulate that Edward I's military architect, Master Bertram, was a military incompetent, who failed to notice this detail. It is easier to accept that, at the time of the castle's construction there was no military disadvantage to being overlooked, and that the Welsh did not have the military capability to threaten the fortification at this distance.

Hillforts were also constructed on the downward side of a slope or hill, with clear ground higher than the outer defences. None of these forts are within the hillfort dominated zone. They are concentrated in Cornwall and mid-south Wales (Forde-Johnston 1976, 83-6). The vast majority of the latter are small (enclosing less than 1 ha.), and many of these may date to the Irish settlement of the sixth century AD (Laing 1975, 92-9). Of the few that have been excavated, Walesland Rath, Pembrokeshire proved to be a multi-phased site enclosing 0.25ha, with occupation lasting from *c*. 100 BC to the Roman period. The site appears to be primarily a defended (enclosed) farmstead (Wainwright 1971). To use these examples as evidence that hillforts in the hillfort dominated zone are not defensive is to deny the regionalism that is such a characteristic of the Iron Age. The hillforts described above may not have served a primarily defensive function, but this is impossible to infer, without relating them to the potential regional social conditions that prevailed when they were constructed.

Despite what appears to be a significant function of defence, hillforts within the hillfort dominated zone pose further problems in relation to their use. Despite their size, they do not appear to have been heavily occupied (Cunliffe 1984b, 560; 1995, 89-90). Many enclosed areas that would be unsuitable for the construction of habitable structures. Examples include Caer Caradoc, Shropshire, where a natural outcrop of rock containing a cave has been enclosed, leaving little level ground for the construction of buildings; or the Breddin where the area available for occupation has been calculated to be only 20% of the total enclosed (Musson 1991, 7). Where reasonable areas of the interior have been excavated, structures that can parsimoniously be interpreted as storage facilities, either pits or 'four-post structures', occupied a large percentage of the space available (Gent 1983). Where occupation has left visible traces on the surface (e.g. Conway Mountain, Griffiths and Hogg 1956), this has been interpreted as evidence of relatively intensive use. However, since the excavation of many of these sites is limited or non-existent, the chronological relationship between the individual structures cannot be reconstructed, and any estimate of the level of occupation must be seen as hypothetical.

At Danebury, the estimated population varies between 100-300 (Cunliffe 1984b, 560), and it has been suggested that the fort was only used on a seasonal basis (Stopford 1987). Population estimates are of course notoriously difficult, but a comparison between Danebury and the Pucara of the Mantaro Valley in Peru highlights the potential low level of occupation of Middle Iron Age hillforts. In the Mantaro Valley, during the Wanka II (AD 1300-1450) phase of occupation (essentially a Neolithic or Copper Age technology), a number of sites were enclosed with strong defensive walls against the raids of adjacent polities. Because of the relatively recent nature of the occupation, reasonable estimates of population can be made. Enclosures of a similar size to Danebury (i.e. between 4.1ha. and 6ha.) have estimated populations of between 369-540 persons and 615-900 persons (D'Altroy 2001, 89, Table 4.2). Even the minimum estimates are greater than the maximum suggested for Danebury, and the maximum estimates are beyond any suggested for hillfort occupation. This would suggest that hillforts were not intensively occupied, though some long term occupation almost certainly took place, at least within the Wessex region. Another defining feature of hillforts is that the low levels of occupancy do not seem to relate to individuals who had access to a wider variety or better quality of material goods than those living outside the hillfort. There appears to be no significant gap in access to archaeologically identifiable sources of wealth. One exception to this appears to be storage. Hillforts, especially those on the Welsh Marches, do seem to have vast storage capacity (Stanford 1974; 1981). This may represent access to power through the use of staple finance to support a stable, if relatively politically weak, elite (D'Altroy and Earle 1985, 188; Earle 2002, 342-3).

Using large areas of a hillfort interior for storage would reduce their effectiveness as places of refuge during times of stress. The space available for people and, perhaps more importantly, livestock would have been limited, particularly as some storage pits seem to have been left open to partly silt up naturally (Poole 1995, 250-1; Smith 1977, 45). These would have represented a real hazard to both people and livestock.

The lack of water storage, or the enclosing of natural water supplies, would also tend to suggest that hillforts were not designed to withstand any period of 'siege'.

Humans can subsist with limited water supplies for considerable periods, but animals, and in particular cattle, need significant quantities of water every day. Modern dairy cattle require up to 180 litres every day (Stansfield 1990, 56). Clearly the smaller, almost dwarf, Iron Age cattle (Clutton-Brock 1987, 68) would have required significantly less water but, even if they needed say 20 litres a day, a small herd would still require large quantities of water, and there is no available evidence that such provision was made within the vast majority of hillforts, in the form of dew ponds, cisterns or the like. A rare exception to this is the Buckbean Pond within The Breiddin. However, the cistern here, created in approximately 350 BC, shows no signs of its steep edges being eroded or damaged by cattle (Musson 1991, 83-9). Another exception is Burrow Hill Camp, Shropshire (Musson and Watson 1993, 28), as its banks enclose a natural spring.

Outside of the hillforts, the landscape appears to have been relatively densely occupied with two main types of settlement, consisting of open or unenclosed 'villages' and enclosed farmsteads (Hamilton and Gregory 2000, 61; Hingley 1984, 78-9). It would appear that these two settlement types occupied differing ecological zones (Hingley 1984, 78-9). Open settlements were generally in low lying valley bottoms whereas enclosed settlements tended to be in upland areas. These distributions are not mutually exclusive and both types of settlement pattern occur within both broad farming zones, though in significantly reduced numbers. Surveys would tend to suggest that these settlement types were widespread throughout the hillfort dominated zone. Given the longevity of the majority of excavated examples, often in the region of five hundred years, it would be parsimonious to suggest that many of the known settlements were occupied contemporaneously. If this was the case, there is no real evidence for areas between polities being unoccupied and left open as a 'buffer zone', as suggested by Tacitus for the early Germans (*Germania*, 16) and by archaeological evidence from the Cuzco Valley, Peru (Bauer 2004, 76-7, Map 8.3), or through natural features as in Peru (Burger 1992, 12), Greece (Hanson 1998, 88-9) or Southeast Asia (Gibson 1990, 126

Open Villages

Open villages, as their name suggests, seem to have no defensive perimeters or were only partly enclosed by a ditch and bank. This has clear implications in relationship to how warfare functioned in the Middle Iron Age. Most of these open settlements appear to have been continuously occupied from the Early Iron Age to the Late Iron Age. Though there is evidence of rebuilding, there is no evidence of violent destruction.

How much of the population lived in such settlements is impossible to estimate. Most open villages contained a number of houses, but the chronological relationship between these is often problematic, so an estimate even of the total occupation is often difficult. At Winnall Down, nine houses were constructed during Phase 4 of the site (the Middle Iron Age), a maximum of seven of which could be contemporary (Fasham 1985, 18). The population at its maximum has been estimated to be about 92 persons (Fasham 1985, 130). Winnall Down was originally an enclosed Early Iron Age settlement, approximately 0.4 ha. in area (Fasham 1985, 11). The original ditch was V-shaped with a depth varying between 1.3 and 3.35m, the shallowest section to the eastern side where the ditch was also particularly wide (*ibid,* 9). The ditch showed no signs of recutting and appears to have been allowed to silt up naturally, including some deposits of 'rubbish' (*ibid,* 11, see also Hill, 1995a, Chapter 8). By the Middle Iron Age, the ditch had silted so completely that a house was constructed across its course. Once the ditch had filled there was no clear definition of the enclosure, though a concentration of pits to the east of the house may have provided some form of demarcation in this direction. The presence of neonatal lambs among the faunal remains indicates that sheep were reared close the settlement (Hambleton 1999, 87-8; Maltby 1985, 138; 1995, 85).

Evidence that open settlements could achieve material wealth, and should not be seen as being at the bottom of a social hierarchy, is shown by Slonk Hill (Hartridge 1977). During Phase 2 of the settlement (Middle Iron Age), evidence for iron working in the shape of slag, and of bronze working in the shape of moulds, parts of two crucibles and copper/bronze slag (*ibid,* 91) was recovered. This indicates that its occupants not only had access to such goods, but also were able to manufacture them themselves, which tends to weaken any arguments for authority based on manipulating access to 'prestige' goods. The evidence from Eldon's Seat, Dorset (Cunliffe 1968), also seems to support the above conclusion. Here, an open village was occupied (though not continuously) from the Late Bronze Age to the second century AD. The area excavated constituted part of a larger site, and included an area where Kimmeridge shale was worked during the Middle Iron Age (*ibid,* 225-26, Plate VI, Figure 21). Manufacturing of prestige items in an open settlement would certainly seem to infer that the lack of clear defensive structures was not a deterrent to the accumulation of significant quantities of a relatively rare material.

At both Boscombe Down East (Stone 1936), and West (Richardson 1951), partly enclosed settlements were constructed. At Boscombe Down East the V-shaped ditch enclosed three sides; its depth varying from 1.2m to 1.25m, with a width of 1.8m to 2.1m, from the existing ground surface. Although the excavator believed that the bank had been ploughed away (Stone 1936, 466), the ditch section (*ibid,* 471) is clearly symmetrical suggesting that either there was no bank, or that it was positioned far enough away to have no effect on the natural silting of the ditch. At Boscombe Down East the site was only enclosed (on its northern and southern sides) during the Late Iron Age (Richardson 1951, 134). Throughout the Early and Middle Iron Age the settlement seems to have had no defensive barrier. It is possible that both these settlements could have been fully enclosed by hedges or the like, but the use of ditches and possibly banks seems to indicate that there was a desire to define a space rather than to actively defend one.

Enclosed Farmsteads

Enclosed farmsteads tend to cluster in upland or areas of higher land. They also tend to occupy slopes facing south or west. They are normally surrounded with a continuous ditch and presumably a bank. Some, such as Collfryn, Powys, (Britnell 1989) have multiple banks and ditches. Many also have elaborate entrance ways, with antennae stretching away from the single entrance (e.g. Little Woodbury (Bersu 1940) and Gussage All Saints (Wainwright 1979)). The variety of shape and form is really only conflated by the single feature of an enclosing ditch(es). Within this section, only very limited reference will be made to site typologies, as they have little bearing on the military value of these enclosures.

Despite the apparently formidable nature of the enclosing ditch and banks of many of these sites there is a contradiction in their construction. At Little Woodbury (Bersu 1940) the ditch is described as being *nowhere too wide for a human being to jump, from the more gently inclined upper part over the steeper lower part* (*ibid,* 35). The ditch also displayed a feature that is common among enclosed settlements. It was, at least in part, was rapidly backfilled following its excavation. The material for this work seems to have been supplied by dumps on either side of the ditch (*ibid,* 39), potentially indicating that the provision of a defensive bank was not considered part of the original construction. From observations relating to the weathering of exposed chalk, Bersu 'guesstimated' that this backfilling could have occurred within eight weeks of the ditch being dug (*ibid,* 39), yet the settlement has a long chronological occupation following its definition by the ditch (at least 300-100 BC – *ibid,* 99). Such backfilling seems to have been a regular feature in the social processes relating to enclosed settlements. At Owslebury (a banjo enclosure), the ditch varied in depth from 1.5m to 2.1m (Collis 1970, 248), yet the western part of the ditch was backfilled *not long after it had been dug* (*ibid,* 248). The same pattern is observed at Bishopstone. After an initial period of silting, the ditch was in parts deliberately backfilled using large lumps of chalk (Bell 1977, 55-6). The excavator records that there was no evidence of a bank, though some of the fills suggest that it stood on the inside of the ditch. This suggests that the material used for backfilling the ditch was derived from the presumed bank. At Pimperne (Harding, Blake and Reynolds 1993) the entrance feature at the eastern gate was particularly elaborate, possibly consisting of a double wattle fence capped with flints (*ibid,* 11). However, the northern side of the enclosure showed very different activity – the ditch here was shallower than on the eastern side and, following a short period of natural silting, the ditch was once again deliberately backfilled with densely packed chalk lumps (*ibid,* 19). There is also some evidence of ritual activity during this backfilling as fragments of human skull(s) were recovered from the upper levels of the fill (*ibid,* 20). This ritualistic theme may also been seen at Winnall Down (Hill 1995, Chapter 8).

Although deliberate backfilling of ditches seems to have been a widespread practice, it was by no means universal. At Mingies Ditches, Oxon. (Allen and Robinson 1993), the ditches that surrounded the settlement showed no definite signs of recutting, or any traces of more than one phase of construction and upcast (*ibid,* 26). The ditches here appear to have been excavated and then left to silt up naturally with no attempt at maintenance. Given the settlement's long period of occupation (400-10 BC), this would probably have led to the ditches being little more than depressions by the time the settlement was abandoned, if they were visible at all. At Old Down Farm, Andover (Davies 1981), the silting of the initial ditch was so complete by the Middle Iron Age that the settlement was effectively open (*ibid,* 122). This process appears to be repeated at Meon Hill (Liddell 1933), where the ditches show no signs of recutting or cleaning and appear to have been allowed to silt up naturally (*ibid,* 132). The cross section drawn by Liddell (*ibid,* 140) is symmetrical, suggesting that there was no bank surrounding the site. Even the triple-ditched and banked enclosure at Collfryn, Powys, (Britnell 1989) shows no signs of maintenance from its first phase of construction during the third century BC until the ditches were recut in the first century BC. The ditches appear to have been allowed to silt up naturally despite their apparent use as drainage features for the site (*ibid,* 109). The banks appear to have been wide (4-5m) but not high (1-1.5m) (*ibid,* 92). The excavator feels that the lack of maintenance would have rendered the defences more or less useless after one generation (*ibid,* 92). Assuming 25 years per generation, this means that the site would have been effectively indefensible for approximately seven successive generations.

At Fisherwick, north of Birmingham (Smith 1979), though there is evidence of one recut at some point during the third century BC (*ibid,* 22), this was the only evidence for maintenance during the next three hundred years. There is no clear evidence for either an internal or external bank (*ibid,* 22), and the creation of major features close to the edge of the ditch would tend to concur with this view (*ibid,* 28, Figure 9). There was some potential evidence that the site may have been surrounded by an hedge. The waterlogged remains recovered from the ditch include species that are suitable as hedging plants (e.g. blackthorn, hawthorn, willow, oak and elder). The remains of the blackthorn showed cut marks that would be consistent with either hedge trimming or laying (*ibid,* 24). Such a hedge would be a suitable deterrent for cattle or potential predatory wild animals, but would not in all probability have acted as a reasonable defensive feature against human beings. Such hedges tend to have a maximum height of 1.5m (personal observation) and, though defendable, offer as much a hindrance to the defenders as attackers.

At Gussage All Saints, the non-defensive message is reinforced by the use of an external bank, to the 'rear' of the settlement, away from the elaborate entrance features (Wainwright 1979, 3). The existence of an external bank is suggested by the silting of the ditch (*ibid,* 3), and the position of features near to its internal edge. The ditch sides were unweathered, and the section showed no signs of subsequent cleaning following its initial construction (*ibid,* 3). The ditch also seems initially to have been causewayed (the entrances were simple gaps with no evidence of gates or the like). Although some of these were removed during the partial recutting of the ditch in the Middle Iron Age, the bank appears to have remained external.

The settlement evidence from the Middle Iron Age of the hillfort dominated zone seems to suggest a polarised society. On the one hand, the hillforts appear to be evidence of an ongoing inter-polity aggression, against which the community or parts of it needed considerable maintained defence. The ditches at Danebury were cleaned between twelve and seventeen times (Cunliffe 1995, 23) and still present a formidable obstacle. The potential lack of material difference between hillfort and non-hillfort sites, at least in type if not in actual quantity (Hill 1995a; see Cunliffe 1995, 93-4 for a differing view relating to the quantification types), does not necessarily encourage the ideology that hillforts were the top of the social scale. It suggests they were rather part of a relatively socially level whole, perhaps based on staple finance. The other settlement evidence, with its long term occupation of both open and enclosed sites, suggests a relatively peaceful landscape where the requirements of defence played very much second place to social expression. This is perhaps best illustrated by the deliberate backfilling of the ditches that surrounded many of the enclosed farmsteads. This would have rendered them useless in terms of defence, yet the elaborate entrances to many of these sites with their outstretched antennae are suggestive of a social concept, where defence was a valued ideology. The lack of ditch maintenance surrounding these settlements suggests that it was the definition of space was all-important.

Morphological Evidence

At present there is little archaeologically identifiable evidence of the funerary rites for the majority of people in the Middle Iron Age of the hillfort dominated zone. There is some evidence for a potentially local tradition of socially levelling inhumation (Hey *et al.* 1991, 560). Other evidence from hillforts does not obviously constitute a normative burial practice (*pace* Fitzpatrick 1997, 82). This means that any evidence for trauma on the bodies recovered from hillforts is unlikely to be representative of the mass of the population, and may represent special cases, buried in a particular way with unusual or unique rites (Carr *et al.* 2005).

The evidence for the Middle Iron Age of the hillfort dominated zone comes primarily from Danebury. Evidence from Cadbury (Alcock 1972b, 105-6) and Bredon Hill (Hencken 1939, 55-9) dates to the Late Iron Age, as does the majority from Maiden Castle (Wheeler 1943, 348-56). Given the fact that no known Middle Iron Age skeletal remains, have been found in a context that would suggest they had been buried after a battle or similar violent engagement (e.g. mass burials showing high levels of trauma), morphological evidence can, at best, be a guide to the use of a particular weapon type against a particular individual, rather than confirmation that these were used in warfare.

From Danebury there is evidence for the use of spear (Hooper 1991, 430, Table 8.1) and, at least nine individuals suffered trauma that is most parsimoniously interpreted as sword wounds (Carr and Knusel *pers. comm.* 2005; Hooper 1991 428-9), and two more have recently been identified (Carr *et al.* 2005, 171). The only other potential evidence for sword use is from Old Down Farm, Andover (Davies 1981). One trauma that appears to be missing is the characteristic cranial depression caused by a sling stone (Wells 1964, 49). However, as Wells observes (from Peruvian examples) these wounds are almost always well-healed, indicating perhaps that sling wounds as such were rarely fatal, and thus may not have occasioned the 'special treatment' of the body that seems to be associated with skeletal remains from the Middle Iron Age of the hillfort dominated zone.

CHAPTER THREE

THE SOCIAL SIGNIFICANCE OF WEAPONRY

> *'Indeed, as a rule, he compelled all the subject population who had been disarmed to practice the use of the sling: it was, he considered, a weapon for slaves* (Xenophon *The Education of Cyrus VII*, Chapter 4,15)'.

That artefacts carry particular messages and levels of social significance has long been recognised not only by ethnographers and anthropologists but also by archaeologists (Hodder 1982a; 1982b). In the latter case, obviously, much of the direct meaning is lost, particularly in the prehistoric periods. We will never know the true depth of significance relating to the deposition of a single glacial pebble beneath a broken potsherd recovered from pit 2476 at Danebury (Poole 1995, 262).

Weapons also carry messages, but as with any artefact, they are not inherent, but must relate to the society within which they existed. Swords for example are often uncritically seen as representations of elite power. However, we must be wary of applying our own ideological preconceptions to artefacts from the past.

To attempt any reconstruction of the potential social significance of weaponry, there must be a reliance on ethnographic and historical records that portray these weapons within their own cultures. This will enable the exploration of these artefacts within a social context radically different from our own, and should present complex and competing ideologies of weapon use. In an attempt to produce an interpretation of the detailed nature of the social significance of weaponry within specific cultural contexts, four ethnographic/historical[2] studies have been undertaken. If at all possible, these descriptions should be internalised and should not be written by outside observers who bring their own prejudices and ideologies to bear. This was not always possible. The descriptions provided by classical authors of 'Celtic' peoples have been deliberately excluded from the case studies. The stereotypical imagery applied to the descriptions of the 'barbarian' has long been recognised (Hill 1989; James 1999) and uncritical application of these descriptions to European prehistory has, in many ways, hampered the exploration and development of alternative models relating to these societies. In particular the use of Roman and Greek descriptions of warfare have been generally applied to warfare models of the Iron Age across Europe both temporally and spatially (Cunliffe 1991, 488-97; Jope 1961; Harding 1974, 70-2; 1977; Piggott 1950; Pleiner 1993, Chapter 2). This has led to a situation where modelling of warfare has become stagnant.

This methodology essentially follows that developed by Earle (1997) in his exploration of the politic of 'chiefdoms'. By studying in some detail a series of ethnographic and historical case studies, Earle hoped to demonstrate patterns of social behaviour that are cross-cultural. By limiting the case studies relating to the social significance of weaponry to pre-state sedentary agriculturists, many of the pitfalls presented by the cross-cultural generalisations of processual archaeology should be avoided. This, by its very nature, excluded those cultures or historical periods that used mechanical weapons. The employment of mechanical weapons tends to make warfare a technical pursuit where generals become significantly more important, not as direct leaders of men in battle showing their own bravery, but as strategists, using intellect and experience to conduct their armies' movements (Wheeler 1991; Hunt 1998, 195).

This is not, of course, to deny the possibility of other interpretations based on other evidence, but the application of specific ethnographic/historical analogy should allow a reasonable reconstruction of warfare in the Middle Iron Age of the hillfort dominated zone to be attempted. As with any archaeological interpretation, there is, of course, a level of generalisation. It is not possible to create a model without such a tool (Dark 1995, 60-1), and it is within this framework, that certain assumptions relating to the nature of warfare in the Middle Iron Age will be made.

The Case Studies

The four cultures chosen to provide evidence for the initial generation of a model for the social significance of weaponry, are: early Mediaeval Ireland (AD 500-1000), Late Geometric and Early Archaic Greece (850-700 BC), Early Iron Age Israel (1000-750 BC) and Late Intermediate Peru (AD 1100-1450). This may appear a 'rag-bag' of cultures, chosen at random to produce a certain set of evidence. However, they should provide not only conflated evidence but also contrasting ideologies that should illuminate the potential social significance of weaponry, while highlighting the dangers of projecting this evidence too literally into the past.

Early Mediaeval Ireland, Late Geometric and Early Archaic Greece and Early Iron Age Israel all have one significant feature in common. They all present (through iconography, literature, etc.), internal representations of societies at pre- (or emergent) state level. These sources are often contained within a narrative sequence, and were developed for internal consumption by a specific audience. This audience was in all probability, initially at least, an elite who may have been attempting to

[2] Ethnographic and ethnography are used here in a very loose sense, as it will be clear that the texts used are primarily of a mytho-legendary nature.

consolidate an extension of their power, at the expense of older more established systems of control. As the development of elites in all these societies appears to have been rapid, the messages contained within the narratives may relate to the establishment of an ideology of exclusion. The narratives seem to contain specific messages, which appear to illustrate the distinct difference between the elite and non-elite segments of society. It would not be an unreasonable inference that this level of message was required within these specific cultures, as their elites were creating their own identities without significant reference to existing models or to an established past; in many respects they were creating themselves in their own image. Further, it would appear that, in opposition to this ideology, the narratives also show the social identifiers that were to be associated with those sections of society that were being excluded from power. It is probable that these represented an older form of polity and it would appear from the archaeological evidence from all three cultures that this was a less hierarchical society, and one where wealth and, presumably power, was on a more even plane.

The attempts at providing a social meaning or temporal locale for the literary construct as a whole are beyond the scope of this discussion. However, the view expressed by Raaflaub (1998, 170), that tradition only survives if it has relevance to the audience that it is being presented to, is one of the primary theoretical premises behind the case studies. Therefore, any metaphors expressed within the Táin Bó Cuailnge, the Iliad, the Odyssey and other mythological 'histories' must have been not only understandable to the audience that they were performed for but also of relevance to their world. Tradition has the ability to change the outcome of any element, from major political events to the details of everyday life, to suit the existing political or social atmosphere, or to create an idealised past to allow the present social structure to achieve a depth of existence that it never in reality had. The use of traditional stories in an attempt to understand a past society is inherently dangerous. However, if the distance from an authentic past is accepted, certain elements can be used to reveal what the audience, to which the tradition is being presented, thought concerning its relationship not only with the past, but also to present conflicts within that society. By presenting certain elements of society as projected from a distant past, the creators of tradition enable the changes that are in reality occurring at the time to be legitimised.

The older structures of society would of course have reacted to the growth of the elite and changed relative to their expansion and the concentration of power within a specific and exclusive group. The representations of the non-elite by the elites may have been a caricature of what they wished to project onto the excluded members of society.

The elite appears to have attempted to distinguish itself from the mass of the population by utilising specific artefacts. It may be possible therefore to identify specific weapons with specific hierarchical structures. By their very nature, these artefacts must have had a limited circulation or must have required the user to have gained specific skills that necessitated long-term training, which would have been incompatible with a subsistence economy. The use of the sling with any degree of accuracy requires considerable practice. This, however, can be gained as part of everyday agricultural pursuits. The use of a throwing spear or javelin also requires considerable levels of practice. This is more likely to have been incompatible with normal farming activities, as the use of a throwing spear or javelin is limited in essence to killing. The use of weapons that required specific training and practice outside the subsistence economy would in itself have removed them from mass use. The context of the social significance of weaponry is illustrated in these three cultures. In fact, in early Mediaeval Ireland and Archaic Greece (Finley 1979, 113-8; O'Rahilly 1976,143; van Wees 1998), it would appear from the literature that has survived that warrior status was of the utmost importance, and clearly defined the elite from the rest of the population.

In the context of Middle Iron Age warfare, these societies also display the range of weapons that, from the archaeological record, were available to Middle Iron Age cultures. They do not, however, contain weaponry that would have been outside the technology of Middle Iron Age peoples to manufacture. Early Mediaeval Ireland has particular relevance, as this has often been used by other authors as a source of information for Middle Iron Age warfare (Cunliffe 1986, 81, 86; Hamilton 1968; Harding 1980, 8; Jackson 1964, 15-18; Pleiner 1993, 31-3).

Late Intermediate Peru is, of course, the odd one out. The written record relating to this period is almost entirely European, except the writings of Huamán Poma, a native curaca (lord) who, in the late sixteenth/early seventeenth centuries, composed a letter to Philip III of Spain (1978 [1613]). In his letter, he explained the ills of Peru and outlined a brief history of the country. Perhaps most important, though, are his series of ink drawings showing the Incas and pre-Inca peoples in unacculturated dress, using pre-Hispanic weapons. The lens of sixteenth century Spanish ideologies is, of course, a significant factor in any attempt to interpret this material. However, it has become clear recently that although working to their own agenda, the Spanish chroniclers recorded much information that they did not fully understand within a native context, which allows the careful reconstruction of many pre-Hispanic practices and ideologies (Salomon 1986; Urton 1990; Wachtel 1977; Zamora 1988). Late Intermediate Peru has one specific advantage in attempting to construct a model for Middle Iron Age warfare, as alone of the case studies these societies used hillforts (Earle *et al.* 1980; Earle *et al.* 1987, 10-11; Earle 1997, 56-61; Hastorf 1993, 77; Poma 1978 [1613], 27-8). Often achieving proto-urban levels of occupation, these settlements may at certain levels be analogous with the hillforts of southern England. Also unlike the three Old World case studies, in Late Intermediate Peru the social structure was dominated by an ideology of community (Hastorf 1993, 213-6; Mayer 1985, 74-7; Isbell 1997, 98-100). Zinchis (elected war leaders) were extremely weak and had little or no coercive power (Brundage 1963, 13; De Gamboa 1999 [1572], 38-9; Earle 1997, 114-7). This would appear to be directly comparable with the archaeological evidence for the Middle Iron Age of the hillfort dominated zone, which would strongly suggest that there is no significant difference in access to material goods for any settlement type (Cunliffe 1995, 90; Hill 1996, 104-7). This in turn has been interpreted as representing a culture where power was not vested in the hands of a few individuals, but was more accessible by an extensive group(s) within that society.

The cultures in the case studies display a reasonable distribution of social contexts. Late Intermediate Peru represents a weak elite where individual power was limited and contextual. The social context of Early Mediaeval Ireland presents a situation where individual power was becoming more embedded within a particular social group. Late Geometric/Early Archaic Greece and Iron Age Israel I & II are most probably incipient states. This range will allow the social context of weaponry to be compared with the two current competing models of society within the Middle Iron Age of the hillfort dominated zone. The traditional model assumes that there was some from of control vested in an individual, normally described as a chief. The primary source of evidence for this interpretation is the sense that hillforts could not have been built without the use of coercive force and that those individuals living within the fortifications represented an elite. The other interpretation assumes that, since the material differentiation between those living within hillforts and those occupying the surrounding countryside is minimal, the social differentiation must also have been limited if not non-existent and that society was therefore more egalitarian (Hill 1995b; 1996). The expression of warfare is also strongly associated with particular types of society and, in turn, the types of weapons that are chosen to be used. The results of the exploration of the social significance of the weaponry used within the Middle Iron Age of the hillfort dominated zone should reflect the prevailing social structure of the polities that constructed and maintained hillforts.

The tacit assumption that runs through the case studies is that within Europe, it was not until the end of late Middle Ages that weapons were chosen for their mechanical efficiency, before this time frame weapons were primarily a social or cultural expression. Had the former been the case, a particular weapon or weapon type would rapidly have dominated the arena of war. Clearly, this occurred during the Renaissance, when, during the two centuries 1500-1700, gunpowder weapons effectively dominated the battlefield, replacing all mechanical and muscle-powered distance weapons. The pike became the principal shock weapon (Gush 1975, 10) only to be succeeded by the bayonet attached to the musket (in effect a spear) (Stone 1961, 107).

Nearly all forms of weapons found in the Bronze Age of Europe, the sword (Brown 1982; Colquhoun and Burgess 1988), spear (Ehrenberg 1977; Greenwell and Brewis 1909), bow (Clark 1963; Mercer 1970), sling (Vutiropulos 1991), halberd (Harbison 1969), axe (Roe 1966) and shield (Coles 1962) continued to be manufactured, and used in war until *c*. 1450 AD, albeit in new materials and forms. By *c*. 300 BC, with the development of ring mail (Randsborg 1995, 26-7; Zirra 1993, 383) and torsion weapons (Marsden 1969, 12), the full panoply that would be available in Western Europe for the next 1500 years had evolved. As important as the lack of new weapon types, no weapon was abandoned during this period. It is only towards the end of the fourteenth century that the sling seems finally to have vanished from war. This is probably more to do with the rise of semi-professional armies than with its effectiveness (though the increasing use of plate armour amongst a growing proportion of troops may equally have been one reason for the sling's demise).

This is sufficient evidence to suggest that no single weapon was capable of total dominance during this period. Even the longbow, that often espoused 'super weapon' of the mediaeval world (Hardy 1995), did not achieve widespread use in Europe, despite its success at Crècy, Agincourt and Poitiers. It must also be remembered that, despite the longbow's victories, the French did win the Hundred Years War without widely adopting the weapon. The aristocratic nature of mediaeval warfare is likely to have had a severe limiting factor relating to the adoption of the longbow by other nations. No such limitations were evident when firearms evolved into practical field weapons. This in turn would strongly suggest that the choice of weapon, before the introduction of effective gunpowder weapons, was not a matter of pure mechanical efficiency, but one of social expression. The introduction of muscle-powered mechanical weapons in the form of the crossbow, and some siege weapons (Marsden 1969, 5-12; Ober 1987), obviously had an impact on the nature of war, as did the increasing sophistication of armour and defences, but no innovation was universally adopted until the introduction of gunpowder weapons.

Weapons, far from representing merely changes in warfare as a direct reaction to increasing effectiveness, or as part of some prehistoric 'arms race' (*pace* Osgood, Monks and Toms 2000, Chapter 7), are a component of a people's world view. This may be both a conscious expression of elitism or an unconscious expression of exclusivity or inclusivism, depending on the situation. In either case, weaponry in all its combinations or its general absence, should relate to a particular culture's identity. Like all artefacts, weapons are, and in all probability always have been, poly-representational. They are not simply weapons nor can they simply be described as having a dual function, practical and expression of power. They can easily express gender, age associations, social standing, as well as power and prestige and many of these meanings will be in operation at one and the same time and/or only be visible during certain situations or to different groups.

In each of the four case studies, weaponry is explored to see if there are any underlying trends that may articulate a commonalty of expression. As stated, the objective is to avoid the pitfalls of generalisation by relying on a core group of cultures that share a similar economic basis. As with any analogy, there are inherent dangers with this approach. The material remains of a culture could represent a unique and utterly unrecoverable social expression. However, such pessimism would all but destroy any attempt at understanding the prehistoric past. Without attempting a detailed examination of the social significance of weaponry, we are forced either to use typology as a descriptive base or attempt to understand weapons and war using the experiences generated in modern times. The latter situation is untenable, and the former negates any in-depth exploration of the past.

Case Study 1

Early Mediaeval Ireland (*c.* AD 500-1000)

Early Mediaeval Ireland has often been used as a primary source of information for the creation of models relating not only to prehistoric Iron Age warfare, but also to society in general (Alcock 1972b, 34; Jackson 1964; Megaw and Simpson 1979, 490; Ross 1967). The most famous of these social explorations is Professor K. Jackson's *A Window on the Iron Age* (1964). These ethnographic correlations have been heavily criticised, and this has led to many of the concepts being removed from more modern considerations of the Iron Age in Britain. However, Early Mediaeval Ireland has remained one of the core sources relating to the study of warfare (Cunliffe 1986, 81, 1991, Chapter 18; Pleiner 1993, 24-33). This use of Early Mediaeval Ireland as one of the core analogies for Iron Age Britain makes this case study of particular importance. Although Early Mediaeval Ireland carries strong messages relating to the social significance of weaponry in a pre-state context, its social context makes it an unreliable source for analogy relating to the Middle Iron Age of the hillfort dominated zone, in general.

Early Mediaeval Ireland has an abundant literature relating to mythological characters who inhabit a concrete landscape, and the main mythological texts, the Táin Bó Cuailnge and other heroic or kingly cycles, have an extensive critical literature attached to them. Much of this critique revolves around the internal meaning of the Táin Bó Cuailnge and how it relates either to earlier periods of prehistory, both in Ireland and elsewhere within the British Isles, or how it is structured to relay a particular political viewpoint within its own context (Aitchison 1994; O'Riain 1994; Kelleher 1971). The question of its validity to earlier periods has been most vigorously challenged by Mallory (1981; 1986), who, by using particularist methods, has identified various elements (the use of silver, size of swords, different clothing etc.) that seem to date from the eighth or ninth centuries, and certainly seem more akin to the Norse world of that period than that recovered archaeologically from earlier periods in Ireland.

Political and Social Structure

Early Mediaeval Ireland was divided into numerous competing polities (Byrne 1987; Mac Niocaill 1972, Chapter 2), numbering perhaps as many as one hundred and fifty at any one time (Byrne 1987, 7). These petty kingships were arranged into provincial hierarchies within the five provinces of Ireland. There was a mythological assumption that there should be a high king operating as an overarching authority over all the kingships, with his seat at the mythological centre of Ireland, Tara. Although this position had been claimed by the Ui Neill as early as the seventh century (Byrne 1987, 53), it only become a short lived political reality in the eleventh century, under Brian Boruma mac Cennetig of the Dal Cais kings of Munster (Ó Corráin 1972, 125).

The archaeology of Iron Age (pre-Christian) Ireland is extremely fragmentary, leading one eminent Irish archaeologist to entitle them 'the invisible people' (Raftery 1994, Chapter 6). Although no detailed picture of the society that existed before the Early Mediaeval period is possible, certain trends can be seen. There appears to have been a significant reduction in human activity, if not population, between *c.* 300 BC and *c.* AD 300 (the precise dates vary from site to site; for a tabulated comparison see Stout 1997, Figure 8). So much so, that the palynological records recovered from the extensive peat bogs show an almost universal regeneration of hazel scrubland (Jelicic and O'Connell 1992, 119-40; Weir 1994, 96-7). This is followed by a rapid upsurge in human activity and the clearance of this initial forest regeneration, resulting in the reappearance of the typical weeds of cultivation (Weir 1994, 91-2).

It is during this period of increased human activity that ringforts, one of the most numerous archaeological features in the Irish landscape first appeared (Stout 1997, Chapter 2). Ringforts (normally a single-ditched enclosure about 30m in diameter) tend to concentrate in areas with good to average soils on the higher ground, but always avoiding lower, poorly drained, riverine areas. Few, if any, are in clearly defendable locations. The ringfort is normally associated with the rise of the free farmer, often referred to in the law texts (Kelly 1998, 364).

In contrast to the ringfort, hillforts in Ireland, like those elsewhere in Britain, tend to occupy clearly defendable locations. The apparent similarity between hillforts in Britain and Ireland has led to those in Eire being tentatively dated to the Iron Age on purely typological grounds (Raftery 1972, 51). However, unlike the British examples, there are no hillforts that can be dated unequivocally to this period in Ireland. Although a relatively common feature in the landscape, only a limited number have been excavated (Cotter 1994; Condit 1995; Grogan and Condit 1994; Mallory 1991; Raftery 1969).

The majority of these appear to have been constructed during the late Bronze Age. Even the massive stone promontory forts that are such a significant feature of the west coast, have produced radiocarbon dates that indicate that they too were commenced towards the end of this period, although there is some evidence of later occupation (Cotter 1994, 28). The majority of hillforts appear to have been abandoned before or immediately after the onset of the Iron Age (*c.*400 BC). Only Freestone Hill, Co. Kilkenny (Raftery 1969, 101) has produced clear evidence of occupation dating to a later period (*c.* AD 450), and, even here, there is evidence that the original construction may date to the Bronze Age (Raftery 1994, 59). Evidence from Haughey's Fort suggests that hillforts were occupied by elites (Mallory 1991). Small fragments of gold leaf, beads and a single gold stud measuring only 4.5mm in diameter, seem to indicate the support of craft specialisation that is normally associated with elite social group activity (*ibid,* 20, Figure 15).

It is possible that the introduction of Christianity was primarily responsible for the major changes that appear to occur between AD 600 and 900 (Mytum 1992), although similar social change must have been occurring well before its arrival, as the construction of ringforts as early as AD 250 demonstrates. The hierarchical nature of the Church and the construction of proto-urban centres like Clonmacnoise and Armagh (Hodges 1982, 47-9) would have served to accelerate the acquisition and legitimisation of power in the hands of an emergent elite (Mytum 1992, 105-8). The construction of a new landscape-defining feature (the ringfort) represents a real break with the traditions of the immediate pagan Iron Age past.

There is only limited evidence for elite activity in much of Iron Age Ireland. The La Tène metalwork that has been discovered has a distribution primarily limited to the North and West, and can be tightly dated to either side of *c.* 100 BC–*c.* AD 100 (Raftery 1998, 22). It is almost totally absent from Munster (Warner 1998).

Although there is a general acceptance of the existence of an elite within the Iron Age of Ireland (Mytum 1992, 47; Raftery 1994, Chapter 4), none of the major landscape features dated to the Iron Age (O Riordáin 1960; Waddell 1983; Wailes 1976; 1990; Waterman 1997) appear to have any strong archaeological evidence for secular group activity. Their construction may not necessarily have been beyond the capabilities of a non-elite based society (Burger 1992, 37-8). It would be difficult to argue that the construction of such landscape features as the Mound of Hostages at Emain Macha really required the input of specialist elite individuals (*pace* Lynn 1992, 37).[3] The evidence of a self-reproducing elite in Ireland before the late fifth to early sixth centuries AD has yet to be clearly demonstrated.

The basic political unit of Early Mediaeval Ireland was the *tuath*, led by a *ri* or king (Mac Niocaill 1972, 44). It was fully hierarchical (Byrne 1987, 28). The personage of the king, by the time of written records, had achieved a sacred status. This may represent an attempt by a newly empowered office to telescope itself into a mythological past and thus legitimise its recent accumulation of power.

In the *tuath*, the basic social unit was the *derbfine,* a related group descended from a common great-great-grandfather (Mac Niocaill 1972, 49-50). This probably represents the longest period of descent that can be accurately remembered without reference to either written records or a professional body of specialists trained in long term mnemotechny. This kin group may represent a survival from the late prehistoric period, or it may be a reaction by the populus to the rise of an elite. In either case, this type of social complexity is unlikely to have remained static for any length of time, and would continually have evolved in response to new pressures.

There appears to be a strong underlying principle that land was held by the defined kin group, sharing a 'common ancestor'. Though normally limited to the *derbfine,* further kin layers existed. The deepest of these was the *indfine*, extending back six generations.

The ancestor of the *indfine* may have been a mythological creation before the arrival of Christianity. The tomb marker found at Wroxeter, dating to the early fifth century, appears to contain a mythological formula for claiming descent from a non-human ancestor, CUNORIX MARCUS MAQUICOLINE which, in partly Latinised early Irish, translates as 'Hound-king, Son of the Holly' (Wright and Jackson 1968). The designation 'Son of the Holly' would certainly associate the owner of the first name with a divine identity, as the holly was sacred among the pre-Christian Irish (Ross 1967, 64).

The power of the community ideal can be seen within the law texts of the seventh/eighth centuries (much of the work in translating these difficult texts has been undertaken by Kelly (1988)). The vast majority of farmland and, therefore, the underlying unit of wealth in

3. One of Lynn's arguments is that in order to lay out the structure under the mound at Emain Macha, the builders required a knowledge of Pi. The apparent use of Pi is more likely to be a mathematical product of the use of a 'standard' form of measurement (i.e. the pace, a cut length of timber etc.) and a desire for the internal divisions of the structure to be the same as the external divisions rather than an understanding of Pi.

Early Christian Ireland, was known as *fintiu* or kin land. Even though an individual controlled aspects of the land within his ownership, his *fine* or kin had certain powers limiting his actions. He could not, for instance, sell any of the *fintiu* without the permission of his kin. Not only was the individual constrained in his actions but the *fine* also bore a level of responsibility to the land as a whole, including rights of sharing, relating to any extra produce that a member of their *fine* produced (Kelly 1988, 100).

As late as the seventh/eighth centuries, the community as represented by a defined kin group exerted considerable control over the actions of its members. This control extended even into the elites, and to kingship. It is possible that there had been an erosion of these powers through the period of rapid social change that Ireland experienced in the late fifth to early seventh centuries.

It is against this background of apparently rapid landscape change, presumably associated with equally rapid social transformation, that the Táin Bó Cuailnge and the other mythological cycles were first written down. The surviving manuscripts cover a range of dates from the late tenth to the early fourteenth century. Wherever possible the earliest manuscript has been referenced, assuming that these have been altered least, and thus represent Early Mediaeval Ireland before any significant non-insular social change had occurred.

Within the literature of Early Mediaeval Ireland, three forms of manufactured weapon predominate: the sword, the spear (both throwing and thrusting), and the sling. Archery equipment is totally absent. Occasionally, heroes are compared to arrows for their speed, showing at least a passing knowledge of archery, but heroes never use the bow; nor do any other characters within the literature.

The primary source data will be the Táin Bó Cuailnge Recession 1 (TBC1), translated and edited by Cecile O'Rahilly (1976). The Recession I dates to *c.* AD 1100 (O'Rahilly 1976, vii). Historically, it can be shown to date to *c.* AD 809 (O'Riain 1994, 32). Linguistically, it can be dated to the eighth-ninth centuries (O'hUiginn 1992, 29) and may even have its earliest roots in the seventh (Olmsted 1992, 5). A synopsis of the Táin can be found in Mallory (1992, 9-27). The use of the Finnan Tales has been deliberately skewed. The earliest of these date to the twelfth century, but the majority were not written down until the fourteenth or fifteenth centuries, by which time Ireland had continual contact, not only with Europe, but in particular with the state level bureaucracy of England. It is likely that they portray the social reality of the time and all the concepts that this carries, which would have been outside that of a pre-state polity.

The sling

At first glance, the literature of Early Mediaeval Ireland seems to provide a context for elite use of the sling. Not only does Cu Chulainn, the most celebrated hero of Ulster, use the sling, but so does Conare, the most respected mythological 'High King' and even Lugh, the solar deity, uses the weapon during his battle with the Fomor. Yet not all is not what it seems.

The Destruction of Da Derga's Hostel

The story of Conare's succession to the throne of the High King is first mentioned as being included with a now lost manuscript entitled the Book of Druimm Snechtai, which is dated to the first part of the eighth century (Gantz 1981, 20). Fortunately, an almost complete version survives in the Lebor na hUidre (The Book of the Dun Cow) which dates to the twelfth century (Gantz 1981, 21). It represents an early survival for an Irish manuscript, and the listing of the Destruction of Da Derga's Hostel in the first part of the eighth century makes the work comparable to the earliest presumed dates for the Táin.

Within the narrative, the succession of Conare is predicted following the undertaking of a bull feast, at which the bull feaster (presumably a member of the Driu, although this is not stated) predicts that the true heir to Tara (the seat of the High Kings) will appear naked at dawn, carrying a sling with a stone in its pocket. At this point in the story, Conare is unaware that he is the son of the last High King. While Conare is on his way to Tara for the inauguration of the new High King, he pursues a flock of birds, a magical troop that had defended his father. As part of the chase, he enters the sea, which necessitates him removing his clothes, thus fulfilling part of the prophecy.

The emphasis on nudity may have a direct bearing on the sacredness of the moment. The recent recovery of three well preserved bodies, two of which show signs of violent death, are clear indicators that human sacrifice was, if not common, at least widespread among the pre-Christian Irish. The naked lower half of a human was discovered in association with a series of wooden stakes, which were uncovered during peat extraction at Lemanaghan Bog. Although not securely dated, the general context would indicate an 'Iron Age' date (Archaeology Ireland 13, No 48: 4). The other two, one from Croghans Bog, Co. Offaly (dating to *c.* 392-201 BC), the other from Clonycavan Bog, Co. Meath (dating to 362-175 BC), both shows evidence of extreme soft tissue trauma (Grice 2006, 19-21), similar in practice to the presumed sacrifice of an adult male found in Lindow Moss (Stead *et al.* 1986). Although the dating of the Lemanaghan Bog body is yet to be confirmed, and a full palaeopathological study has not yet been undertaken, all three may represent a graphic connection between the 'naked and the sacred' (Marinatos 2000). That the concept of such sacrifices was known in a more clearly Early Mediaeval Irish context is demonstrated by the appearance of the three naked sacrificial victims recorded at the end of Conare's story, the Destruction of the Da Derga.

> *I behold a trio, naked on the ridge pole of the house: their jets of blood coming through them, and the ropes of slaughter on their necks* (Stokes 1901, 314).

It is a consequence of Conare's nudity, although achieved by literary conceit, that results in his acquisition of 'divine/other world' status. Although his nudity strips him literally of direct cultural connections, it allows him to enter a clearly defined alternative position.

Where does this leave the inclusion of the sling and stone? The most obvious interpretation is that the sling is used as a symbolic connection with the biblical David, who also achieves his kingship using this weapon. However, the very nature of that achievement may have made such an association unacceptable to the Early Christian Irish mind. David not only killed the champion of the Philistines using a sling, but had refused the use of 'proper' warriors' weapons and armour (Samuel 1:17). The lack of kin relationship between Saul and David would no doubt also have disturbed the audience's mind. In many respects, David's assumption of power represents the antithesis of Irish ideology.

Conare is clearly envisaged as a youthful individual who has yet to achieve full rights within society. He was indulging in playful activity with his foster brother before going to Tara. The sling, as will be discussed below, appears to have strong associations with youth in Early Christian Ireland. However, within the context of Conare, the sling may have other symbolic connotations.

Had Conare appeared carrying a sword or spear, not only would this have placed him in the adult world but also removed his anonymity, which is achieved by him being fostered in secret. The sling, by the nature of its use, requires little physical strength. Thus the image of Conare as a youth, not yet a fully mature man, is reinforced still further.

The sling also allows Conare to indulge in two of the most important 'elite' occupations, hunting and war. By creating the bird troop, the story allows Conare literally to kill two birds with one stone. He would rather indulge in hunting for sport than travel directly to Tara for the inauguration. By his pursuit of the bird troop he indirectly undertakes the actions of war, as the disguised birdmen are the warriors of his father's bodyguard. Their entry into the sea and Conare's pursuit into the waves prevents him from further using his sling. Entering the sea prevents Conare from possibly killing an identified warrior with the sling.

The sling also represents a known pastoral tool, a herding weapon. Conare's appearance naked, and therefore associated with the divine, carrying a loaded sling by implication presents him as a driver of cattle. There is a clear connection between the health of cattle (and thus the wealth of the elite) and the righteousness of the king (Lucas 1989, 12-3). Conare is viewed as one of the most noble and righteous kings ever to have held the High Kingship. This accords with the proposition, laid out above, that, at his moment of succeeding to power, he may have been envisaged as a divine leader of men. The sling plays an important role within this construct but, unlike the potential of a sword or spear for primarily warlike activities, the sling allows Conare to be armed but not directly aggressive.

Táin Bó Cuailnge

In order to find fame and allow his name to be remembered, the young Cu Chulainn (aged seventeen) sets out to encounter and fight his first warrior. He approaches one of the watch points on the borders of Ulster. There he meets Conall Cernach, who is willing to accompany him on his quest. However, Cu Chulainn wishes that all the glory of his first encounter with an enemy should be his alone and, in order to stop Conall from accompanying him, he casts a stone from his sling that shatters Conall's chariot pole. This forces Conall to return to Emain, thus leaving Cu Chulainn to patrol the borderlands alone (TBC1 669-703).

The sling is used here to perform a disreputable act. Cu Chulainn, anxious to gain renown, has to disable Conall in a manner that is not belittling for either hero. He achieves this with the sling stone. He does not threaten Conall, as would be required if he used a spear or sword, but makes it unacceptable, through the use of custom, for Conall to follow him.

Later in the narrative Cu Chulainn defends Ulster single-handed as the other men of Ulster (Cu Chulainn is not a man of Ulster by virtue of his birth) are suffering the pains of childbirth, as they were cursed to do by Mancha (Gantz 1981, 128-9). During his defence, he comes across the charioteer of Orlam, the son of Medb and Ailill. He kills Orlam and decapitates him. Orlam's charioteer fears for his life but Cu Chulainn says he does not kill charioteers; but he does threaten to cast a sling stone at him and break his skull unless he carries Orlam's head back to the camp of Medb (TBC1 880-915).

There is a direct contrast between the treatment of the charioteer (who remains anonymous throughout the episode) and the member of the warrior elite. Cu Chulainn kills Orlam (though the method is not specifically stated, decapitation implies a sword). The charioteer is treated very differently: despite originally stating that he did not kill charioteers, Cu Chulainn immediately threatens to kill him with his sling. Although it could be argued that the sling is the only distance weapon available to Cu Chulainn, he is earlier recorded as being exceptionally skilled in the use of the javelin (TBC1 415-17). This would indicate that in this case the sling is specifically being associated with dishonourable actions.

The reference to breaking of the charioteer's skull also tends to reinforce the concept that the sling is seen within a context of dishonour. It is evident throughout the Táin that the most honourable death for a warrior is to be decapitated by his victorious enemy. The head had strong magico-sacred overtones within a broad Western European 'Celtic' tradition (Ross 1958; 1967, Chapter 2). Evidence from the tale cycles and archaeology make it clear that such beliefs also existed in late prehistoric and

Early Mediaeval Ireland (Rynne 1972). The proposed destruction of the head (even the charioteer's) may therefore represent the destruction of an element of post-mortem social interaction. This may have been considered as a complete destruction of the individual's personality.

As the cattle raid continues, Cu Chulainn attempts to kill Medb herself, not through personal combat but by casting a stone at her whenever he could. Although there is no direct reference to the sling, the use of the verb to cast in O'Rahilly's (1976) translation would tend to indicate that a mechanical method of propelling the stone is being used.

Although Medb's position has often been interpreted as a representation of the empowerment of women in Early Mediaeval society, recent work has pointed towards a more misogynistic interpretation (Kelly 1992, 77-84; Ni Bhrolchain 1994). Within this context, Medb can be seen as a troublemaking woman, the instigator of all the death and destruction that the Táin caused. As part of this contextual construct, it would appear that Cu Chulainn singles her out for specific attack, using a common unheroic weapon. No other named member of the elite is so treated, not even Ailill, who could easily be seen as a co-conspirator in the raid. Death from a distance, with no opportunity for defence or corresponding reaction, is a direct contradiction of the heroic ideal of personalised combat. It would seem that the sling is being portrayed as a weapon suitable for killing a woman, who has not only dared to venture into the masculine world of war but, in the process, has brought destruction in her wake. Once again Cu Chulainn aims specifically at Medb's head, potentially leading to the ultimate destruction of the self. Cu Chulainn's tactic also leads to Medb having to be protected by the men of the army, as though she were incapable of such defence herself. Once again, the belittling of a woman in what is primarily a male preserve is projected.

In an attempt to prevent Cu Chulainn from harrying Medb's army, Ailill decides to give Cu Chulainn his daughter Finnabair. In order to avoid risk, Ailill's jester is disguised as the king; Cu Chulainn spots the deception and kills the jester with his sling by driving out his brains (TBC1 1605-07).

The killing of the jester seems almost casual. Cu Chulainn merely happened to have the sling stone in his hand; he makes no recourse to sword or spear. Death by a thrown stone, not only of a deceiver, but a man identifiable by his coarse speech, again places such weapons within the context of the common people, distancing the elite from such acts through the form of death, as well as through the manner of killing. The destruction of the jester's head once again places him outside the realm of the hero, as he would be disfigured and thus become an anonymous corpse with no identifiable personality or history.

As the narrative continues Cu Chulainn is approached by the Morrigan, the Irish goddess of war, and she appears to have a marriage proposal for him, as she is impressed by Cu Chulainn's martial ability. This he rejects and as a result, the Morrigan threatens him. Her threats consist of two attempts to stampede cattle at Cu Chulainn. In response, Cu Chulainn states that he will stop her by using his sling, once to smash her eye, when the Morrigan is in the form of a wolf, and once to break her leg when the Morrigan is in the form of a heifer (TBC1 1859-74).

The sling is clearly placed within a true pastoral context, described as a herding weapon or tool. It can wound and drive away the most conspicuous threat to a herd, the wolf. It was the legal duty of a client to carry out a foray against wolves on a weekly basis and there are many other references within the law codes to the perceived threat that wolves presented (Kelly 1998, 186-7). The sling is also seen as able to stop stampeding cattle by again wounding the lead animal. It is significant that in no case does Cu Chulainn claim that he would be able to kill the Morrigan, even though he is unaware of the divine status of his opponent. This would indicate that, in this pastoral context, the sling is portrayed as a wounding weapon and not truly deadly, a weapon inherently unsuitable for a hero. That it is acceptable to be used against a woman (albeit divine) may carry overtones of Christian misogyny.

Tiring of Cu Chulainn's ability to continually defeat her champions, Medb breaks the terms of fair (heroic) play (that only one warrior is to come at Cu Chulainn at any one time), and sends five to attack him at once. Cu Chulainn slays them all single-handed. In his rage, *Cu Chulainn pelted (with sling-stones) from Delga so that no living creature, neither man nor beast, could get past him south between Delga and the sea.* (TBC1 2036-39)

After breaking the terms of fair play, Cu Chulainn is under greater pressure than before. It is under these circumstances that he resorts to using the sling to prevent the advance of Medb's army. The sling may be being identified as a weapon of the weary, and by correlation, of the weak. It is only a hero of Cu Chulainn's stature that would be able to wield such a weapon with this effect. Even so, unlike the weapons of the elite, the spear and the sword, the sling in this instance can be seen as a disrupter of the narrative sequence; it prevents the story from progressing by hindering the continued single combats that are central to the Tàin.

The Death of Aile's Only Son

This is linguistically as well as documentarily a late text. The surviving manuscript dates from the fourteenth century (Gantz 1981, 21). Cu Chulainn was, in earlier texts, described as being only a boy when he attended the 'warrior school' of Scathach (mother of Aile). It is unlikely that he would have sired a son in these stories. In this later text, Conla is described as his son and, having been trained by Scathach in all aspects of weaponry except the *Gae Bolga* (see below on the spear), sets off at the age of seven to find his father. He approaches the coast of Ulster in a bronze ship with golden oars. As he comes, he fires his sling at a flock of birds on the sea, knocking them out (but not killing them) and then releasing them only to once again knock them from the sky with his 'jaw feat'. The exact nature of the latter is not given, but the etymology does appear to have some connection with the use of sound (Kinsella 1969, 39).

Clearly, in this context the sling is a weapon strongly associated with youth, here indeed with childhood. That Conla is able to knock birds from the sky with his sling without killing them places the sling within the realm of less lethal weapons. The use of phrases relating to sound is of particular interest. The analogy of humming sling stones with the tuning of his voice seems to be a metaphor for the second stunning feat carried out by Conla. The expression that the feat was undertaken in a manner that is in some form *'faster than the eye could follow'* (Kinsella 1969, 39) strongly suggests a comparison between this act and the flight of a sling stone. Initially a stone cast from a sling is almost impossible to follow with the naked eye. Under certain circumstances, the stones flight can produce an audible 'hum' or 'buzz'. Whether the analogue of the voice and the sling is a representation of the sling being as weak as a shout, all noise and no danger, or if the conjunction of the two terms should be regarded as an expression of the power of Conla's voice, is problematical. However, taking the other evidence presented above, the former case is likely to be more appropriate. This would again place the sling within the context of a less than effective weapon, a mere noisy toy.

As Conla approaches Ulster, Conchubar is concerned that he will ravage the province and decides to send a warrior, Condere, to find out what the boy wanted. Conla, however, refuses to give either his name or allow Condere to pass. Conla challenges the Ulaid to combat, and Conall Cernach comes forward to fight the boy. Conall Cernach specifically describes the feats that Conla has undertaken as *'pretty games'*. Conla responds that they will work just as well on him and, using his sling, he sends a '*stunning-shot into the sky*' which successfully disabled Conall Cernach long enough to allow Conla to bind him and send him back to the men of Ulster in disgrace (Kinsella 1969, 42).

In this passage, the damaging effect is not the sling stone but the noise that the stone makes as it accelerates into the sky. By employing the noise of the sling stone to produce the defeat of Conall, the narrative escapes the problem of having a named individual being overcome by such a weapon. There is no direct contact between the hero and the sling. It is perhaps no coincidence that Conall as a personal name means 'strong as a wolf' (O'Corráin and Maguire 1981, 56). The defeat or disablement of Conall, both by Conla and Cu Chulainn (O'Rahilly 1976, 144) would perhaps be of great relevance to an audience familiar with the meaning of Conall's name. As discussed above, the hunting of wolves was a requirement of a lord's clients. If the supposition that the sling is associated with youth and the non-elite is correct, the inference that Conall is metaphorically being hunted by both Cu Chulainn and his child is given potential meaning within a mythic/social context.

Aided Meidbe: *The Death of Medb*

The earliest extant manuscript relating to the above narrative dates to the twelfth century and is contained within the Book of Leinster (Hull 1938, 52). A later document comprises several independent fragments that are chronologically disparate. Both these are written in Middle Irish. The linguistic evidence tends to support the concept that both documents are copied from an earlier manuscript originally composed in Old Irish (*ibid,* 53). Whereas the existence of the presumed earlier document cannot be proved, this story may be chronologically comparable to the Tàin.

The Aided Meidbe relates how Medb's brothers rebelled against their father. In an attempt to prevent this, one of her sisters, Clothru, allowed each of them to have sexual intercourse with her. She hoped that this incestuous act would be disgraceful enough to prevent them from attacking their father. The three brothers were, however, not dissuaded and they were subsequently killed in battle. Clothru was killed by Medb because she usurped the sovereignty of Connaught. During the attack on Clothru, Medb's sword blows were responsible for the birth of Furbaide mac Conchobair.

When Furbaide mac Conchobair achieves adulthood, he seeks revenge on Medb for the death of his mother. His is specifically described as practising *casting* at two stakes set in the ground, and not stopping until he had struck the apple that was set on the top of one of the stakes. The Old Irish word 'taibleth' is the denominative form of 'taball' and means 'slinging, casting' (Hull 1938, 60, Footnote 25). So, although the sling is not directly mentioned, the action of casting is not likely to be associated with the use of a spear or javelin.

Furbaide is finally able to kill Medb during an assembly between Ulster and Connaught. Medb was bathing in Loch Ri when Furbaide saw her and, being told who she was, cast a piece of hard cheese at her and, striking her on the crown of her head, killed her. The text is specific that Furbaide did not seek a stone in order to attack Medb.

At first sight this narrative appears to present clear evidence of the sling being used, not only by a named individual, but also for killing another named individual who may be a euphemised goddess (Kelly 1992, 77). The act, however, although potentially planned, gives the impression of being impulsive, almost recklessly, casual. This appears to be a method of obscuring the type of death that is being perpetrated. Furbaide is killing his maternal aunt, and is therefore performing the act of *fingal* or kin slaying. Kelly (1988, 127-8) describes *fingal* as one of the most horrendous crimes possible in Early Christian Ireland, as it struck at the core of kin-based society. It was impossible for the *fine* to revenge itself without indulging in further acts of kin slaying. Nor would the normal payment of an honour price be sufficient, as this merely recycled the wealth within a defined group who were related not only to the slain but also the slayer. By using the sling, Furbaide not only negates the use of a heroic weapon for such a socially unacceptable act as *fingal* but also places such actions within a broader context of correlation. Acts undertaken with the sling are not worthy of a true hero, and thus, by association, those who use the sling as their only, or primary, weapon are not worthy to be classified among the warrior elite.

Throughout the narrative, Medb is shown as a meddling or troublemaking woman, who not only kills her sister (another act of *fingal*) but also usurps the sovereignty of Connaught from her father. That she is killed with a sling shot while bathing denies her a heroic or warrior's death in combat. She is likely to be, if not naked, then at least only partly clothed and therefore unable to defend herself. This vulnerability clearly places her within a perceived female role. Her death by a sling and a piece of hard cheese appears to reinforce the femaleness of the context. The vast majority of ethnographic examples of the sling are manufactured using some form of textile or weaving process. The manufacture of textiles in Early Christian Ireland was firmly within the female domestic realm, as was the production of cheese (Kelly 1988, 327-30). The association of such female elements with the death of Medb would disassociate her further from the male realm of warriorhood, and could be seen as the final insult to a troublesome woman who dared to venture into the male dominated world of politics.

The Great Battle of Magh Tuireadh

Another narrative where a named individual is killed using the sling is contained within the mythological cycle. Here Lugh, champion of the Tuatha De Dannan, leads them into battle against the Formor, who have been taxing Ireland heavily for some time. Lugh slays Balor, the King of the Formor, by casting a sling stone into his magical eye (Heaney 1994, 20).

As with the Death of Medb (Hull 1938), this narrative appears to present clear evidence of the sling being used not only by a named individual but an individual who holds a position of primacy within the 'gods' of the pre-Christian Irish, who kills another named individual. The context also appears to be in the mould of heroic combat, personalised and individual. However, Lugh is Balor's grandson and therefore, by killing him, is undertaking the act of *fingal* that correlates with that described above for Medb; the creators of the myth of the Battle of Moytura enabled Lugh to escape the consequences of his actions, primarily by describing Balor as evil; his killing was therefore justified. Also Balor's death is prophesied, so Lugh had no option other than to follow the tides of fate. By using the sling, Lugh once again removes the taint of *fingal* from the truly heroic weapons of the sword or spear. Lugh drives the sling stone through Balor's head and, although he is later described as decapitating Balor, the disfigurement of his head by the sling stone should be seen within the context of 'non-heroic' acts.

The Spear

From its first mention in the Tàin the spear is given a special place. It is placed directly in the realm of Cu Chulainn, not only as the hero of Ulster, but as a beautiful member of the elite, who is performing great feats of battle with the *gae bolga*, sword and spears (TBC1 81-3). The association of Cu Chulainn to various weapons continues later in the text: Fergus compares him with the spear point, he is sharper, keener, quicker (TBC1 386-7), each of the descriptive terms has the potential for clarity of action.

However, Cu Chulainn's association with the spear goes back to his childhood deeds when he is described as coming to Emain Macha (the capital of Ulster) with his toy javelin, which he casts and then runs after, catching it before it falls to the ground. Even here, the spear is given a special place. It is seen as fast and difficult to catch. It is a mark of Cu Chulainn's special status as a future hero that allows him to undertake this feat (TBC1 412-3).

The expression of the elite nature of the spear is more clearly demonstrated when the young Cu Chulainn takes up arms. In his attempt to find weapons suitable for his prodigious strength, he destroys fifteen panoplies before Conchobar, King of Ulster, offers him his own, which even Cu Chulainn cannot break. Cu Chulainn brandishes his newly acquired status symbols while Conchobar declares *'Happy the people and race over whom reigns the owner of these arms!'* (TBC1 624-5). The direct association with particular weapons, in this case, the spear and sword, is clear. No other form of weapon is so obviously connected with the right to rule.

In an initial attempt to prove his worth, Cu Chulainn sets off to fight the enemies of Ulster single-handed. His first combat encounter is with three unusually named warriors, Foill (Sly), Fannall (Swallow) and Tuachell (Cunning) (TBC1 732-59). Cu Chulainn kills each of these warriors in turn. The names of these warriors appear to hold special significance for the type of warfare that the Early Mediaeval Irish aspired to (even if they never achieved it). Both Foill and Tuachell carry names that signify unacceptable types of combat within an 'Heroic' context. Slyness and cunning have no place in the open field of battle that is personified in the Tàin. Even Fannall could be seen as having an epitaph that would be less than acceptable, namely the ability to skim over water like a swallow. Cu Chulainn especially refers to this trick as an attribute that is outside the concept of 'Heroic' warfare (TBC1 736). Although Cu Chulainn kills all three warriors by casting his spear at them, there is no mention of the weapons they use. It is as though there is a wish to eschew the association of particular weapons with the characteristics displayed in the warrior's names.

As the Táin continues, Cu Chulainn is defending Ulster single-handedly, as a series of heroes from Medb's camp come to fight him. One of the heroes, Nad Crantail, approaches Cu Chulainn and casts nine fire-hardened stakes of holly wood (in effect javelins) at him. Cu Chulainn, who is hunting birds at the time, avoids them all by running over the top of them as he pursues his prey (TBC1 1419-33). That the stakes are described as only fire-hardened, i.e. without a man-made metal tip, implies that the society that is projected in the Táin was deliberately distancing itself from the use of such weapons. Although the holly is one of the sacred trees of Early Mediaeval Ireland (Ross 1967, 64), it would be difficult to argue that the use of holly for spears is being discouraged by the Christian writers of the Táin, when there is so much of a blatantly pagan origin included within the text. The inference would appear to be that the use of such simply manufactured weapons is associated with the non-elite. They are thus seen as not only unsuitable weapons for a hero to use but are simply not seen as weapons at all by 'true warriors'. This motif is repeated later in the Táin when Fer Baeth, Cu Chulainn's foster brother (sharing a common foster mother) comes to fight him (TBC1 1778-83). Cu Chulainn is unwilling to fight due to their ties of kinship, but Fer is insistent. Cu Chulainn leaves their meeting in anger and, as he does so, treads on a holly shoot that penetrates up to his knee. Having extracted the shoot, Cu Chulainn shouts after Fer, saying that he must not go away until he has seen what he has taken from his leg. Fer foolishly asks Cu Chulainn to throw it to him, and Cu Chulainn kills him by casting the holly like a javelin. Although Fer and Cu Chulainn were not related genetically, the bonds of fostership were as strong as those of kin in Early Mediaeval Ireland, and the killing of a foster brother was the equivalent of *fingal,* as discussed above. The use of the natural weapon of the holly shoot, like the use of the sling, removed this act from the realm of the elite.

The worth of a warrior's weapons and, in particular, his spear is illustrated in the Death of Redg the Satirist. In an attempt to disarm Cu Chulainn, Ailill sends Redg to ask Cu Chulainn for his javelin. Cu Chulainn refuses to give him the spear and Redg threatens to satirise him. Satire was a significant threat in Early Mediaeval Ireland: not only could it lead to loss of honour or face, it has been recorded as being fatal (Kelly 1988, 44). Such reports may not be pure fantasy, as the psychological damage done could be considerable which, in turn, may lead to acute harm and even death. In order to avoid this fate Cu Chulainn casts his javelin at Redg and impales him through the head (TBC1 1518-20). Cu Chulainn does not take Redg's head as a trophy. This may have significance, as the satirist was one of the lowest grades of poet in Early Mediaeval Ireland, and was often regarded with fear and hatred and, despite his powers, was classified among the lower classes.

Spears could also carry magico-mystical associations of their own. This is highlighted by the weapon carried by Cuscraid Menn Macha, son of Chonchobar king of the Ulaid. A silver band runs up and down the shaft of the spear, when there will be victory for the Ulaid (TBC1 3631-3). This direct correlation between the spear and the ability to predict the future moves the spear into the realm of the 'other world', and all the power with which it is associated.

The strangest weapon that Cu Chulainn wields is the *gae bolga*; it is separated from other weapons, not only by its name (O'Rahilly 1984, 61-2), but also by the nature of its use and the inferred style of its manufacture. The *gae bolga* is no ordinary spear. Only Cu Chulainn has been taught to use it and, on the three occasions that he does, it is used specifically to kill his own son (Kinsella 1969, 44), his foster brother (TBC1 3099-3105), and Loch (TBC1 2022-6), a warrior who would not fight Cu Chulainn as he had no beard (the mark of maturity in Early Mediaeval Ireland). The first two are clearly killings within the context of *fingal* and, as such, would be unacceptable to the audience of the Táin. This would suggest that the *gae bolga* is being used to deflect these acts away from the elite. It is clear that none of them ever did or could hope to own a weapon like the *gae bolga.* With the killing of Loch, we see an honourable man trapped; he has no wish to fight Cu Chulainn as a beardless boy, it is below his honour. He will only fight true men. This marks him out as a hero; he does not fight or kill those he sees as weaker than himself. Even though Cu Chulainn killed his brother, Loch will not fight him until Cu Chulainn uses blackberry juice to give himself false stubble.

The etymology of the *gae bolga* raises issues relating to the social context of the weapon. *Gae* is the Old Irish for spear. This does not differentiate between a throwing or thrusting weapon. The word *bolga,* however, carries a potentially deeper significance. The Fer Bolga were, in mythological terms, the original inhabitants of Ireland. Their supremacy was destroyed by the Tuatha De Dannan when they invaded and conquered Ireland. The Fer Bolga were given the province of Connaught as their homeland (Gregory [1904] 1970, 31). Medb comes from Connaught and this would suggest that the weapon that Cu Chulainn uses in the act of *fingal* is associated with two women, Medb and Scathach.

Scathach teaches Cu Chulainn alone to use the *gae bolga.* This, given the often misogynistic nature of the Tàin (Sessle 1994), would suggest a connection between the use of the *gae bolga* and the treachery of women as portrayed in the narrative. The *gae bolga* may therefore have been associated with the involvement of women in warfare, something that the Tàin is often at pains to present as unnatural.

This would potentially extend to Cu Chulainn's famous spear, the *gae bolga.* By its name it would appear to be associated with women and therefore in the authentic social context of the audience, the non-elite. As women were in most ways excluded from the legal process of Early Mediaeval Ireland (Kelly 1988, 75-6), the use of the *gae bolga* in acts of *fingal* and other unacceptable killings would be palatable to the audience of the Tàin, as their social context is not threatened by the use of the 'proper' heroic weapons available to them. In this context, it is perhaps easier to see why each time Cu Chulainn uses the *gae bolga,* he kills his opponent in a manner unbecoming to a hero. In the case of Fer Diad[4] and Loch, the spear pierces them though the anus, surely the foulest of deaths; in the case of Conall, the spear is described as '*bringing his bowels down around his feet*' (Kinsella 1969, 44). In each case, the death of the hero is unbecoming. Cu Chulainn almost 'cheats' as he kills them. If this is the case, the use of the *gae bolga* in unacceptable acts could be transferred onto an excluded group (and therefore by definition all excluded groups?), reaffirming the cohesion of the emergent elite.

Throughout the Táin, spears are associated with particular warriors, and are given qualities that may have been desirable for a hero to have; for example: even Medb's javelins are described as being *'keen, sharp-edged and light'* (TBC1 3212-14). At no time are spears used to describe qualities that would be undesirable to a warrior. There appears to be a direct and strong correlation between the weapon and its social context.

There is a direct relationship of spears and, in particular, throwing spears, to the elite. Although the descriptions within the Tàin often do not describe exactly how a spear is being used, of those that do, two cases are thrusting weapons; the remaining ten cases are cast or thrown. This would certainly imply that throwing spears were seen as suitable for a hero and, by adoption, the elite of Early Mediaeval Ireland.

The Sword

The sword is not as frequently mentioned within the Táin as the spear. It is never associated with non-elite individuals and, despite the lack of direct reference, the sword would appear to be the primary way of decapitating an enemy, which, as discussed above, would appear to be the most culturally acceptable death for a hero. That Cu Chulainn is able to cut a branch from a tree with a single blow, and impale four heads on its branches, would seem to be a narrative link between the two concepts (TBC1 332-8).

This association with the elite is most clearly demonstrated by the continual reference to swords in the possession of heroes. The portrayal of a hero is incomplete without a description of their sword. These descriptions usually concentrate on the hilt: ivory (TBC1 3660-2); silver (TBC1 3795-6); or gold (TBC1 3777), all materials of high prestige, and all, except for 'ivory' (Hencken 1950, 88, Figure 24D) (or other ivory-like materials, such as antler), missing from the archaeological record of Early Mediaeval Ireland. The materials described correlate more closely with the period after the Norse invasions. This has led some authors to question the validity of the sword within the context of the Táin (Mallory 1992, 135-7). It is probable that although the type of sword may have changed, the social context of the weapon remained fairly stable. The archaeological recovery of swords from high status sites such as Lagore Crannog (Hencken 1950, 91, Figures 25-6) would support the view that the sword held a significant place within Early Mediaeval Ireland. In the Táin itself, Cu Chulainn is described as performing feats and great deeds with four small swords by the prophetess O Feidelm (TBC1 79). This would certainly infer that, even if the smaller swords of the pre-Norse period had lost some of their importance by the eighth century, when the Táin was first written down, potentially the smaller 'earlier' swords maintained their social position.

Swords go beyond being the identifiers of the social elite; they become directly identified with them and seem in part to represent them within an ideological framework. During the raid, Ulster's cattle are driven off, the women taken captive and the men slain (TBC1 1214). These three elements are described to Conchobar as he lies stricken with the wasting of Ulster. It is only in the context of killing men that a weapon is mentioned, and then Fergus mac Roig is described as *'the brave one who wields a sword'* (TBC1 1219). This clearly identifies Fergus with the sword and with the hero's quality of bravery. Earlier, during the raid, Fergus mac Roig has an affair with Medb. In order to prove this, Cuillius, Ailill's charioteer, removes Fergus' sword and takes it to Ailill. Ailill ensures that the sword is kept in good condition, as it will be required later. When Fergus discovers that the weapon is missing, he takes his charioteer's sword and uses it to carve himself another from wood. He does not, it should be noted, merely take another man's sword, but fashions himself another 'fake' weapon from timber (TBC1 1044-1103).

It is as though the identity of Fergus is partly encapsulated within his sword and another's will not suffice. However, this would seem to place Fergus in a liminal situation. The narrative seems to suggest that the use of wooden weapons, no matter how large (TBC1 1305, for Fergus' wooden sword being as large as a boat's rudder), are associated with youth and fine feats (TBC1 1328-9). This in a sense emasculates Fergus, making him a boy, by the removal of his social identifier, the sword. The association with youth seems to be reflected onto the non-elite sections of society. They hold no power, are reliant on their social masters for legal or physical defence and can do nothing without their permission (Kelly 1988, 81-2).

Deference to the sword's social position can be seen in the confrontation between Cu Chulainn and Etarcomal (TBC1 1323-87). Etarcomal comes to view Cu Chulainn under the guard of Fergus mac Roig. Cu Chulainn will not attack Etarcomal while he is under Fergus's protection, despite the fact that Etarcomal challenges Cu Chulainn to a duel at the ford. In an attempt to dissuade Etarcomal, Cu Chulainn first cuts a sod of earth from

[4] Although Fer Diad is wounded by the *gae bolga,* it is not the weapon that kills him. Cu Chulainn casts his spear over Fer's fallen shield and penetrates his heart. It is this heroic action that kills Fer Diad, not the sneak attack of the *gae bolga.*

under his feet, so he falls to the ground with the sod on his belly. There is no mention of the weapon used. This may be because there was an association between the cut sods of soil and the non-elite (Lynn 1992, 52) or that no weapons should be directly associated with the potential denigration of Etarcomal to the non-elite class. Etarcomal will not retreat and still demands to fight Cu Chulainn, so he removes his clothes by cutting them away with his sword. Still, Etarcomal will not leave. Cu Chulainn shaves him with his sword without breaking the skin. Still, Etarcomal insists he wants to fight Cu Chulainn. Finally Cu Chulainn splits the youth in half with his sword. He does not therefore take his head, denying him a true hero's death.

The misuse of the sword in a non-heroic manner is demonstrated by the death of Nad Crantail (TBC1 1481-2). Following his mortal wounding with a spear by Cu Chulainn, Nad returns to the fight and casts his sword at him. Cu Chulainn avoids the cast, and the sword strikes a pillar of stone and is broken in two. This may, of course, be seen as merely a 'realistic' response to throwing a sword against a large rock. However, later in the Tàin, Cethern, an Ulster hero, thinking, that he is attacking Ailill drives his sword and fist though a stone pillar, upon which Ailill's crown had been placed (TBC1 3315-19). Towards the end of the Tàin, Cu Chulainn overtakes the fleeing Medb, who pleads for her life (TBC1 4118-9). He takes her (presumably as a captive) to Ath Luain, where he strikes a flagstone three times with his sword. The exact meaning of the latter act is now lost to us but, in each case, a sword is driven into or strikes rock without any noticeable damage. The acts are both within the ken of heroes, one seeking combat with the leader of the enemy force, and the other sparing the life of a non-heroic individual. In both these cases the sword is not broken, even though 'practicality' would demand that it was. The inference is that when used correctly a sword is 'unbreakable'. This may represent a reflection of the elites' visualisation of themselves.

The ultimate expression of the sword's elite status within the Táin comes when, at the last great battle, Fergus regains his sword from Ailill and attacks the Ulster host (TBC1 4016-24). Fergus seizes his sword and, using it in a two-handed manner *'cleared a passage for a hundred through the line of battle'* (TBC1 4037-8). Fergus finally cuts his way to Conchobar and aims a blow at him. However, he is restrained and, through honour, is dissuaded from attacking the men of Ulster further. Instead, he attacks three hills and strikes off their summits leaving them flat-topped. This act pushes the sword distinctly into the realm of the divine, placing the users of the sword, the elite, within, if not that category, at least within its sphere.

Conclusions

The sling

The sling exists within a clearly defined context. Cu Chulainn is described as being only seventeen in the Tàin. Conla is only seven when he sets out on his fateful voyage to find his father and, although no direct age is given for Conare, it is manifest that he is still in his youth. Again, no direct age is given for Lugh and, as far as it is possible to age divinities, it would appear that he is to be perceived as not yet 'mature'. Furbaide is also not directly aged but, by inference, it may be assumed that he is again seen as young. His practice of casting his sling shot at a stake is comparable with the youthful games of the other boys of Ulster (Hull 1938, 60), placing such activities within a non-adult context.

Irish law codes have a strict division of age grades (Kelly 1988, 81-2). Conla being below the age of fourteen has no legal capacity beyond exchanging goods of equal value. He has the same legal responsibilities as the base clientage (*ibid,* 81). He cannot make any legal contract without the support of his legal superior, in this case his father or guardian rather than his lord. Cu Chulainn, by contrast, would have been classed among the *fer midboth,* who appear to have been aged between seventeen and twenty. He would have greater legal responsibility than Conla but, even though he was able to accept a fief from a lord, he still lacked the complete legal capacity of a fully-grown adult (*ibid*, 82). Cu Chulainn would, by the nature of his divine and, therefore immortal, father, never truly achieve adulthood, as this could only be achieved on the death of the child's male parent, and the assumption of full responsibilities within the kin group or *fintiu.* Cu Chulainn is forever in this liminal state, neither fully adult, nor completely a child. There is no direct age given for Furbaide or Conare but, as Conare assumes the kingship of Tara, it would perhaps be safe to assume that he is also a *fer midboth.* He is able to assume adulthood because his father has already died, although within the context of the narrative he is unaware of this. The descriptions of these three characters as being in their minority would imply strong associations between them and the mass of the population, who had to rely upon their lords for most forms of legal and personal protection. Lugh may be the only character seen as a legally integrated male described as using the sling. However, as discussed above, it seems that Lugh is using the sling in a clear attempt to deflect or obscure the dreadful crime of *fingal.*

It would be a reasonable inference, from the above evidence, that the social position of the sling is being described. It is not associated with adult males, who have been integrated into elite society, but is associated with youths, who are less responsible for their actions, and must be 'protected' by a social superior. Although holding a clearly defined legal position within Early Mediaeval society, these individuals that had not yet achieved adulthood had more in common with the mass of the population than they did with the elite.

The sling can been see as a liminal weapon and as a metaphor for the condition of the individuals using it. The sling is both a tool and therefore below the contempt of the elite, and also a deadly weapon, the sole preserve of the elite. It is, in part, this contradiction that allows it a place within Early Irish 'heroic' literature at all. The sling is seen as a boy's or youth's weapon, requiring no great strength, and/or requiring continual 'playful practice' to achieve the dexterity necessary for effective use. Heroes within the Tàin throw stones to kill or maim their opponents. This requires greater strength and may be an indicator of physical power.

Throughout the surviving literature from Early Christian Ireland, the sling is shown to be a deadly weapon. However, its lethal potential is only used against those who are themselves weak either physically or mentally (as in the case of Ailill's fool) or socially inferior. Even when the Morrigan is wounded she is in the form of an animal, a hornless heifer, which by its nature is not only defenceless but virginal, requiring protection and thus, by analogy, weak. The death of Medb is perhaps the most obvious example; not only is she a woman but, at the time of her death, naked and bathing; ultimately vulnerable. Even Balor with his deadly eye can be cast into the mould of the enfeebled. He cannot raise the lid of his evil eye without assistance (Heaney 1994, 22). The use of the sling against the weak would place it in their arena, or at least those that the elite would wish to make weak, not necessarily physically but certainly socially and economically, the unfree.

By correlation, it may be assumed that the sling is seen as a weapon of the common people unworthy of mythological heroic use, and thus below the contempt of the new elite that sought to distinguish itself from the rest of the population. The sling is clearly used for disreputable acts, not limited to merely preventing another warrior accompanying Cu Chulainn on his initial quest for heroic recognition, but even being responsible for *fingal*, perhaps the ultimate crime within the kin-based society of Early Christian Ireland. Other than for the act of *fingal* described above, the sling does not kill a single named individual. It is used entirely for harrying the unknown faceless mass of Medb's army or for killing pets or jesters. The sling is not a hero's weapon in the terms of the sword with its near magical properties (TBC1 4077-82) or the spear, that continual killer of foes, friend and enemy alike (TBC1 3065-3113). If the sling represents the weapon of the community, the references to its use for socially undesirable acts could be seen as an indirect attack by the elite on the structure of the *fine*, which still held considerable and presumably restraining power over the individual in the seventh and eighth centuries.

The spear

The spear has distinct forms, casting and thrusting. Only when cast as a stake and not as a full spear does it undertake fully unheroic deeds. Such simple and effective weapons would have been widely available to all sections of Early Mediaeval Irish society, and their manufacture impossible to control. The same cannot be said of metal-tipped weapons. Here, control of production would have been easier and their association with the elite is stronger. Spears are used to attack Cu Chulainn when the rules of fair play have been breached, but these attacks are inevitably frontal, so at least Cu Chulainn would know he was being engaged in combat.

Throwing spears are used in most combat situations. Only the *gae bolga* appears to be a specifically designed thrusting weapon and, as explored above, this is used only in connection with acts that the elite would find socially unacceptable. This leaves the throwing spear as one of the exemplifiers of elite status. Due to its nature, the spear could be appropriated by non-elite sections of the community, so it is not free from the taint of non-heroic action. The narrative allows for this by including both the *gae bolga* (associated with women) and the use of stakes as weapons (natural spears). The inclusion of stakes in the boyhood deeds of Cu Chulainn also constrains these weapons to members of the elite, who have not yet become adult and thus, by implication to the non-elite, who will never gain full legal recognition. This ploy is similar to that for the sling.

The sword

The sword represents the ultimate expression of exclusion between the elite and non-elite of Early Mediaeval Ireland, and the ultimate expression of inclusiveness within the elite group. With the mythological power to change the very geography of the land itself, the sword appears to have almost divine status. Swords also seem to have the power to act as delimiting social markers. During the Táin, Iliach, an old warrior, attacks the host of Medb. He is finally killed by Docha, who decapitates him. Later, Docha takes Iliach's head and sword to Iliach's grandson. They make peace and, as a mark of this, Docha keeps Iliach's sword.

The social distinction between the elite and non-elite is expressed during the fighting leading up to the final conflict of the Táin: Medb's servants attack the Ulster force not with spears or swords but with clubs. These are not just natural weapons that would be readily available to most people, but are specifically described as being made of iron and thus requiring to be forged. It is that they are wielded by members excluded from the elite that draws them into such sharp contrast with the sword.

Causes of Conflict in Táin Bó Cuailnge.

The most obvious cause of conflict in the Táin would appear to be a single large cattle raid undertaken between two significant polities. There may be another deeper cause – that of honour. It is the honour of Medb that is at stake, because her husband Ailill, her equal in all other matters, has a bull greater than hers. Given the overriding importance of cattle in Early Mediaeval Ireland (Lucas 1989), such a disparity between equals would surely have been keenly felt. It is Medb's desire to satisfy her honour, also a matter of great concern in Early Mediaeval Ireland, and be the true equal of Ailill that is the root cause of the Táin.

The honour of Ulster is also at stake. Originally the owner of the brown bull of Cuailnge had agreed to lend Medb his bull. This would have been a satisfactory outcome, except that one of Daire mac Fiachna's servants heard Medb boast that she would have taken the bull had it not been offered. From this point forward recovery of the stolen bull becomes a matter of honour for the men of Ulster.

As the raid continues, Medb's army does not limit itself to taking cattle and other movable goods but also takes the women of Ulster as slaves. That the standard unit of measurement was a *cumal* or single female slave, within Early Mediaeval Irish legal documents, and not a male slave or *mug*, is potentially an indication of how widespread this type of action may have been. If this was the case, raiding for women may have been as acceptable as raiding for cattle.

Of course, it could be argued that the use of honour is merely a cover for the true economic rationale behind raiding, the taking of cattle and possibly female slaves. However, Lucas (1990, 174-5) has shown, through a detailed investigation of the historical records dating from the twelfth to sixteenth centuries, that the overwhelming number of raids were unsuccessful and often resulted in the death of the leader of the raid. This would tend to infer that the economic gain was limited and that honour was the one of the prime motives for raiding.

Case Study 2

Iron Age Israel (*c.* 1000-750 BC)

> *Like tying a stone in a sling is the giving of honour to a fool* (Proverbs 26:8)

The Hebrew Bible is littered with references to weaponry and its use within a social context. Rather than attempt a synthesis of this extensive material (which would be well beyond the scope of this study), a single narrative will be used to explore the social significance of weapons at a particular moment of the 'history' of Israel and Judea. The story of David will be used, from his initial appearance in the Nevi'im, where he defeats the Philistine Goliath, until he succeeds to the Kingship of Israel following the death of Saul (Samuel 1:17-31 and Samuel 2:1-6). Although there are at least two other stories of how David entered Saul's service, David's defeat of Goliath is the best known and best illustrates the social significance of weaponry. This narrative is also one of the few within the Hebrew Bible where a detailed description of the use of weaponry is given, although, as with much of the Bible, there is no real description of the organisation and tactics used. Therefore it is almost impossible to reconstruct a full picture of warfare (Carroll 1995, 28).

The context within which the storyline is presented may represent the first major shift towards a centralised state level society, from a less centralised system of sheiks, and rule by a 'theocracy of prophets'. The complex narrative appears to contain both elements that an emergent elite may have wished to denigrate, and an attempt to contain that power within a clear religious context where some secular power was still vested in the priesthood. This conflict is epitomised as being between Saul, representing the rise of a more 'secular' state, and Samuel who, as a prophet, represents the older more 'egalitarian' mode of government. The last phrase in Judges, the books before Samuel, '*In those days there was no king in Israel: everyone did as he pleased'* (Judges 21:25), is surely an echo of a perceived period of loose non-centralised government. Another advantage of investigating the story of David before he assumes the kingship is that the full range of weapons available to Middle Iron Age peoples is represented and their potential social context is illustrated. These weapons are fully represented in the archaeological record for the period (Yardin 1963).

The Hebrew Bible[5] has often been used as an historical text. The narrative has been used uncritically and has been assumed to represent an unbiased historical record (Kenyon 1969; Wright 1962). This ideology has been seriously challenged (Thompson 1992) and the Bible is now widely recognised as a product of centuries of evolution, reflecting the changing political and religious patterns of Israel. However, there is no consensus of opinion among Biblical scholars. It is of no concern within the context of this study if, in historical reality David, Samuel, or Saul existed. It is the social context of the weapons they wield that is of interest.

It is probable that the majority of Biblical texts were first written during the seventh or eighth centuries BC and substantially edited, if not rewritten, following the exile of a proportion of the Hebrew population in Babylon around the year 587 BC (Blenkinsopp 1995, 119; Thompson 1992). The general outline of the political situation is accepted as reliable (Miller 1991; Laughlin 2000, Chapter 8) but details of conflicts etc. are difficult, if not impossible, to correlate with actual events. It is

[5] Throughout this section, the Prophets or Nevi'im, as translated by the Jewish Publication Society of America, will be used. This publication uses the oldest surviving texts as its basis and includes details which are not found in the official King James version of the Bible.

possible that the coarse details of earlier social institutions are embedded within the narrative. It is often difficult to reshape such material without significantly rewriting the story as a whole. As will be explored further in the section relating to Geometric Greece, it is probable that, without some form of written record, the memory of events is quickly and continuously moulded to fit and explain the existing political situation. The events described for the early life of David do not directly correlate with the political situation existing in the Babylonian period. It is probable that the writers of the Nevi'im were drawing on earlier texts. There is, of course, no way of knowing what level of editing took place during this period. In any event, there is strong potential for the story of David and Goliath to illustrate a social pattern of the emergent Hebrew state during the Iron Age IIA-B (1000–750 BC). It probably represents a worldview of its now literate elite. It was during this period that Israel moved from being an acephalous, segmented, society to a more centralised state (Holladay 1995, 372).

The Sling

The story of David and Goliath is perhaps the best-known narrative in the Western World relating to the use of the sling. The narrative is not necessarily a simple tale of the boyhood deeds of the future King David. It may contain other messages relating to the nature of conflict or competition during the early secondary state formation of Israel (Holladay 1995, 376).

Firstly, David is the youngest son of Jesse and does not accompany his elder brothers in joining the army called together against the Philistines by Saul (Samuel 17:13-14). It is only by delivering food to his brothers that David enters the camp of the Israelite army (Samuel 17:17). This would indicate that David only has status within a 'domestic' context in comparison with his brothers, who are fully grown men, who would presumably have been integrated into the society of the early Hebrews.

David hears Saul's declaration that he will give his daughter in marriage to any man who will face and kill Goliath, who, up to this point, had remained unchallenged due to his sheer size and terrifying appearance. Saul also declares that he will make the victor's father's house free in Israel (Samuel 17:25). This related to the royal levies that were required as part of the apparatus of state level government. The necessity to provide such levies sets David's household within a clearly subservient position, potentially relating him to the new position of the mass of the population, which in a state must have surrendered much of its independent power to the emergent elite.

In the absence of any other volunteer, Saul accepts David's offer (Samuel 17:33-37). Initially, Saul provides David with his own battledress of a warrior; a sword, helmet and 'breastplate'. As David is clothed in the armour, this probably would have been some form of scale, which had a wide distribution throughout the Levant during and before the Iron Age (Yardin 1963, 196-7, 354). These combined elements separate the world of warriors and, in particular, the elite from the rest of the population. None of these artefacts serve any purpose other than the cultural expression of war. Even the spear, otherwise the ubiquitous weapon, is excluded. The text shows the exclusiveness of war as an expression of elite power, as it would have been only the elite who could command the resources to manufacture or purchase such uni-purpose artefacts (Samuel 17:38-9). However, the armour is too large for David and he is unable to even walk in it. This infers there is a significant step in separating David from the warrior elite. As a boy, he is unable to operate fully within this adult masculine world, and has to rely on a weapon that should be restricted to the non-warrior world, of the domestic i.e. the sling. By placing David squarely within the context of boyhood and thus separate from the male world, he is denied even the use of the spear. This would indicate that this is an attempt to exclude the following narrative events from the male world of war and thus from the elite. The only other 'weapon' David carries is a stick (which would be an invaluable aid in any mixed agricultural economy for controlling domestic stock). Again this places him firmly outside the realm of the warrior and squarely into the realm of the domestic or, at least, the non-military.

Before facing Goliath, David selects five river-worn pebbles from a nearby gully. The pebble represents the very essence of the simple weapon drawn from the earth itself. The only knowledge required is a basic understanding of shape and an instinctive feel for the weight of each pebble to ensure as close a possible match between them. This contrasts directly with the forged, manufactured weapons of the warriors. To manufacture these, there would have been significant input of time on the part of specialists, who, in all probability, would have been supported by the elite and would have been at least temporarily, if not permanently, separated from the majority of the agrarian community. There is also a contrast in the materials used: the pebbles are, of course, easily available and access to them impossible to control. Their very abundance makes them worthless as expressions of difference. They can only be representations of sameness, almost of blandness, the common lot. However, manufactured weapons, particularly those of iron (which does not occur naturally in Israel), can be controlled and used as a display of distinction from those segments of society unable to acquire sufficient raw materials or the time of specialists to manufacture goods beyond the domestic, multipurpose tool.

David then proceeds to challenge Goliath. A series of formalistic insults follow (Samuel 17:43-44, 46). They have echoes in the Iliad (Redfield 1994, 168-9) and other Eastern Mediterranean texts (Wyatt 1998, 219, Footnote 195). They appear to reduce Goliath, despite his size and reputation, to the same level as a wild animal, which David has specifically described himself as conquering earlier in the narrative, using just his sling (Samuel 1:35-36). This section of the narrative would also place David in the context of protector of a flock, a common analogy for a ruler throughout the Near East (Haubold 2000, 17).

As David approaches Goliath, he declares that he comes to face him, not so much with weapons, but in the name of the Lord (Samuel 17:46). It appears there is a deliberate attempt to remove success from the sling. David's ultimate victory is directly attributed to Yahweh: it would not have mattered with what weapon David faced Goliath. His fate was already sealed by divine intervention. This reduces David's youthful nature and the sling's social context. From here on David is the chosen of God. This places him directly in the context of the older pre-state society, which may have been resisting state formation under the early kings. There appears to be a compromise between religious power and secular power, though David becomes king, his power is seen as coming directly from Yahweh so he is still part of the older order.

The power of Yahweh in this context is further shown in Samuel (1:25, 29), where his ability to destroy Saul's enemies is directly related to the use of the sling, as Yahweh is able to cast away their lives *'as from the pocket of a sling'*. This statement, given the nature of other references to the sling (particularly the later Proverbs (26:8)), where tying a stone in a sling is correlated with honouring a fool, does not lessen the power of Yahweh as a war god but increases it, as he is able to defend the king even with a sling.

David then proceeds to cast at Goliath, striking him on the forehead with such force that Goliath falls to the ground (Samuel 17: 48-51). Despite the fact that the text describes David as striking Goliath down and killing him, it would appear that David is unable to slay Goliath with a sling stone alone. David has to use Goliath's own sword to '*dispatch him and cut off his head*' (Samuel 17:51). The verses stress that he did not have a sword in his hand (Samuel 17:50). This reiterates the lack of elite context for David; he is outside this element of society at this moment. David's use of Goliath's own sword projects him into an elite context (Samuel 17:51). This appears to be an attempt to remove the sling from the world of lethal warrior weapons and requires David to slay the enemy of Israel with the appropriate elite weapon *par excellence,* the sword. This could be seen as a metaphor; although the common people could harm their enemies, it required the intervention of the warrior elite (the sword) to destroy them.

The sling is further used as a signifier of social context in Chronicles (1:12,2), where the slingers listed in the army are described as being Benjamites and the kin of Saul. The writers of the Hebrew Bible associate ethnic origin with particular weapon types during the period of the united monarchy (Chronicles 1:12,2), when the various tribes of Israel are described each with their own unique weapon class. Clearly, this is a literary fabrication, as it is unlikely each tribe would have used only those weapons designated to it within the list. However, in the minds of a contemporary audience, such classification would potentially have had great resonance.

In particular, Saul is described as being a member of the Benjamite tribe (Chronicles 1:12,2). The Benjamites are the progenitors of a 'civil war' in Israel immediately before the rise of the kings (Judges 20-21). As part of their punishment, they were not allowed to marry women of the other tribes of Israel and had to take women from outside (by force). The Benjamites were ethnically associated with the sling (Judges 20:16). This association belittles the social position of the sling further; not only is it a domestic weapon, but one directly correlated with internal strife and troublemaking.

This would suggest that the association of David killing Goliath with a sling may be related to Saul as king. In some ways this may be an attempt to blur the social boundaries that the emergent elite may have been creating. In particular, it may have been a reminder of the ancestry of Saul (the first king) and the taint of internal strife that this carried. By associating Saul with the Benjamites, and thus with David, his position as king is drawn into the social relationships of earlier patterns.

The Bow

The comparison of Saul's son Jonathan with David, is of particular interest in the context of the social representation of weapons. Jonathan is a fully integrated member of the elite. As such, does not use the sling, but is incorporated into the wider field of Middle Eastern elites with his use of the bow. Archery had acquired widespread elite connotations by the late Bronze Age, throughout the Levant, which it retained until at least the end of the Achaemenid Persian period. Jonathan's association with the bow may been seen as joining him, and therefore the elite of Iron Age Israel, to a wider, more cosmopolitan world outside the close knit tribes that appear to have characterised the earlier Hebrew period. Following David's defeat of Goliath, Jonathan presents him with his own tunic, sword, bow and belt (Samuel 1:18:4). This act draws David from the 'egalitarian' world of his father into the elite world of Saul and Jonathan, and directly connects the latter group with the defeat of Goliath and the Philistines.

The type of bow being represented would, in all probability, not be a simple self-bow manufactured from a single piece of timber, but the more complex composite bow. These artefacts not only required access to a range of materials (e.g. good timber, horn or bone, sinew, and some form of adhesive) but also required the employment of specialists in their manufacture (Rausing 1967, 88). Such bows were widespread across the Middle East and represented one method of display employed by the elites of the area.

The power of the composite bow, as a symbol of elitism, is demonstrated by the earlier narrative of The Story of Aqhat from the late Bronze Age city of Ugarit (Wyatt 1998, 246-313). Within the narrative, a composite bow is manufactured, which is desired by the goddess Aqhat. The resulting tragedy leads to the death of the main human characters. This narrative clearly shows the desirability of archery equipment in display and as a signifier of rank or position. It would appear from textual evidence from Ugarit that certain family names (e.g.; *Bib Anat* and *Abdlabi*) were associated not only with the goddess of war, Aqhat, but also specialist military archery families (Milik 1956, 5-6).

Within the development of the state in Israel, the power of 'mighty men' was of particular importance (Holladay 1995, 378-9). It may be inferred that these individuals gained and maintained power primarily through the exercise of military power, and that the society within which they operated had not yet developed to where the status of an individual was ascribed rather than archived. It is clear from the story relating to David and Jonathan that Jonathan's succession was not guaranteed, and that the acquisition of a loyal following of retainers was one basis for securing secular power. Identification of individuals with distance weapons is highlighted by the discovery of 'arrowheads' manufactured during the Israeli Iron Age IIA-B. These are inscribed with individuals' names (Cross 1992; Milik 1956). Although usually classified as 'arrowheads', their size (98.5-105mm) and therefore their weight (this is not given, but must be significant, considering their size) would suggest that they were javelins or darts rather than arrows cast from a bow.

The inscriptions cut into the projectile heads often refer to an individual as 'the man of X' (Cross 1992, 22), which would strongly indicate that a trend towards a hierarchical structure of dependence was developing. The inscribing of personal names onto weapons would suggest that in the society where they were used the individual had become more prominent, with the lessening of group identity. It has been suggested that the use of individually identified 'arrowheads' is associated with their use in archery competitions (Mitchell 1986, 151; Cross 1992, 26). However, arrows used for competition can easily be identified by individual archers, not only due to the variation in shaft length, but also by any decoration that the shaft or fletchings may have carried. The use of the simple tapering insertion tang as the hafting method for these projectile points creates an inherently weak joint and requires the use of a hollow shaft. This, most likely, would have been reed for arrows or, for larger javelins or darts, *Arundo donax,* (a giant reed growing to 6m high and common throughout the region). The inherent weakness of an insertion tang would suggest that the heads could easily become detached in the violence of battle or the hunt, making identification of the individual responsible for the death of either another warrior or prey difficult.

The potential need to identify casualties to a particular individual may have a direct bearing on the level and type of warfare engaged in during the Iron Age I-IIB period. If the description in the Book of Samuel was a reality, then single combat between 'heroes' would appear to have been commonly accepted. Here may be one of the rationales behind the exclusion of the sling from this type of warfare, river-worn pebbles being not only readily available but, in their essence, anonymous, the antithesis of heroic warfare.

The final acts associated with archery in the story of David and Saul demonstrate the fluid position that the bow held in early Hebrew society. During Saul's last battle against the Philistines, he is surrounded by enemy archers and severely wounded. All three of Saul's sons are killed in the rout following the Battle of Mount Gilboa, though there is no reference as to how. It is clear, however, that the Philistine archers do not actually kill Saul. He attempts suicide (itself considered an act against the divine laws of God) using his own sword, and is finally killed by an Amalekite using a spear. This suggests that although the bow was a powerful weapon, the death of Saul removed the bow from the act of regicide, allowing it to maintain its position in the elite. This ideology is carried further as the dirge of Saul's death, sung by David, is entitled the 'Song of the Bow'. This directly associates archery with the defence of Israel against its enemies, in particular the Philistines. Jonathan's exploits with the bow are particularly highlighted: he is described as never turning from a fight while using his bow:

> *From the blood of the slain*
> *From the fat of warriors*
> *The bow of Jonathan*
> *Never turned back* (Samuel 2:1:22)

Jonathan, and thus the bow, is placed within the context of a true warrior, one who does not flee when faced by the enemy. Jonathan's direct link with the elite also connects the bow with this group. The last verse of the 'Song of the Bow', specifically describes both Jonathan and Saul as the weapons of war. This expresses a strong association between the bow and war, the latter being one of the main expressions of the elite.

The Spear

From its first appearance in the Book of Samuel, the spear is categorised into two distinct forms. Goliath is described as having *'a bronze javelin (slung) from his shoulders. The shaft of his spear was like a weaver's bar and the iron head of his spear weighed six hundred shekels'* (Samuel 17:6-7). Clearly there is a distinction not only of size but also of manufacture associated with the throwing javelin and the thrusting spear. The javelin is specifically described as bronze. This would correlate well with the archaeological evidence discussed above relating to projectile heads with inscribed names. It is Goliath's thrusting spear that is specifically compared to a weaver's beam. This may associate such weapons with a more domestic context, as weaving was undertaken by women within the home (Samuel 2:5:29). This context may at first glance appear inappropriate for a weapon of war. However, if spears and, in particular, thrusting spears represented the most common form of weapon available to the average tribal levy within the Israeli 'army', then the context begins to assume more relevance. When Saul arms David before his duel with Goliath, the spear is excluded from the weapons given to him. Although Saul later in the text uses a spear, this is under quite specific circumstances that will be discussed below.

Certainly, from the iconographic evidence, the spear would appear to be the most widely used weapon within contemporary Levant (Yardin 1963, 352, 368, 420-1). Although the arming of troops of major states such as Babylon, Egypt and Assyria would have been carried out from centralised stores, it is more probable that the weapons of a tribal levy would have been the individual possessions of each warrior, as is clearly the case in Dark Age Greece, which is roughly contemporary with the period of David and represents a non-state level society. The association with the domestic realm therefore is clearer. Each warrior would have maintained his spear at home. Only those weapons directly associated with the elite, and therefore outside the normal household sphere of influence, are presented to David. More importantly, the throwing javelin of Goliath is not included within this reference, and this may indicate that this is a deliberate attempt to differentiate such weapons from the common place 'domestic' thrusting spear of the tribal levy.

There is a clear association with Saul and the use of the throwing spear. Following a further defeat of the Philistines, David's growing power appears to threaten Saul who, while listening to David playing the lyre, attempts to kill him by throwing his spear at him (Samuel 18:11). Saul tries to kill David in exactly the same way in the next chapter (Samuel 19:10). He also attempts to kill his own son, Jonathan, with a throwing spear, after he has raised his anger by defending the actions of David (Samuel 21:33). It is clear that the carrying of or, at least, the possession of a throwing spear by the king at 'court' was not considered unusual. In fact, it would appear to be part of the symbolism associated with kingship, at least that expressed by Saul (Samuel 22:6). He is described as both carrying such a weapon during the day and sleeping with a spear stuck in the ground near his head (Samuel 26:7).

The death of Saul may also serve to illustrate the liminal nature of the spear. Although he falls on his own sword having seen his sons killed by the Philistines (Samuel 1:31:2), it is clear from the next passage that Saul is not dead, as an anonymous Amalekite (a group defiled by the Nevi'im) tells David that he killed Saul, after he found him *'leaning on his spear'* (Samuel 2:1:6).

The potential for the spear to be the instrument of discord is demonstrated in the tale relating to Abner and Asahel. During the aftermath of a battle between the men of Israel, of David's army, Asahel pursues Abner, leader of the Benjamite forces, across the battlefield, despite warnings to desist and an apparent offer to provide captives in his place. '*Seize one of our boys and strip off his tunic'* (Samuel 2:2:21), Abner shouts to his pursuer. Asahel refuses. Even after threats of death, Asahel continues the pursuit. Eventually, Abner kills him by thrusting his spear backwards right through him. The Hebrew relating to the actual incident is difficult (Jewish Publication Society 1978, 170, Footnote b) and would suggest, given the overall context of the pursuit, that both runners were *'Swift of foot like a gazelle in the open field'*, that the fatal blow is more likely to have been dealt using a throwing spear, and in particular, as it is described as a 'backward thrust'.

The spear then is given something of a liminal position within the Nevi'im and, by analogy, in Iron Age I - IIb Israel. Although the evidence rests on the interpretation of one reference, it seems possible, that the thrusting spear was ideologically associated with the domestic arena. This does not necessarily imply a feminisation of this particular weapon, merely an association with the everyday running of an agriculturally based society. This is understandable if we accept that the mass of Israeli troops were a tribal levy who would have maintained their own primary weapons. This may go some way to explaining the desire of the Nevi'im's authors to classify weapons ethnically. These groups were perhaps supposed to have these weapons; the reality of the situation was immaterial. This would separate the thrusting spear from the elite and place it firmly within the context of the mass of the population. The Nevi'im does not degrade this weapon, it essentially ignores it. It has no dangerous social context, and any attempt to lessen it would potentially have been seen by the majority as an attack on their status as warriors, within the war-defined society that the Nevi'im presents to us.

The throwing spear in contrast is placed within a liminal position. It is clearly associated with the elite. In fact, it is shown to personify the ultimate member of that class – the king. Yet, it is this weapon that is used in attempts to kill both David and Jonathan. In both cases, the guilt for such actions is directly placed on Saul, but it is noteworthy that he never tries to kill David or Jonathan using a sword, dagger, or arrow. Each time he uses a throwing spear. This could be read simply; it was the only weapon available on each occasion. However, this concept is eroded when considering the killing of Abner by Joab (Samuel 2:3:27). Joab avenges the death of his brother, Asahel, by striking Abner in the belly. The same scenario is used for the death of Ish-bosheth. On neither occasion is a weapon type mentioned, although with the death of Ish-bosheth we are told that he was stabbed to death, after having been struck in the belly. The stabbing may indicate the use of a sword or dagger, but could equally apply to the use of a spear; no clear definition is given, suggesting that specific internal Israeli killings are not directly associated with any particular weapon type. This gives greater meaning to the attempted killings undertaken by Saul, as the throwing spear is specifically described in all three cases.

There appears to be a strong distinction between the thrusting spear, with its potential connotations of 'domestic' stability and internal continuity, and the throwing spear, with its connections with attempted kin slaying and disruptive kingship. The duality of their associations has distinct social significance: on the one hand, the stability of the 'older' pre-kingship society and, on the other, the potential disruption and internal strife presented by the rising elite. Of particular interest is the fact that David is rarely associated with the spear. Although he asks for a spear or sword (Samuel 1:21:9), he is offered only the sword of the defeated Goliath. The only other time David is in possession of a spear is when he has taken it from Saul, in order to show his loyalty to the monarch (Samuel 1:26:22). This lack of identification with the presumptive heir to the throne of a united monarchy is of interest. It is as if the growing power of the monarchy should no longer be connected in any form with the older tribal political structures. Once David has slain Goliath, if he is associated with weaponry at all (and in most cases he is not), he is associated with the sword.

The Sword

It is sometimes difficult to see, when authors of the Nevi'im describe the destruction of a city or town by the sword (e.g. Samuel 1:22:19), that this should be viewed within the context of righteous conquest. Such acts, far from being repulsive to the contemporary audience, were part of the expression of their might over the other peoples in the area (the historical reality of such matters is not at issue here, as the narratives were believed within the context of their time). Within this context, it is not so much that might is right, but that the right have (or are given) the might forcibly to impose their wishes on others.

The use of the sword as an instrument of 'mass' destruction is strongly associated with the Israelites within the Nevi'im. The destructions of towns and cities by non-Israelites are recorded but they invariably eschew the use of the sword. Instead, the enemy is described as 'storming' (Samuel 1:30:1), burning or raiding (Samuel 1:23:1). No direct mention of the weapons used within these contexts is made. Yet the response of David and his troops to such attacks is to *'Gird on his sword: David too girded on his sword'* (Samuel 1:25:13). This disassociation may, in the first place, represent an attempt to reduce the association of the sword with acts that would be unacceptable to the Hebrew elite, in particular, the destruction of their own settlements. Secondly, it represents a meshing of the revenge of David against the destroyers of Israel, symbolised by the righteous use of the sword.

The elite nature of the sword is confirmed within this context throughout the books of Samuel. From the first instance, when the boy David sets out to fight Goliath, the sword alone is placed within the context of the ruling elite. David is given Saul's arms and armour; the only weapon presented to him is the sword. As has been explored above, it would appear as though David, following his defeat of Goliath, is unable to kill the Philistine other than by using the defeated giant's own sword. Later Jonathan integrates David into the ruling group by providing him with (amongst other things) his sword.

The sword's position within the ideology of the elite is further strengthened when David, fleeing from Saul, comes to the priest of Nob, Ahimelech. He specifically asks for a spear or sword, as he claims to have left his behind in his hurry to take a message for Saul. Ahimelech is unable to offer David any weapon save the sword of Goliath, and David says the same as *'there is none like it'* (Samuel 1:21:9). No other weapon is described in such terms; the sword is clearly seen as incomparable. It would also appear that Goliath's sword is the only weapon of his panoply preserved as a form of trophy within a religious context. This would in turn, suggest that the sword held a special place within the ideology of 'historical' trophies.

Even within the context of single combat (perhaps one of the prime signifiers of elite warfare), the use of the sword can be denied. During the internal conflict for power between David's and Saul's sons, following their fathers' deaths, it is decided that, before battle should be joined, twelve young men from each side should *'come forward and sport'* before the assembled armies (Samuel 2:2:14). It is clear from the context that this sport is single combat. None of the combatants uses swords. Instead, each kills the other using a dagger. The dagger, an evolved form of knife, possibly has similar connotations to the thrusting spear. This shift away from the sword as the killing instrument would strongly suggest an attempt to deflect any blame for such internal strife away from the sword and its strong links with the elite.

The use of the sword in the context of Saul's attempt to commit suicide is unusual (Samuel 1:31:4). The Nevi'im is not solely an elite-dominated text. There are strong connotations of conflict between the secular and the divine (suicide being against Hebrew law). This may infer that the death of Saul is an expression of this conflict. Saul tries to kill himself, not only by his own hand, but also by the ultimate expression of elite status, the sword. This would appear to be a closed circle of paradoxes. Saul the king, the leader of the new elite, is destroyed by the very symbol that defines him and his class. It is as though the elite contain the roots of their own destruction. The sword, however, later escapes this deed; the killing of Saul is described as being undertaken by an Amalekite using a spear (Samuel 1:9-11). It would appear that this is an attempt not only to reduce the impact of Saul's suicide but also to shift the 'blame' for his death onto a group that had threatened Israel at this time. It also successfully reduces the sword's involvement in a crime against Yahweh.

Conclusions

If the proposal that Israel's state formation is not an internally driven evolution, but results primarily from secondary sources (e.g. the Phoenicians) is correct, and that this transformation occurred within less than one generation (Holladay 1995, 372), the David narrative may have particular significance. It may be that the message (at least in part) is that the mass of the population, under the older elder-controlled system, although capable of defending itself against outside aggressors (in the shape of the Philistines), is unable to defeat these enemies without the direct intervention of the new elite.

The use of weapons as indicators of social inclusion or exclusion is particularly well drawn within the books of Samuel, and their use underlines the conflict that appears to be occurring between the secular and the religious elements of Iron Age I and II Israel. Each weapon type is given a context. The two most prominent exemplars of social exclusion (to the mass of the population), and inclusion (within the elite) are the sword and the bow. The spear appears to cross social boundaries between the elite and the populous. The sling, by contrast, appears to be firmly embedded in the culture of the non-elite and is excluded from the elite's worldview.

The sword is never given negative connotations. Even the potential 'regicide' in the attempted suicide of Saul is deflected by the actual killing being committed by an enemy of the Hebrews. Alone, the sword stands for the elite of Iron Age I and II Israel. It is given, to the exclusion of all other weapons, to David by Saul when he first faces Goliath. Goliath's sword is the only weapon kept from his panoply in a religious context. Whole 'cities' are put to the sword by the Hebrews. War itself is metaphorically identified with the sword (Samuel 2.2.26). *The sword of Saul never withdrew empty* (Samuel 2:1:22): in this line of the Song of the Bow, Saul's bravery in battle is directly associated with the sword.

The bow, the other symbol of exclusivity, is strangely denied a context in relationship to combat when used by Jonathan but it is given context when used by the Philistines (Samuel 1:31:1). This may be the result of omissions by later historiographers or may be a requirement of the type of heroic warfare that Goliath's challenges engender.

It would be easy to strike and potentially kill an enemy at a distance. However, only a true warrior would dare to face a hero in physical combat. The bow, though directly associated with the elite through Jonathan, could be seen as a weapon of the coward and this context is specifically denied in the 'Song of the Bow' (Samuel 2:1:22). The bow also has strong associations with the more cosmopolitan world of the Middle East, and the dangers that its polytheistic religion may have posed to the emergent monotheism of the Hebrews. It would therefore be in the interests of the priests to play down the importance of the bow, and thus erode connections with the outside world.

The use of the sling by David would, by correlation, appear to carry connotations of, if not actual cowardice, at least a reluctance to be fully involved in war that would be associated with the elite. In conjunction, this may strongly associate the sling with the relatively unstratified elder-led society, where all (male) members, at least, had the potential to exercise power in the future (Holladay 1995, 378). The sling is clearly placed in the category of a peasant weapon, incapable of overriding success against Israel's traditional enemies. This could only be achieved by the continued existence of the elite, and the social stratification that this entailed.

The spear straddles these two worlds. As a thrusting weapon, there appears to be an attempt to place it within a more 'domestic' context. As a throwing weapon, it is the choice of Saul. With it he twice tries to kill David, and once his own son. The spear is also the weapon that eventually kills Saul at Mount Gilboa. It would appear that the elite could not make redundant such a useful weapon as the spear; it would simply not be practical. Therefore they created a liminal social context for it. The spear is neither elite nor non-elite. It allows them to carry out disreputable deeds, without sullying the sword or bow, but it is still retained as a principal fighting weapon of the mass of the population. Unlike the sling, which can be replaced by the bow as an expensive distance weapon demonstrating the power of the elites, the spear has no such replacement, and must, due to its ubiquitous nature, be given a place, however grudgingly, in the world of the elite.

Using the narrative of David, it has been possible to illustrate the potential use of weapons as social indicators in a fully sedentary agricultural society that had only recently moved from an egalitarian style of government to one that would soon be dominated by the unified monarchy of David. In conclusion, it is probable that in the period of early state formation of Israel and Judea, weapons carried particular messages. Elitism is expressed through the use of 'expensive' (in terms of materials and the requirements of time in their manufacture) weapons, the bow and sword. The older, less centralised order is also represented through the use of weapons that can be more directly related to a domestic or self-sufficient context, where contact with a more cosmopolitan world is not required and, in the case of early Iron Age Israel may have been considered undesirable. It is in this context that the spear, and, more particularly the sling, stand out as social indicators.

Causes of conflict in Iron Age I and II Israel

Carroll (1999) is correct to state that it is impossible to construct a full picture of early Hebrew warfare from the information provided by the Bible. However, it may be possible to see some of the motivation behind the conflicts, or at least those presented within the context of the narrative.

The most obvious, and perhaps the most contentious, is the ideology of conquest. The Nevi'im contains an historicised account of the conquest of the Levant by the Hebrews. The historical nature of this conquest has been seriously questioned in recent years. Within the context of the contest between David and Goliath, it is clear that the concept of individual combat allies itself to this ideology of conquest. Quite specifically, Goliath states that, if David defeats him, the Philistines will become the slaves of the Israelites, and *vice versa* (Samuel 1:17:9) This would strongly infer that it is the conquest of people as much as of land that was idealised within the Nevi'im.

Outside this grand scheme, there are distinct references to other forms of warfare, in particular to raids against enemy tribes, in an attempt to drive off cattle, asses, camels and other goods (Samuel 1:23:5, 1:23:27, 1:30:20). The passage associated with Nabal (Samuel 1:25:2-25) would certainly suggest that the wealth of an individual was measured by the number of livestock he owned. Thus the raiding by David would no doubt have increased his personal wealth and, in turn, his social standing. This type of raiding appears to be typical of an emergent elite, where prestige is confirmed by the ability to give gifts to retainers. Raiding also confers prestige upon the raider, by his continual acts of bravery against the enemies of his group. David's raids are typical of a leader gaining greater and greater prestige among not only his followers but also among the population at large. This made his transition into kingship easier.

One type of raiding that appears in the Nevi'im in a disguised context is the taking of women as captives for wives and possibly slaves. The latter case appears in Samuel (1:30:1-2), when the Amalekites had taken the town of Ziklag and carried off all the women *lowborn and highborn alike: they did not kill any* (Samuel 1:30:2). This would certainly imply that the captives are to be sold into slavery, rather than taken as wives. The clearest example of wife taking is in Judges, when the Benjamites, having fought against their kin, are refused wives by the other Israelite tribes. They therefore raid the inhabitants of Jabesh-Gilead and Shiloh. The fact that the Benjamites are instructed to kill the men and women of Jabesh-Gilead, and take only the remaining virgins, and that the captives taken at Shiloh are described as girls, infers that the captives are not merely slave women or concubines, but will be married into the Hebrew community. That this type of specific wife taking takes place in a context when the writers of the Nevi'im believed that there were no kings in Israel is of particular importance, and may indicate that raiding for marriageable women formed part of the social fabric of the Hebrew community, before the rise of a more centralised authority.

Case Study 3

Late Geometric and Early Archaic Attica (*c.* 750-*c.* 650 BC)

'No bows will be stretched in numbers, nor slings in multitude, when Ares joins the struggle in the plain' (Archilochos Fragment 3)

Introduction

The Late Geometric and Archaic periods in Greece encompass an era of significant social change (Coldstream 1977; Fisher and van Wees 1998; Osborne 1996; Tandy 1997). Unlike many other emergent complex societies, in Late Geometric and Early Archaic Attica, there appears to be a move away from hereditary stratified elites towards a form of democracy where a large (specifically the male citizen) section of the population empowered itself, and thus eroded or removed power from the existing aristocracy (Osborne 1996, Chapter 9; Seaford 1994). Many of these changes appear to have their roots in the very period when Late Geometric and Early Archaic Greece was entering the literate world of the Western Mediterranean (Burket 1992; Coldstream 1977, Chapter 11).

The surviving literature provides a detailed, though often complex and contradictory, view of what this changing society was like, and how it reacted to the various stimuli that were in force during this period. The type of weapons, and in particular their use, as with any social representation, mirrored these developments, and often expressed the changing ideologies of the elite and the *domos* who often opposed them.

The exploration of the social significance of weaponry is complicated by the fact that Greece was neither uniform, nor unified, during this period. It comprised a series of competing city state polities, which, although sharing a common language, did not share a common culture. It has become apparent that although our knowledge relating to warfare in and between classical Athens and Sparta is reasonably developed (Hanson 1991; 1998; Pritchett 1971-1985; Munn 1993), outside these areas our knowledge even for the Classical period itself becomes poor (Hanson 2000, 201). There is evidence that certain areas of Greece did not develop hoplite warfare until into the Classical period and that the Macedonian North retained much of its warrior ideology until the start of the Hellenistic period. This lack of knowledge is even deeper in the 'Dark Ages' that followed the collapse of the Mycenaean Palatial system. It is, therefore, impossible to study Late Geometric and Early Archaic Greece as a single element, and the strong regionalisation must be taken into consideration. Fortunately the city state of Attica provides a wealth of detail, both archaeological and iconographic (Coldstream 1977; Snodgrass 2000). It is also fortunate that Homer composed his works in a dialect that was a fusion between the closely related Aeolic and Ionic (Parry 1971, 5). Ionic was the dialect spoken in Archaic and Classical Athens. Mythologically, Athens is presented as a significant starting point for many of the Ionic migrations to Asia Minor (Herodotus 1.145; Pausanias 7.1.4), which is the location of Smyrna, the traditional birthplace of Homer (Coldstream 1977, 341). There is consensus among Homeric scholars that neither the Iliad nor the Odyssey relate to a particular polity at a particular time, but seem to have been created to appeal to a pan-Hellenic audience. One other significant author provides an insight into the society of Archaic Greece: Apollodorus wrote a handbook of the mythology of Greece, which, although written in the second century AD, can be shown to include many elements that are drawn from earlier documents (Gantz 1993, 2). This evidence will be used, where the narratives presented by Apollodorus supplement those of Homer or Hesiod, with the knowledge that it may contain many centuries of corruption.

It was during the inception of the Archaic Period that the works of Homer[6] and Hesiod were first written down. Archaic Greece also produces a number of other poets whose surviving works vary significantly in length and completeness. Homer and Hesiod both present a world where the hereditary elites control the mass of the population either through gift giving (Finley 1979, 120-3; Haubold 2000, 37-40; van Wees 1992, 169-72), or by violence (Jackson 1995; Tandy 1997; van Wees 1992, 172-82). The works of Homer plus the archaeological record will provide the main avenue for investigating the social significance of weaponry and how this changes in response to the democratisation of Attica.

Before proceeding any further, the difficult issue of how the early literature relates to the palatial system of the Mycenaeans must be addressed. There is a vast corpus of work attempting to prove or disprove a connection between the two. The most widely accepted interpretations of the Iliad and Odyssey are those by Finley (1979) and van Wees (1986). Whereas Finley places the works of Homer in the ninth or tenth centuries (i.e. firmly within the Dark Ages), van Wees opts for a later date at the start of the Archaic Period. The social interpretation of van Wees (1986) will be used as the background for this study. This has particular pivotal importance in relationship to the social significance of weaponry. There is considerable evidence to indicate that all the elements of warfare as illustrated by Homer can be directly derived from the immediate and contemporary environment (both political and religious) of Late Geometric and Early Archaic Greece.

It has long been recognised that the Late Geometric and Early Archaic Greeks had open trading contacts with the more complex societies of the Levant, in particular the Phoenicians. This interaction is widely recognised as being one of the prime motivators in the internal changes that occurred in Greece during the eighth century BC. However, an ideology persists that many elements within the Iliad and Odyssey must be remnants of earlier Mycenaean stories. This theory is not supported by the available evidence.

A classic example of this ideology of survival is the war chariot. As recently as 1995 (Jackson 1995, 66), it has been stated that the chariot in painted scenes of Late Geometric Greece must be a legacy of the Bronze Age. This ignores the fact that the dominant state of the Levant in the seventh and eighth centuries BC (Assyria) used war chariots as a principal arm of its offensive armies. It is of course during this very period when the works of Homer are first written down (Osborne 1996, 6). There is even some evidence that some Late Geometric Greeks may have fought the Assyrians (Luckenbil 1926, paragraphs 286-8) and may have had direct first hand experience of such combat. Certainly, Cyprus was incorporated into the Assyrian Empire in 709 BC (Karageorghis 1976, 109). There were large-scale Greek enclaves on the island, and adoption of symbols of power from the Assyrians could easily have taken place here. It is much more likely that the use of war chariots by the Greeks is an emulation of their powerful neighbour. The use of such symbols of power would have buttressed the elites in Greece, and provided them with a visual connection to the dominant state of the period. The description by Nestor (Iliad 4: 347-56) of how to use chariots seems to fit the use of 'heavy' Assyrian chariots better that the 'lighter' chariots apparently shown on the Dipylon painter's vases. Finley's (1979, 45) statement that Homer did not understand chariot warfare is flawed, as it assumes that there is a general and recognised model for chariot warfare. This is simply not the case (see Greenhalgh 1973 and Littauer and Crouwell 1983 for opposing views). The method of fighting in the Iliad is echoed not only in Caesar's description of the British in the 1st Century BC, but would appear to be paralleled in the Táin Bó Cuailnge from Early Mediaeval Ireland. This would suggest that Late Geometric and Early Archaic chariots were as much an expression of elite exclusivity as they were vehicles of war.

The same can be said for the use of bronze weapons by the heroes. At the end of the Geometric Period, a number of Mycenaean tholos tombs were opened. They represented a focal point for hero worship (de Polignac 1995; Morris 1988) which, in turn, acted as a powerful connection by the elite (who claimed ancestral descent from these heroes) to a deep mythological past (Antonaccio 1993; Whitney 1988). It has been claimed that such use of Bronze Age tombs is a direct response to the widespread dissemination of surviving Bronze Age poems (Coldstream 1976, 349-52). However, there is no distinct chronological correlation between the two (Snodgrass 2000, 391), and it is equally likely that the opening of the tombs stimulated the poets. The desire of the 'Dark Age' elites to provide themselves with a deep 'historical' past may be a direct response to significant contact with the Levant, where such history was more commonplace and allowed the dominant elites to justify their position through the use of time and tradition.

Even the famous boar's tusk helmet worn by Odysseus is an element that is not recapitulated within the societies of the seventh/eighth centuries, when the epics were first written down, and can be shown to be a late literary interpolation (Fagles 1990, 14). It presumably represents an ancient archaeological find placed within a pseudo-historical framework. It is against this background of assuming that the Homeric literature is a recent invention (as oral history has a great capacity to change in the face

[6] Homer will be used to identify the author/authors of The Iliad and The Odyssey.

of differing social needs) that the exploration of the social significance of weaponry will be undertaken. During this period, Greece can perhaps be seen as a 'peripheral' culture that emulates the symbols of its more complex neighbours without necessarily understanding or incorporating their meaning totally within their ideology. The evidence that any of the institutions or ideologies of the state level Mycenaean period survived to the seventh/eighth centuries BC is contradictory. Such evidence can be demonstrated to be within the purview of the Late Geometric and Early Archaic Greeks without reference to earlier periods. Therefore, the significance placed upon weaponry within Homer and other archaic authors is taken to be derived solely from a pre-state sedentary agricultural context.

The model for Late Geometric and Early Archaic Greece presented by both Finley and van Wees posits a small dominant warrior elite (literally in this context an aristocracy – *the best of men*), controlling the rest of the population, which comprises a mass of free (to varying levels) peasants and a number of traders, whose social standing seems to be ambiguous, somewhere between pirates and commercial agents. The warrior elite did not engage in trade, and looked down on those who did.

Warfare is expressed by a series of single combats between these elite members with a general 'free for all' also taking place (van Wees 1994, 2-9). Despite this apparent level of continual violence, fortifications in the form of upstanding walls are rare, confined to the Greek colonial areas on the western coast of Turkey (Coldstream 1977, 303). These are more likely to be a response to an external native threat rather than a Greek internal one, particularly at Smyrna (Nicholls 1958) where the Greek settlement is only some 20km from the border with the expansive polity of Lydia, which eventually in the sixth century BC conquered the town (*ibid,* 128-9).

The full panoply of weapons used by the middle Iron Age peoples of Britain was used by the military forces of Late Geometric and Early Archaic Greece. The use of chariots by the elite will not be taken into account as they present too many interpretative problems. Indeed, many authors question the use of chariots in the warfare of this period, and see them as either funerary 'carts', sporting vehicles (used for racing), or merely iconographic representations of a conceptual ideal (Coldstream 1977, 352; Finley 1979, 149; Snodgrass 1964, 169).

The Sling

Archaeological evidence for the sling in the Late Geometric and Early Archaic periods is slight. Rounded pebbles seem to have been the norm (Nicholls 1958, Note 116), the use of lead shot not being rediscovered until Classical times.

The sling is mentioned twice within the Iliad. Firstly, it is used to bind the wounded hand of Helenus by his retainer Agenor (Iliad 13:689-92). The sling is specifically described as being made of tightly twisted wool, and is therefore a fabric which, within the Greek world, places it firmly within the field of the feminine (Blundell 1995, 141), an attribute wholly unacceptable to the Greek warrior ideal (see Ovid *Metamorphoses* xiii and Hyginus *Fabula* 96, for Achilles dressed as a woman to avoid going to Troy). The Agenor passage is a demonstration of the social position of the sling within the heroic world of Homer. In brief, the following action takes place: Helenus is wounded by a spear thrown by Menelaos, Helenus having himself cast an arrow at Menelaos, which had bounced off his breastplate with no damage. The wounded archer is now tended by his retainer, who appears to be a slinger. Both Helenus and Agenor are Trojans and already in a socially inferior position to the Achaeans. The obvious connotation is that an archer, as will be shown below, was not necessarily regarded as a true warrior, and is tended to by a man of lower social status.

To more fully explore the potential social context of the sling in Late Geometric and Early Archaic Greece, it will be useful to quote the second reference in full (Iliad 13:822-33):

> *But the Locrians followed little Ajax.*
> *They have no love for stand-and-fight encounters-*
> *had no crested bronze helmets to guard their heads*
> *no balanced shields in their grasp, no ashen spears*
> *only their bows and slings of springy, twisted wool.*
> *Trusting to these, they followed their chief to Troy,*
> *shooting these, salvo on pelting salvo,*
> *they tore the Trojan battle line to pieces.*
> *So the men in heavy armour fought at the front*
> *they grappled Trojans and Hector helmed in bronze*
> *while Locrians slung from the rear, safe, out of range*

The passage shows that the Locrians are the retained troops of little Ajax, and that they did not attend the siege of Troy armed with the proper panoply of the warrior elite but, instead, carried both slings and bows, despite the fact that these weapons were of great importance in conducting siege operations among the city states of the Levant coast. They never appear to have been used in the assault of the walls of Troy. Instead they play their part on the open field of battle.

It is clear that the main thrust of the above passage deals with the sling and not the bow, *'salvo on pelting salvo'* appears to refer to the casting of sling stones rather than arrows. What is also apparent is that the Locrians were incapable of 'proper' combat as portrayed by the Iliad. They could only sling into the Trojan lines while safe and out of range. This associates the use of the sling with desire for safety and avoidance of the close combat that was becoming a feature of Greek warfare, and would become its defining attribute (Hanson 1991, 74-5; 1998, 180). Slingers are seen as deadly, that is clear, but also as inferior, avoiding the real test of a man, hand-to-hand combat. The concentration of military power in the hands of the hoplites at the end of this period (Hunt 1998, 9-11) only served to enforce this concept for the sling and the bow. An inscription at the shrine of Artemis at Chalcis, Euboea, records an agreement '*not to use long distance missiles*' (Murray 1993, 78-9) and the poet Archilochos (Fragment 3) in his poem of Enyalios says, '*No bows will be stretched in numbers, nor slings in multitude, when Ares joins the struggle in the plain'*. There is a clear desire to remove or denigrate the use of distance weapons and the sling in particular. This may be a reaction by the 'Hoplite class' to their assumption of power in Attica during the sixth century BC. As they had removed the aristocracy from its position of power through the use of military force, they may have feared that the remaining underclass of landless peasants would do the same to them if offered the opportunity. By degrading the sling, by reflection they degraded its user, thus creating the ideal of an ineffective peasant military force that would have buttressed their ideology of military superiority. There was a fear of the uprising, not only of free disowned Greeks but by the large slave population of Attica as well (Hunt 1998, 51-2).

In the Iliad and the Odyssey there is a distinct difference between the elite (*arisitoi)* and the people *(laoi)*. The *laoi* are seen as being derived from the stones in the ground, in effect an autochthonic group. The analogy that the *laoi* are literally the stone-people allows them to be portrayed as requiring protection, as they form the basis of society stretching back into mythical times, before the rise of the elites (Haubold 2000, 42-3). The association of the *laoi* with stones (see, Apollodorus i, 7.2; Ovid *Metamorphoses* I, 260-415, for a mythological explanation) may also place the use of the sling into this category. After all, stones (pebbles) are its very ammunition and therefore may have been regarded as a primeval weapon, in use before the rise of the elites and not part of their exclusive world. When the emergent democracy of Athens was first formed, the method of voting was to place black or white pebbles in a pot, voting for or against a proposal. This underlines the autochthonic concepts of the *laoi*. The pebbles, of course, also give anonymity. They, like the people themselves, are anonymous, the very antithesis of the heroic ideal. This anonymity associates the sling with this antithesis as well.

The iconographic evidence for sling use in Late Geometric and Early Archaic Greece is very limited. It is not shown on any of the Late Geometric pots that illustrate a variety of combat scenes (Ahlberg 1971). One of the earliest representations is from an Early Corinthian alabastron (Dugas 1928, 137, Figure 3). Here a slinger is shown preparing to cast while standing behind a fully armoured hoplite. This depiction has been interpreted as hoplite use of the sling, as part of a panoply of helmet, spear and shield are standing near the slinger. However, given the evidence above, this would be unlikely and it is preferable to see this naked figure (who has no other weaponry) as a *gymnete* (*pace* van Wees 2000, 152; Rawlings 2000, 240), so degraded by the Spartan poet Tyrtaeus (F 11.35-8) as being '*here and there, squatting under the 'hoplites' shields'*, an action that separates the *gymnete* from the warrior ideal either, of the single combat of the Homeric hero or the co-operative combat of the hoplite line (Hanson 1991, 2000).

The Bow

The bow holds an unusual position in Late Geometric and Early Archaic Greece. The evidence from the later Archaic period, when the bow became associated with the East and in particular with the ever present threat of the Persian Empire, will be ignored within this case study, as this represents a state level reaction to a specific threat and not an internal social context. The iconographic evidence of Late Geometric Attic ware commonly represents archery in scenes of combat (Alhberg 1971, 44-5; Boardman 1998, Figures 41.1-41.2, 50, 60; Coldstream 1977, Figure 33(b)). The interpretation of these images is, of course, inherently difficult, as we have no direct access to the ideology that produced them (Boardman 1983; Coldstream 1991; Snodgrass 1980). However, the majority were either grave markers (i.e. above ground), or included within the burial itself (Coldstream 1977, 109; Snodgrass 2000, 149-50). It would be unlikely that images of an unacceptable nature would have been included on such memorials or offerings. Therefore, it may be inferred that the bow was acceptable in combat and may have been used by the elites of this period, as arrowheads have been recovered from graves both in Attica and outside the region (Nicholls 1958, 131; Snodgrass 2000, 233, 263, 267).

Bows are also used by the divine twins Apollo and Artemis. However, Artemis and her use of archery can be seen within the context of inappropriate use. Artemis is a perpetual virgin and therefore excluded from the proper realm of Greek women, to be a wife and childbearer (Blundell 1995, 71-3). She also operates within a male context, being a huntress, a protectress of young warriors (Marinatos 2000, 92-3). She is also specifically described as being 'manly', a contradiction to the Greek ideal of womanhood (Apollodorus iii. 5.6; Pausanias v. 16.3: viii). This could place the bow within a similar context, a weapon outside the normal range of elite property, one that does not operate within a proper context; it becomes ambiguous. This is sharply drawn in Book 21 of the Iliad where Artemis insults her brother Apollo for his inaction in defending the Trojans against Zeus. At this, Hera flies into a rage and, in her outburst, she taunts Artemis: *'How do you have the gall, you shameless bitch, to stand and fight me here, you and your archery!'* (Iliad 21:547-8). In order to teach her a lesson in warfare, Hera boxes Artemis' ears with her own archery equipment. The sobbing Artemis withdraws from the Trojan Plain and seeks her father Zeus, who takes her on his lap and comforts her as though she was a child (Iliad 21:560-90). This would certainly seem to place archery equipment in the hands of the weak and childlike. The image is reinforced by Teucer, who in the midst of battle would, *'mark a target, shoot through the lines – the man he hit dropped dead, on the spot – and, as a youngster ducking under his mother's skirts, he'd duck under Ajax's shield, and the gleaming shield would protect him head to toe'*

(Iliad 8:308-12). These images, in turn, suggest that archery equipment is outside the realm of proper warriors or, perhaps more importantly, outside the realm of the citizen hoplite. The ambiguity is particularly germane when relating the use of the bow to Apollo. Clearly, he is male and a god, so his actions should fall within the range of acceptability for the elite. This is to misunderstand the nature of the Greek gods. Unlike the monotheist religions of the Near East, the Greek gods did not create the world or mankind, and therefore, in many respects, they operate outside the bounds of human acceptability (Zaidman 1992). That said, they are, like all other divinities, a product of the culture that created them and must operate within its confines, however loose these may be (Vernant 1990, Chapter V). In a literate society, gods will often carry forward old ideologies that perhaps have less direct relevance to the society that continues to worship them. This may mean that the use of the bow by Apollo may have been socially acceptable, even desirable, during the Late Geometric period. Such an association may have become less appropriate through time but, due to the historical nature of the divinity, he was unable to lose this particular attribute. Therefore, another more subtle way of eroding the social context of the bow may have developed. Apollo uses the bow in a particular fashion at the beginning of the Iliad. In order to punish Agamemnon, in response to the pleas of his priest Chryes, Apollo causes a plague to affect the Achaean army (Iliad 1:24-64). This he does by firing his arrows into the troops; each man he strikes (invisibly) falls ill and dies. This narrative associates the bow with disease and death, not from heroic action but by corruption and decay.

Artemis and Apollo also kill all the children, save two, of Niobe, who (foolishly) boasted that she was superior to Lato (Artemis' and Apollo's mother) because she had many children (either seven or twelve sons and daughters) and Lato only two (Apollodorus iii. 5.6; Pausanias v. 16.3: viii). They slay them with their arrows (Carpenter 1998, Figure 71), as if to remove any possibility that such spiteful murder should be associated with the proper heroic weapons of the spear or sword. Once again, the bow is placed within a context of deceit and shamefulness, characteristics that do not accord with the ideals of the *aristoi*.

Even Herakles, perhaps the greatest hero of all those within the mythological sequence, is actually in a subservient position (classified as a slave (Apollodorus I.9, 19)) when he undertakes his twelve great labours. His use of the bow may therefore have been correlated with this position. It should be noted that Herakles also uses a club as opposed to the spear or sword, which again seems to confirm his temporary servility. It was Herakles who accidentally shot the immortal Centaur Cheiron through the knee, resulting in an agonising and incurable wound (Apollodorus II.5, 4; Lucian *Dialogues of the Dead* 26). Cheiron enters into a liminal state, wounded yet unable to die. Archery, by association, would perhaps have shared this liminal state and could be placed in this category.

The liminal position of archery is further reinforced in both the Iliad and the Odyssey. The great exponent of archery in the forces of the Achaeans is Odysseus, yet he is invariably described as being the man with the quickest wits, the most cunning and treacherous (Finley 1979, 69-70). Even Odysseus does not use the bow in open conflict: he is armed as a true hero, with spear and shield. It is only when undertaking reconnaissance at night that he carries a bow (Iliad 10:304). It is as though the bow is being confined to operations that, although requiring bravery, do not require a man to stand in battle. The bow, as used by Odysseus within a context that would have appeared less than acceptable to the audience of the Odyssey, is illustrated by his slaying of Penelope's suitors on his return to Ithaca (Od. 21). Rather than face the men openly, Odysseus enters his own home in the disguise of a beggar, and it is only during a test of strength that he is able to string his old bow (Od. 21: 452-8) and begins the slaughter of the suitors (Od. 22). The suitors occupy a particular position within the culture of Archaic Greece. They are technically guests and therefore under the protection of their host. In this case Odysseus by their killing breaks this traditional code (Finley 1979, 99-101). It is only after the suitors arm themselves with their own swords (Od. 22:78-110) that Odysseus starts to fight them with the 'normal' weapons of a warrior, taken from his storeroom (Od. 22:111-34). The initial, unacceptable attack is undertaken using the bow.

The denigration of the bow can be illustrated by its use as a weapon of treachery, either explicit or implied. After nine years of siege, the Trojans and Achaeans finally call a truce in an attempt to resolve the kidnapping of Helen without further bloodshed. The competing gods do not wish for the conflict to end until they have decided their own rivalries, using humans as their tools. During the parley, Athena (a supporter of the Achaeans) seeks out the Trojan archer Pandarus and, promising him fame and glory, persuades him to cast an arrow at Menelaos. Athena deflected the arrow from Menelaos, causing it to only to wound him, though enough that blood ran down his legs (Iliad 4:100-60). The Achaeans are outraged, '*look how the men of Troy have laid you low, trampling down our solemn, binding truce!*' (Iliad 4:178-9). Even Zeus is outraged by this breaking of a truce. All this anger is directed towards the archer who fired the arrow, and they surge into battle. The use of the bow here is clearly identified with deceit and treachery. Even the description of the arrow as '*a shaft of black pain*' (Iliad 4:136) appears to carry connotations of an undesirable nature.

The other character who regularly uses the bow is Paris (the abductor of Helen and therefore the main cause of the conflict), who finally kills Achilles by shooting him in the heel (Apollodorus *Epitome* v.3). This directly associates the weapon with an undesirable element in war. One narrative describes Paris as hiding behind a statue of Apollo and shooting Achilles from behind, while he believes that a truce is in operation (Servinus on Virgil's *Aeneid* vi.57). Paris is often portrayed as weak and feeble, preferring the comfort of Helen's bedroom to the rigours of the battlefield (Iliad 6:370-400). His description as a lesser man potentially extends to the bow, placing it once again outside the realm of proper war.

The iconographic evidence from Late Geometric Attica presents a potentially different picture. Here, archers are common in battle scenes, both fighting on land and at sea. They are given a prominent position within the composition, often seeming to lead the action (Ahlberg 1971, Figures 41-3). Certainly, they are not shown as hiding behind a more prominent spear-bearing warrior, as is the case with a Corinthian aryballus dating to about 600 BC (Boardman 1998, Figures 171.1-2). More importantly, all the archers carry swords, which would appear to provide their owners with considerable social standing (van Wees 1998, 344-7). There appears to have been a major shift of emphasis away from the use of archery. This may have occurred as the power of the land-owning citizens increased. As 'proto-hoplites', they would have disliked archery, as it could kill and maim at a distance, which reduced the effectiveness of a single decisive hand-to-hand conflict (Hanson 1998). The manufacture of archery equipment is also expensive and requires continual maintenance (the reference to the use of ibex horn in the Odyssey is taken to be a misunderstanding of composite bow construction (Od. 21:442)). The use of a bow to any level of accuracy requires considerable training, for which time would not necessarily have been available to the citizen farmer hoplites.

The Spear

The spear is the Late Geometric and Early Archaic Greek's weapon of choice. By the Archaic period the poet Archilochos squire of Enyalous could claim that *'by the spear my bread is kneaded. With my spear I win my Ismarian wine, which I drink while I lean on my spear'* (Archilochos 2). It is clear that by this date (*c*. 600 BC), the spear had become synonymous with warfare and the ability to defend or take what the owner desired. It is unlikely that Enyalios' squire was a true mercenary at this early date, as each polity relied primarily on its citizen soldiers (Hanson 2000, 219).

Earlier, the spear seems to have held a similar position. The iconographic evidence from Attica would tend to suggest that the majority of warriors carried at least one and, in most cases, two or three. That would suggest that at least one was a throwing weapon and the evidence from the Iliad would tend to support this argument. The split between thrusting and throwing spears is about 50/50 (both methods are recorded 79 times each, in Fagles' translation). This may be a product of increasing use of 'hoplite' tactics, but the date of the introduction of such primarily thrusting spear warfare is problematic. The best that can be said from the available evidence is that it is unlikely that true hoplite warfare was in operation before 650 BC, when it is first portrayed on the famous Chigi Vase (Boardman 1998, Figures 178.1-178.3). However, as van Wees has recently argued (2001, 155-6), it is probable that the introduction of full hoplite warfare was not a simple single event, but evolved over a significant period of time in the sixth/seventh centuries BC.

The iconographic evidence for spear use tends to show a shift away from a chaotic mêlée where all weapons (except the sling) are shown being used by figures whose other attributes (shield, swords, possible helmet plumes), suggest that they were socially equal (Ahlberg 1971, Figures 33, 41-2), to a more formalised layout where spearmen predominate and always appear to lead the action (Boardman 1998, Figures 171.1-171.2, 175). This would suggest an increase in the importance of the spear as a warrior's weapon.

The archaeological evidence appears to support this view. Attica is rich in Geometric burials (900-700 BC), many of which contain grave goods (Coldstream 1977, Chapter 4; Collis 1984a, 9, Figures 9-10; Snodgrass 2000, 117-51). A high proportion of these appear to indicate the internee held or aspired to warrior status (Strömberg 1993, 81-3). Approximately 41% of male burials contained some form of weapon excluding knives. The cremation of a thirty-year-old male contained within urn D16:4 from the Areopagus cemetery is an exceptional example of this phenomenon. Dating to *c*. 900 BC, it contained a long iron flanged sword, two spear heads, an iron object interpreted either as a chisel (Snodgrass 2000, 233) or a javelin head (Collis 1984a, Figure 9c), an axe and a knife (*ibid*, Figures 9f-g). The sword had been deliberately deformed before being placed in the grave; it was wrapped around the funerary urn. This grave gives a clear indication of the panoply of weapons that potentially were considered as markers of social status (either as a warrior or a member of an elite). However, by 690 BC weapons of all kinds had practically disappeared from burials (van Wees 1998, 340). This may represent a lessening of weaponry as a symbolic marker of status (*ibid,* 344-52).

The Iliad and Odyssey both confirm the status of the spear as the mark of a warrior and a member of the elite. When Telemachus goes to Ithaca to try and remove the suitors from his mother's house, he carries his spear into town (Od. 2:10). Slaves and servants also carry spears to defend themselves and their flocks, but they only carry them in the countryside (and presumably, in the case of slaves, on their masters property) (Od. 14:528-31), they do not carry them in settlements (Od. 17:197-255). It would also appear customary for large households to hold significant quantities of spears as part of their normal inventory (Od. 1:150 and 22:118). These passages strongly link the spear with a demonstration of elite status, at least in the urban environment.

The Iliad provides more detail relating to the spear's social standing. It is invariably described in powerfully dramatic terms: '*your sharp bronze spear* (Iliad 5:147); *if the spear in my hand is mad for bloodshed too* (Iliad 8:130); *and at last he picked up two tough spears tipped in bronze, honed sharp, and the glare flashed off their brazen points and pierced the high skies* (Iliad 11:48-50); *gripping a sturdy spear, bronze edged and sharp* (Iliad 14:13); *full battle gear, take up your slashing bronze lances*' (Iliad 23:892). There is rarely clear differentiation between throwing and thrusting spears, but references to spears being thrown are common, and there appear to have been specific spears for this purpose. It was not until the close of the Archaic period that the use of throwing spears by hoplite forces seems to have ceased. This would indicate that the type of spear use did not necessarily carry particular messages at this time. It was the spear itself that portrayed the status of the owner as a warrior initially of the elite, and later as a citizen solder of a hoplite army and his ability to use it in a battle rather than as a defence against animals and the like.

Such is the power of the spear that the wounded Tydides and Odysseus would rather support themselves on their spears when attending a council called by Achilles (Iliad 19:56), than lean on a staff that may have marked them out as old and hapless. Kings, as they watch the battle from the Achaean boats, lean on their spears (Iliad 14: 46). The spear, it would appear, was a very powerful symbol of a man's position within Late Geometric and Early Archaic Greece. It marked him as free and, potentially at least, a member of the elite.

The most extreme example of the spear's social position is recorded by Pausanias (ix. 40.6), where the spear-sceptre of Pelops is given divine status by the Chaeroneans, and is viewed by them as a representation of their supreme deity. The weapon here transcends its mere human social position and enters a completely different realm. Such a change must surely represent the end of a long process of social evolution, but does demonstrate the potential for such expression within the context of hoplite-orientated Greece.

The Sword

The Athenian historian, Thucydides (I.5.3), states that in the past (i.e. before the fifth century BC, when he is writing), all men would 'bear iron' or carry weapons as part of their normal attire. From the iconographic evidence of the Dipylon painter and others of the Attic school, the weapons carried would appear most commonly to be the spear and the sword (Ahlberg 1971, 12-37; Boardman 1998, Figures 46.1-2, 50, 59-60; Coldstream 1977, Figure 33).

The evidence from Late Geometric and Early Archaic burials would also tend to support the ideology that the sword was seen as part of the panoply of a 'warrior', even if the interred individual never actually engaged in combat. Iconography showing the use of swords in combat as grave markers, and the interment of weapons within the burials themselves are strong indications of the social acceptability of violence as an expression of power. The sword of the Late Geometric and Early Archaic periods, without going into a detailed typological description (see Snodgrass 1964, 93-103), would appear to be a direct descendant of the long 'slashing' sword of the late Bronze Age (Harding 1984, 162-5), even down to the careful manufacture of a flanged hilt in many early examples (Collis 1984a, 42, Figure 9b). This is technically difficult to produce in iron and serves no practical function, which suggests that this feature shows a deliberate conservatism among the smiths producing such weapons for a presumed elite. Such weapons are suitable for the more individualised combat shown on Attic vases, and described in the Iliad and Odyssey. This type of sword would have required considerable amounts of space to be used effectively and does not conform well with the constraints of hoplite warfare. It should therefore come as no surprise that the sword not only became physically smaller, judging from the scenes painted on vases showing these weapons in use (Carpenter 1998, earlier larger swords Figures 222, 224 and 334; smaller later swords, Figures 305, 310, 338; Hurwit 1985; compare Figure 76 with its prominently displayed sword with Figure 122) and the limited archaeological evidence (Anderson 1991, 25-7). It also lost much, if not all, of its social status by the end of the fifth century BC (van Wees 1998, 351), and had become very much the secondary weapon to the hoplite spear (Anderson 1991, 25).

In the world of Homer, the sword appears to operate within a liminal sphere. It is often described in complementary terms as, '*his sword with silver studs* (Iliad 3: 391, 419); *It is solid bronze and the hilt has silver studs* (Od. 8:449); *...the sharp sword drawn from beside my hip*' (Od. 11:53), and yet the sword also seems to carry the sense of being a lesser weapon than the spear. The latter is most often engaged in combat; it is the major killer of heroes on both sides. The sword is at times even used for apparently utilitarian acts such as cutting away the reins of a chariot (Iliad 8:102), where a knife or dagger would have been equally suitable.

The iconography from the Late Geometric period of Attica would seem to support the archaeological evidence relating to the sword as the weapon of the 'warrior' elite. Many of the Dipylon painters' vessels show scenes of combat, where the sword is used extensively both on land and at sea (Ahlberg 1971, 46-8). The sword's predominance in the sphere of social status is perhaps best demonstrated by the fact that the male mourners on the Dipylon vases all carry swords, but, the carrying of spears is less common (van Wees 1998, 352). This would suggest that the interred wished to be seen as operating within a social context where the display of weapons and, in particular, the sword, was a mark of inclusion in an exclusive group.

The sword's shift, from predominance to a secondary, even derogatory, role, is best seen when used in mythological scenes painted in the Archaic period. The suicide of Telamon Ajax at the end of the Trojan War was a common motif and usually shows Ajax falling on his own sword supported in the ground, his shield and spear carefully arranged next to him. The earliest representation of this is on a Corinthian aryballos dating to the beginning of the seventh century. The image becomes more common in the sixth century BC

(Carpenter 1998, 207), just as the ideology of citizen hoplite warfare started to become paramount (Hanson 1991, 77; 2001, 203; van Wees 2001,155-6). Suicide was apparently regarded with some suspicion, as Ajax is initially denied any burial rites, which would condemn his soul to wandering, unable to join the others in Hades (Gantz 1993, 123) and find any rest (see Od. 11:79-88, for Odysseus' companion Elpenor's lament at not having a proper funeral). He is finally buried in a coffin rather than being cremated, as was the normal funeral for a mythological hero (Apollodorus *Epitome* v.7). The use of the sword to commit suicide could be a simple case of practicality, as it would be difficult to fall upon a spear. However, within a mythological context, there are a number of other ways of committing suicide that would have been available to Ajax; hanging (Apollodorus *Asopids III.12, Kings of Athens III 14, Epitome I. 18*), jumping from a cliff (Apollodorus *Epitome I.7*) or self-cremation on a pyre (which would have neatly avoided the problem of burial (Apollodorus *Heracles II.7; Theban Wars III.7*)). That Ajax ignores these other possibilities suggests that the use of the sword, and suicide (as a less than acceptable social act), had become iconographically entwined, and potentially entwined in ideology as well.

The death of Agamemnon offers another potential view relating to the social position of the sword within the Archaic period. Agamemnon is murdered by his wife and her lover (Aigisthos), following his return from the Trojan War. He is persuaded by Clytemnestra to have a bath and put on a new garment that she has woven for him. Unfortunately for Agamemnon, the clothing has no head or armholes, and, as he struggles to put the tunic on, Clytemnestra or Aigisthos kills him by stabbing him to death with a sword. Interestingly, the use of the spear is contrasted on two bronze shield panels both dating to the sixth century BC. One shows Clytemnestra stabbing her husband with a sword (or short dagger); a spear is shown on the scene but is not used in the murder. The other panel shows Orestes, Agamemnon's son, revenging his death by stabbing Aigisthos with a spear, as the latter tries to draw his sword (Carpenter 1998, Figures 350 and 353). Later Archaic images often show Orestes using a sword to kill Aigisthos, as if to imply that the spear should no longer be involved in even this act of aristocratic revenge (Carpenter 1998, Figure 354). The murder of Agamemnon by his wife, while unarmed and incapacitated in his own house, was seen as a deed of great treachery in Greek society. The association with the sword with such acts hints at its reduced social standing. This is further reinforced by scenes showing the rape of Kassandra by Little Ajax during the sack of Troy. What makes this scene of particular interest is that Kassandra attempts to gain the protection of the goddess Athena, who is dressed in the full panoply of a hoplite warrior: spear, shield and helmet. Little Ajax threatens Kassandra not with a spear but with a sword (Carpenter 1998, Figure 336). His action shocked even the Achaeans and they debated later what they should do with him. Again there is a close association with the sword and socially unacceptable acts. As with Orestes, this is contrasted, even highlighted, by the use of the spear in the context of being on the acceptable side of the social situation.

One of the definitive demonstrations of the sword's weakening social position is seen when Achilles, maddened by the death of his friend and companion Patroclus, rampages through the Trojan army as it retreats before him (Iliad 21:1-260). Finally, he reaches the river Xanthus, which many of the Trojans are attempting to cross. Achilles in his rage leaps into the stream and starts killing all the fleeing Trojans he can reach. However, the Iliad makes it clear that he leaves his spear on the bank of the river and does all the killing with his sword (Iliad 21:20-29). None of the Trojans killed in this initial onrush are named. Achilles insanity (Iliad 21:6) is given full reign only later. As he re-enters the river (having taken twelve Trojan princes prisoner for sacrifice at Patroclus' funeral), he meets Lycaon, a son of Priam, king of Troy. Lycaon is unarmed: he carries no shield, helmet, sword or spear (Iliad 21:59). In an attempt to save his life Lycaon grasps Achilles' knees as a supplicant (Iliad 21: 79, 86), and is therefore under the protection of Zeus. Achilles tries to kill him as he approaches with his spear but misses. It is only after Lycaon has pleaded for his life that Achilles kills him with his sword (Iliad 21:132). Next, Achilles meets Asteropaeus (Iliad 21:185-230) and they cast spears at each other. Asteropaeus wounds Achilles in the arm and again Achilles misses with his spear, which embeds itself in the riverbank. Achilles now draws his sword and springs at Asteropaeus, who is struggling to remove Achilles' spear from the bank. Despite three attempts, he fails, and Achilles kills him with his sword (Iliad 21:203-4). Achilles casts both the bodies into the river, where they are consumed by fish (Iliad 21:228-30), which denies them the proper funerary rights of a hero, and thus potentially the right to join the other souls in Hades (Gantz 1993, 123).

Conclusion

The Late Geometric and Early Archaic periods in Greece were periods of great social change. The aristocratic elites were being challenged by the ideologies of democracy based on the ideal of the citizen farmer. It is against this background that the major literary works were produced and they potentially reflect social identifiers.

Although the ideology of farmer citizens was the main political force within many of the Greek city states (it must be remembered that the aristocracy of the major polities of Sparta, Boetia and Thessaly never lost power), the use of 'peasant' weapons never achieved significant recognition, despite the suitability of Greece's mountainous terrain for skirmish tactics (Xenophon *Memorabila* 3.5:25-7). In fact, the minority that constituted the citizen soldiers was always fearful of the potential power of the landless peasantry and, as a result, it would appear that they degraded the natural weapon of this group, the sling. The bow suffers a similar fate. If the interpretation of the iconographic evidence from Attica is correct, it would appear that archery played a significant role in Late Geometric warfare and was associated with the individual warfare of an aristocratic elite. This position radically changed with the rise of the hoplite. The bow, like the sling, began to acquire overtones of fear and cowardice, as it could kill at a distance. This degradation was accelerated after the Persian wars, when archery began to be associated with Medism, eastern decadence and defeat (Hall 1995, 111).

The sword again, if the iconographic and archaeological evidence is representative of social reality, seems to be associated with individualised warfare involving 'single combat'. The wealth of the graves containing swords excavated in Athens would suggest that this weapon in the Late Geometric period was a signifier of elite status. With the rise of the hoplite, the social position of the sword declines and it is eventually relegated to a secondary weapon, much reduced in size.

The spear, however, becomes the very symbol of the hoplite citizen. In time this recognition steadily becomes associated with the thrusting spear, and the use of the throwing spear becomes rare and eventually non-existent. The thrusting spear is, of course, ideally suited to hoplite warfare, where the clash of heavily armed warriors became the classical ideal (Hanson 1991, 3).

The change of weapon types from sword, throwing or thrusting spear, and bow, to reliance on the long thrusting spear alone could be seen as a mechanistic reaction to the strictures of hoplite warfare. However, the rise of the hoplite citizen is itself a social phenomenon, and would suggest that the change in the status of weapons is part of this process. The sword and bow, with their individualised aristocratic connotations, were deliberately eschewed in favour of the spear (both throwing and thrusting), whose use required little training (unlike the sword or bow). The manufacture of the spear, though beyond the landless masses, was well within the economic capacity of the landed citizen. By the early Classical period, even the throwing spear had fallen from use and hoplites relied on their thrusting spear almost entirely. The military disadvantages of relying on a single weapon was sharply shown by the defeat of the Spartan force by the peltasts[7] and psiloi (using slings) of Cleon at the battle of Sphacteria in 425 BC (Thuc. 4.30.4-4.37.5). Even after this defeat, the hoplite remained the mainstay of Greek armies until the Macedonian conquests in the fourth century BC.

Causes of Conflict in Late Geometric and Early Archaic Greece.

The motivation for Late Geometric and Early Archaic Greek warfare has been extensively explored by various authors (i.e. Finley 1979, 46-7; Jackson 1995, 68-75; Lendon 2000, 6-11; van Wees 1992, Chapter Four) and the following is a brief synopsis of the major works. They illustrate the broad spectrum of rationales behind conflict as the Greeks themselves saw them. No attempt has been made to see the 'true' meaning underlying these ideals in terms of economic gain, land pressure etc. These were, in all probability, beyond the ken of the Greeks who actually fought and died in battle. Most of the conflicts described by the Early Archaic Greeks appear to stem from the concept of loss of honour between two competing polities (van Wees 1992, 206-7). Although this is described in a personal form, as though the hero had been insulted, there is considerable evidence that this loss affected the whole community, whom the hero represents. As individual polities competed against each other in the increasingly competitive world of the period, the need to maintain the honour of the polity became increasingly important. As the concept of individual defence became less acceptable and eventually was superseded by the homogeny of hoplite warfare, the ideology of insulting a community became a common part of the expression of conflict.

However, there is also a strong element of personal gain in warfare (Finley 1979, 63; van Wees 1992, 87-9, 304-7; Tandy 1997, 73-5). Heroes are stripped of their armour by the victor, spoils of raids are shared, so that each man gains the appropriate share relative to his social position. One of the greatest prizes from a conflict appears to have been women. The whole of the Iliad initially revolved around Achilles' anger at Agamemnon for having taken his prize, a woman called Briseis, captured when Achilles raided a local 'city'. The taking of women does not appear to be limited to sexual desire; they were highly valued as slaves to undertake domestic tasks. There are male slaves but they appear to have been captured as boys.

The potential for personal gain may have been one of the central elements in the continuing strife that appears to have characterised Late Geometric and Early Archaic Greece but this carried a high and dangerous price. Homer makes no apology for describing the full horror of warfare, and the fact that even the greatest of heroes share the same fate of all men, even the meanest slave, in becoming shades in Hades. The greatest pressure for continual warfare is the perceived insult from a rival polity. This level of honour loss, either directed at the elites or at the community as a whole could not be ignored, and drove men to face the terrors of battle so vividly drawn in the Iliad.

7. Peltast tactics in many respects resemble those of the Iliad, demonstrating that the military superiority of the hoplite was not inevitable

Case Study 4:

Late Intermediate and Inka Highland Peru *c*. AD 1400-1530

Introduction

> *'Of how Viracocha Inca shot a fiery stone from his sling against Caitmarca, and how they did him reverence'* (Cieza de León (1959) [1553], opening description to Chapter 69).

Peru is a land of ecological diversity (Burger 1992, 12-25; Hastorf 2001(a), 55-64; Salomon 1986, 22-3). Due to extreme verticality, it is possible in a linear distance of no more than eight kilometres to range from arctic tundra to lush tropical rainforests. Because of this richness and distinctness, each main ecological region has produced a unique cultural expression. This case study deals mainly with those cultures that existed within the Puna, Suni and Quechua regions: the highlands. Reference will be made to coastal societies when appropriate, to illustrate and expand the archaeological data available from the highlands. Limited reference will be made to the Montaña (rainforest) regions. Although these clearly had a significant influence on the early development of Andean culture, so little work has been carried out in the region that the nature and extent of any flow of ideas is almost impossible to gauge.

Of the four case studies, Late Intermediate and Inka Peru is unusual. There is no indigenous recorded pre-Hispanic history or mythology from the region. Strictly speaking, history, in terms of a written record, did not commence until 1532 with the arrival of Francisco Pizarro and his Conquistadors. This is only some ninety years after the traditional date for the start of the expansion of the Inka Empire outside the Cuzco valley (Brundage 1963; Rowe 1946).

It is fortunate, therefore, that several Conquistadors wrote accounts of the Inka realm in the first hundred years following the Spanish conquest. These accounts (not solely histories, they include much ethnographic evidence) date from the middle of the sixteenth century (Betanzos 1996 [1557]; Cieza de León 1959 [1553]), to the early seventeenth century (Huaman Poma 1978 [1613]). The quality, detail and plagiarism of the authors varied. One of the most important documents in relationship to this case study dates to 1572 (De Gamboa 1999 [1572]). The new governor of Peru, Don Francisco de Toledo gave Sarmiento De Gamboa explicit instructions to write a history of the Inkas that portrayed them as latecomers, and thus justified their conquest by the Spanish. This required De Gamboa to interrogate a series of informants, to ascertain what Peru was like before the coming of the Inka, and when they conquered certain areas. It is fortunate that this account was commissioned at this time. It just preceded the demographic collapse of the sixteenth century (Cook 1981), and falls within the time when remembered events can be recalled without significant degradation (Vansina 1985). It is clear from this account, and those written by earlier Conquistadors, that a stable and relatively coherent history of the Inkas was known from AD 1450 onwards. Before this date, significantly more incoherence and confusion appears, primarily concerning the issue of which Inka started the expansion of the empire. This is traditionally set at about 1438 (Rowe 1946) and the archaeological evidence seems to support this (Bauer 1992, 46). Beyond this event, the accounts descend into legend, and then mythology. The evidence provided by the Spaniards must be used carefully, with the recognition that the viewpoint of the writers is that of sixteenth century Europe.

Three accounts are of particular importance: Garcilaso De La Vega El Inca (1987) [1609]) Huaman Poma (1978 [1603]) and Betanzos (1996) [1557]). All three were written by native Peruvians. Huaman Poma was the only one of the above who lived as an adult before the conquest to write an account of Peru as he remembered it, prior to the arrival of the Spaniards. El Inca was a second generation Peruvian (his father was a Conquistador and his mother an Inka 'princess'). Betanzos' account was written using his wife, Cuxirimay Ocllo, as a primary source. She had been betrothed to Atahualpa at the age of ten and therefore had a particular insight into the ideology of the Inka elite.

All these accounts start with a history of Peru before the coming of the Inka. It is evident that most relate to how areas outside the Cuzco Valley were perceived to have been ruled. There is an ideological construct that runs through the native history of Tawantinsuyu,[8] that, before the coming of the Inka, the whole country was in a permanent state of war and no proper social structure existed (El Inca 1987 [1609], 35-6; De Gamboa 1999 [1572], 38-9; Huaman Poma 1978 [1613], 27-8), and that it was the Inka that first created order out of this chaos.

The Inka were in reality the last in a long line of indigenous cultures that first showed complexity at about 2000 BC (Burger 1992; Jennings 1978; Keatinge 1988). After this date, traditionally Peruvian archaeology is divided into five major periods, Early Horizon, Early Intermediate, Middle Horizon, Late Intermediate and Late Horizon (Lanning 1967, 25-7). This case study will primarily be investigating the last two, as they represent the highlands after the collapse of the Middle Horizon at *c*. AD 1000, to the rise of the Inka after AD 1438.

One of the dominant cultural expressions in the highlands is an endogamous social unit – the ayllu. The ayllu

[8] Tawnatinsuyu is the indigenous name for the Inka Empire, and roughly translates as the four quarters together. The Inka divided their empire into four geographical areas Antisuyu (the Montana to the east), Chinchaysuyu (the coastal strip to the south), Collasuyu (the central highlands) and Cuntisuyu (to the north).

attempts to be self-sufficient. By the use of community-based ideologies it resisted the imposition of state level control from outside (Isbell 1997). The fact that it still exists today, albeit in changed form, is a testament to its resilience (Gifford and Hoggarth 1976; Molinié-Fioravanti 1986, 342-56). It has long been a tenet of Peruvian archaeology that the ayllu is an extremely ancient social construct that stretches back to the very beginnings of social complexity (Moseley 1992). This has been challenged by Isbell (1997), who sees the ayllu as a method of resisting state level formation within the Middle Horizon. Whatever the development of the ayllu may have been, it was clearly the dominant social expression in the highlands immediately prior to and after the expansion of the Inka.

The ayllu, as a social tool, limits power within itself, by making that power contextual. Modern evidence (Bauer and Stanish 2001, 75-8) suggests that religious specialists, for example, were able to retain their prominence by ascertaining what the majority of the ayllu required as a response from the other world. This should not be seen as cynical exploitation. The best shaman would be able to understand the wishes of the ancestors through the desires of their descendants, those living members of the ayllu.

The inability of individuals to gain significant power has been recognised as one of the distinguishing features of highland Late Intermediate Peru. Even within the context of warfare, which elsewhere appears to have been a significant method of acquiring and maintaining power (Carneiro 1990; Earle 1978), the ayllu seems to have been able to control and constrain its 'leaders' or Zinchi (a Zinchi was, technically an elected war leader (De Gamboa 1999 [1572], 38-9)). The level of the ayllus' commitment to their Zinchi seems to have extended only to tending his fields and crops. It does not seem to have included any undue reverence or allowed him to express power through other means, beyond the individual household to which he belonged (D'Altroy *et al.* 2002). The expansion of the Inka Empire may have only been possible because their ayllu was challenged by another expanding group. The crushing defeat of the Chanka, in or around 1438, gave the Inka Zinchi (either Viracocha Inka or Pachacuti) significant and sudden power. This created a situation that the Inka ayllu was either unable or unwilling to constrain.

It is clear from the later texts that the position of Zinchi was usually held by a single family, and that the Zinchi's son (normally but not always his eldest) was expected to succeed him. It was of course in the Zinchi's best interests to promote continual warfare, and they are often blamed in the Spanish accounts for the incessant conflict in the highlands before the rise of the Inka. This is probably a stereotypical image derived from Inka propaganda and a remembered past, where war punctuated the normal cycle of agricultural activities and was therefore more easily recalled. However, the requirement for continual conflict was very much a double-edged sword for the Zinchi. He would have been required to be conspicuous in any combat always showing his bravery over the other warriors.

As a result of the continual conflict, one of the most common landscape features of highland Peru was constructed, the pucara. No pucara can be identified in the Cuzco region (the heartland of the Inka) until after the defeat of the Chanka (Dwyer 1971, 146, quoted in Bauer 1992) and the construction of Sacashuaman, instigated by Pachacuti (Cieza de León 1959 [1553], 153; Cobo 1990 [1653], 141). Pucaraconna are in essence hillforts, many of them multi-walled with only the area enclosed by the innermost wall being occupied (D'Altroy 2001, 65-88). There appears to be a correlation between the spacing of these walls and the use of the sling. Pucaraconna are often located in inaccessible places (D'Altroy 2001, 65-88), where their construction did not impinge upon any area that had potential for terracing, thus improving the availability of agricultural land to the ayllu. The pucara was usually heavily occupied. Though inherent with difficulties, population estimates seem to indicate that between six and ten thousand people lived in the larger pucaraconna, and that this may have represented something approaching 72% of the total population of the ayllu (Earle 1997, 119). All the internal space defined by the innermost wall was covered by numerous small houses that clustered around a central courtyard. There was often a single large open plaza where the main religious ceremonials of the ayllu would have taken place. Outside the pucara, a series of small villages occupied other defended locations. These were often within easy reach of the pucara and, in times of significant conflict, their occupants presumably would have been protected by the Zinchi. Recent work in the Mantaro area of highland Peru has shown that although there was a material difference between the larger courtyard houses and the smaller ones, this difference was not great, the presence of differing levels of copper ornaments, tools, higher quality ceramics and cuts of meat etc. has led to the interpretation that they were the residents of an elite (D'Altroy *et al.* 2002). However, both the Spanish accounts and the archaeological evidence suggest that these elites were not strongly formed and exercised no great power.

Despite the ecological differences between the regions of Central Peru and their geographical proximity, there appear to have been no developed systems of trade. Although trade networks do seem to have existed throughout the rest of Peru (Salomon 1986, 97-102), they clearly provided a very limited supply of exotics (e.g. marine shells) to the Central Andean polities (Owen 2001, 265-93). In order to exploit these diverse systems, the indigenous Peruvians appear to have used an institution described as vertical archipelagos. In essence, each ayllu would send some of its members to occupy areas of a particular niche in other ecological zones. They would, by agreement, accept members of other ayllus (from other zones) into their sphere. This method of exploiting various ecological environments meant that the ayllu could more easily achieve self-sufficiency (Murra 1985, 3-15).

It is within this ideology of contextual power and against the background of the community-based ideal of the ayllu that the social significance of not only weaponry but also warfare must be explored. The lack of pre-Hispanic literature means that there must be a greater reliance on the archaeological record than in the preceding case studies. The archaeological record of Peru is varied. Due to the extremely arid conditions that prevail in the coastal regions it can be very detailed, as the survival of organic material is exceptional. Iconography, particularly on polychrome pottery, gives significant context to many weapons used by the indigenous peoples. Many of these pre-date the Late Intermediate and Inka periods. However, the disappearance of some of these iconographic elements is of use in this context, as will be discussed below.

There was a strong tendency by the Spaniards to describe each indigenous Peruvian polity in distinctive ethnic terms. Their dress, marriage customs and burial customs varied. It would appear that a requirement for local distinctiveness was part of Inka policy, a way of controlling the peoples they conquered. This extended to the weapons each group used. It would appear that this choice was primarily social, as many of the groups described would have had access to materials which enabled them to manufacture any of the weapons available in pre-Hispanic Peru (Cieza de León [1553] 1959, 117, 263). For example the Chachapoyas of Ecuador occupied an ecologically rich region, with access to many forms of timber and minerals (Salomon 1986). Despite this, their distinctive weapon was the sling, which was also worn as a headdress. Clearly, there is a social dimension to this choice, the materials for the manufacture of both atl-atl (technically this is only the dart thrower, but the term will be used throughout this study to indicate both the thrower and dart) and archery equipment would have been easily accessible.

Late Intermediate and Inka Peru was technologically Bronze Age. Before the expansion of the Inka the use of copper was widespread, but not universal. After conquest by Tawantinsuyu tinned bronze became the dominant metal (Costin *et al.* 1989, 107-39). Many societies continued to use stone as their principal material for tools etc. Very little large-scale casting was undertaken or has survived the last five hundred years of looting. No swords or spearheads appear to have been manufactured in bronze. During the Late Intermediate and Inka periods, bronze seems (in terms of weaponry) to have been limited to the yauri, a type of halberd used by the Inka elite (Hoyle 1966, Figure 73; Poma 1980 [1613], Figures 5a-c).

Weapons that appear to be widespread were the sling, atl-atl, spear and clubs. The term 'club' will be used for those concussive weapons that are shown being used in a two-handed fashion, and the term 'mace' for those used in one hand. Simple correlations between size and weight will be used to distinguish these when recovered from the archaeological record.

Archery

Throughout much of the Americas archery seems to have been a relatively late introduction, despite being known in the Arctic by 2000 BC (Fagan 1991, 151) it did not reach the Midwest of the United States before about 650 AD (McElrath *et al.* 2000, 12) and the Southwest no earlier than 500 AD (LeBlanc 2003, 270), in both these areas it seems to have relativity quickly replaced the atl-atl as the main distance weapon. However, this replacement did not occur in Mesoamerica despite the bows introduction at much the same time (Weaver 1981, 216). It is therefore a little surprising that it was in the Middle Horizon (i.e. before 1000 AD), that evidence for archery is seen in the Central Andes, where again, it did not replace the atl-atl. Both lithic points interpreted as arrowheads (Hastorf *et al.* 1989, 102), and iconography showing warriors armed with bows have been recovered (Paravicino and Romero 2002). Large, deliberately broken ceramics, recovered at the Conchopata D-shaped ceremonial enclosure, show elaborately dressed and tattooed individuals kneeling on reed boats, armed with both hafted axes (clubs?) and bows and arrows (*ibid,* Figure 8.5). The similarity of other fine decorated pottery found in the same contexts as the figured pieces led the excavators to interpret these as showing a hierarchical military organisation (*ibid,* 240-1), associated with the expansive Wari Empire.

Following the collapse of the Tiwanaku-Wari, that signalled the end of the Middle Horizon, all evidence of archery in the Central Andes disappears. No more arrowheads are recovered from the Mantaro area (just north of Conchopata) and the iconography of archery also vanishes.

The exploitation of vertical archipelagos would have meant that woods suitable for the manufacture of archery equipment would have been available. Archaeobotanical data would suggest that some timber (probably for building) was imported from the Montaña during the Late Intermediate Period (Hastorf and Johannessen 1989, quoted in Earle 2001, 306), and the highlands themselves had considerable woodland cover well into the seventeenth century.

The abandonment of archery and the adoption (or reassertion) of the sling in the Late Intermediate Period may well be a social reaction against the state formation that the Tiwanarku-Wari polity represented. The sling with its connotations of commonality (explored below) is more in keeping with the ideology expressed by the ayllu and, if this social system was developed to resist state level groups (Isbell 1997, 298-9), then the abandonment of archery with its obvious elite connotations may be a reaction against such associations.

Concussive Weapons (Clubs and Maces)

The earliest representation of combat in pre-Hispanic Peru, from Cerro Sechin (Pozorski and Pozorski 1987, 79-82), dates to around 1290 BC, and shows a group of warriors, wearing decorated hats (helmets?) and loincloths, dismembering another series of warriors. The victorious combatants are using what appear to be concussive weapons, though some may have some form of cutting edge. No other weapons are shown. The Cerro Sechin reliefs, although presenting difficulties in relation to dating, seem to have been produced when the Casma Valley (in which they are situated) was undergoing a series of violent upheavals. Other evidence (e.g. ceramics) suggests that the carvings may represent the iconography of an invasive force from the highlands (*ibid,* 82). The representation of the concussive weapons seems strongly associated with a potential expression of conquest and the ideologies with which this correlates.

Later images of club-using warriors occur throughout the Early Horizon, Early Intermediate Period, and into the Middle Horizon. Of these, the best represented are those of the Moche of the North Coast of Peru. Here, a series of spectacular elite burials have been uncovered at Sipan (Alva and Donnan 1994). The sheer volume of interred material marks these burials out as belonging to members of the dominant groups. Contained in these burials are small gold figures of warriors bearing clubs (*ibid,* Figure 87). The club from this region appears to have been a heavy two-handed weapon with a chipped lithic point at the butt end. Its use would best match that of a modern bayonet and rifle butt. This type of club not only appears in the Sipan burials but is a common motif on Moche pottery (e.g. Alva and Donnan 1994, Figures 137-9; Donnan 1976, Figures 20-22, 37; Hoyle 1966, Figures 41, 47; Wilson 1987, Figure 10) and even on plaster frescos (Wilson 1987, Figure 12), where the iconography seems to express the power of this weapon in its own right (Berrin 1997, 154), as would the deposition of such weapons within contexts that strongly suggest ritual (Bourget 2001, 94, Figure 6). On pottery, the club is shown carried by individuals whose elaborate dress would strongly suggest that they were elite. The club, although shown with other weapons on the decorated pottery (the atl-atl, sling and crescent knife), is prominently displayed and appears often to be highly decorated. All the above would strongly suggest that the club was ideologically associated with the elite. Following the collapse of the Tiwanaku-Wari, indigenous coastal style is reasserted, and the Chimu, in many respects the successors to the Moche, continue to represent the distinctive form of 'Moche' club on their modelled black burnished ware (Shimada 1999, Figure 33). The Chimu were an expansive state level society. Their conquest of the coast coincided with, if not slightly predated, that of the Inkas of the Highlands. The conquest of the Chimu Empire by the Inka *c.* AD 1470 ended their independence. The club (as with the Moche) seems to be associated with the elite and can be seen as a demonstration of their power as effective war leaders and conquerors.

The use of the club in the highlands is also represented by iconographic evidence. Certainly it was in use during Chavin times (1250-900 BC) and, although associated with other weapons (sling, atl-atl, knife), it appears to be an expression of power through martial activity (Burger 1983). The archaeological evidence for the club in the highlands as an artefact is poor. A number of doughnut-shaped stones with a carefully drilled hole through their centres have been recovered at Wanka I (AD 600-800) sites in the Mantaro area. Given their lack of wear, it has been suggested that these clubs were used in war. One such stone, from the pre-Inka Chupaychu settlements around Huánuco Pampa, demonstrates the problems of differentiating between 'weapons' and 'domestic' artefacts. As it was found within a residential context, it could be interpreted as either a mace or a clod-breaking maul or, as the excavator suggests, both (Morris and Thompson 1985, 142, Plate 66). This inter-relationship between the 'domestic' and 'military' would not be out of place within the community ideology of the ayllu. During the domination of Tiwanaku-Wari, it would appear that regional peace was established, as the evidence for warfare disappears. Following the collapse of Tiwanaku and the onset of more unsettled times, warfare, or at least the threat of warfare, returned and is expressed by the construction of hilltop pucara, and their occupation by a significant proportion of the population.

It would appear as if the ayllu, as an expression of community ideology, was deliberately eschewing the use of weapons that had been associated with the elites of previous periods. The club, despite its ease of manufacture and efficiency may have been one such weapon. There is certainly no evidence that the Inka, despite their state level society, used the club as an expression of power. Occasionally, the Inka shown in Huaman Poma's drawings (e.g. Poma 1980 [1603], Figures 5(d), 6, 22; Morris and Thompson 1985, Plate 8) holding a 'star mace' made of stone with five or more blunt points. In general, it appears to have been a weapon of the mass of the population. The Inka is shown by Huaman Poma, carrying a mace but, on the rare kero (wooden drinking vessels) (Hoyle 1966, Figure 73; Mason 1969, 40) that survive, they are shown only holding the Yauri which, as discussed below, seems to have been the defining weapon of the Inka elite.

The Atl-atl

The use of the atl-atl, to extend the forearm and increase the range of a projectile (normally called a dart), has an extremely long ancestry in the Americas. Experimental work has shown that Paleo-Indian points (both Clovis and Folsom) were probably used with the atl-atl (Browne 1940; Butler 1975; Howard 1974).

The atl-atl continued to be used throughout the entire pre-Hispanic era. It is clearly associated with Moche elites, not only in warfare, but often in combination with the club, in hunting scenes, where elaborately dressed individuals kill deer (Bankes 1980, 38-9; Donnan 1976, Figure 7b; Kubler 1962, Plate 134(a)), and foxes (Donnan 1976, Figures 41,121). Moche iconography also includes images of elaborately dressed individuals standing on raised platforms that have been interpreted as 'thrones' (Willey 1953, 215-8). These individuals carry either an atl-atl or atl-atl and darts (Donnan 1976, Figure 51). However, the atl-atl does not seem to have been wholly associated with elites. There is significant evidence of its use by the coastal Nazca, where it has been argued that burial patterns suggest a ranked rather than a strictly hierarchical society (Carmichael 1995).

In the highlands, it is the association of the atl-atl with the monumental statues found at Tiwanaku (Middle Horizon), that give the clearest indication of context. On Barrett monolith (Kolata 1983, Figure 6.7), it would appear that the figure depicted is accompanied by attendants carrying an atl-atl and darts in a quiver. The darts appear to be condor-headed, and the atl-atl seems likewise to be decorated with representations of condors. The large figure itself is carrying snuff taking paraphernalia or a sceptre-like object and a kero, all of which are identified with political and religious power (Kolata 2004, 112, 114). Even the famous 'staff god' has been interpreted as carrying, not two staffs, but an atl-atl and darts (Morris and von Hagen 1993, Figure 90). If this interpretation is correct, this has a profound impact on the social significance of weaponry. The staff god represents one of the main iconographic elements of the Middle Horizon Tiwanaku-Wari (Menzel 1964). The archaeological evidence would strongly suggest that Tiwanaku-Wari had achieved a level of significant complexity, in all probability a state. The potential use of the atl-atl, as one of the elements that express this complexity, strongly associates this weapon with an elite, whose power extended throughout the central highlands and possibly onto the coast. The most likely explanation of this expansion is to use the Inka as an analogy of military conquest. If this is the case, a particular weapon is being directly drawn into the ideology of expansion and dominance in a way that would not appear to have been the case before the Middle Horizon (Carmichael 1995).

The interpretation of the monolith showing the 'staff god' is problematic, but the central figure is either a god or a member of the elite, flanked by a series of winged beings, all carrying what appear to be atl-atls. The significance of these winged beings and their relationship to the mythology of the pre-empire Inka, will be explored below. As Earle (1990) has shown, in early complex societies the concepts of gods and elites can be difficult to separate. In either case, the atl-atl is correlated not only to warfare, as with the Nazca and Moche, but directly and forcibly with the divine (Bankes 1996, 205).

This iconography ceases with the collapse of the Tiwanaku-Wari and, following the Spanish conquest, the Colla and Lupaca did not seem to have any memory of who had constructed the major ruins at Tiwanaku, on the shores of Lake Titicaca. These had been incorporated into Inka myth as one of the places where Inti (the sun god) created the first Inkas (Bauer and Stanish 2001). However, the iconography of Tiwanaku was partly incorporated into that of the Inka, indicating that perhaps there was some form of recollection relating to the Tiwanaku-Wari (Julien 1993). Following the Middle Horizon, the use of the atl-atl seems also to have ceased in the central highlands, despite a history going back to Chavin (*c.* 1200 BC). The number of lithic projectile points recovered from the Wanka II (1300-1460 AD) settlements is extremely small (Hastorf and Johannessen 1989, Table five), and these probably were recovered from earlier contexts by the occupants of the later pucara (Earle 1997, 120). Why, then, was the atl-atl apparently abandoned in this area, yet retained on the coast and further north? There would appear to be no significant mechanical reason for its abandonment, although the valleys of the central highlands were becoming deforested, as the possible management of timber would suggest (Hastorf, quoted in Earle 2002, 302). However, little wood is required to manufacture either the atl-atl or darts. These are not complex weapons requiring specialists for their manufacture, so it seems unlikely that the collapse of the Tiwanaku-Wari polities would result in abandonment of the atl-atl. Although perhaps more difficult to interpret, there appears to have been a change in the method of warfare. It is hard to believe that this was so radical as to require the loss of a particular weapon type that had been in use for so long before. If Isbell (1997) is correct that the ayllu was a response to the growing complexity of the Tiwanaku-Wari, and that it specifically evolved to resist that formation, the loss of the atl-atl may represent a deliberate eschewing of this artefact that was strongly associated with the elites and, therefore, with the ideologies of hierarchic societies. The retention of the atl-atl on the coast may be a symptom of Tiwanaku-Wari having less influence in these regions, or the fact the elites never seem to have lost full control in these areas, as appears to have occurred in the central highlands (Heyerdahl *et al.* 1995).

Some of the later post-Hispanic accounts state that, as part of their initiation rites, Inka Pakuyoc (the nobility) had to manufacture weapons with the most basic tools. These included atl-atl and darts (El Inca 1987 [1609], 357). It is also recorded that the Pakcayoc used archery (El Inca 1987 [1609], 356). Why, then, does it appear that the Inka nobility used both atl-atl and bow? Manco Inka is specifically described as using indigenous Antis from the Montaña because of their expertise in archery (Hemming 1983, 214). The sling was used during the siege of Cuzco in 1536 (Cieza de Leon 1959 [1555], 449-50) but there are no references to the atl-atl, which would have been familiar from the Conquistador experiences' in Mexico.

If the description by El Inca, is correct, it may be that the Inka are attempting to integrate themselves into the cultures of Tawantinsuyu as a whole. The very ethnic nature of weaponry would lend itself to such an expression, the Inka had to be competent with the weapons of their conquered peoples so that the ideology that the Inka represented all peoples in a reciprocal relationship could be maintained. It would have only emphasised the Inka's difference from the ayllus' cultural expression of commonality, if they limited themselves to a single weapon type.

The Yauri

From the illustrations of Huaman Poma, this would appear to have been a halberd-shaped weapon, fixed to a long pole (Huaman Poma 1978 [1613], Figures 5(a-c), 25 (a-b)), remarkably similar to European pole weapons of the period. Despite the fact that Huaman Poma wrote his account in the early seventeenth century, and that halberds were used by Spanish bodyguards as markers of rank, confirmation of the pre-Hispanic origin of the yauri is provided by indigenous paintings showing Inka lords carrying such weapons on Inka keros (Hoyle 1966, Figure 73; Mason 1969, 40). Huaman Poma provides the only indigenous account of Peru, as he was an adult before the arrival of the Conquistadors. Huaman Poma illustrated his account with a series of pen and ink drawings showing scenes from Inka ethnography. The accuracy of these has been recently confirmed by the discovery of clothing attachments, shown on the pictures of the inhabitants of the Titicaca basin (Bauer and Stanish 2001, 81-2, Plate 4.8). This level of detail would suggest that the information contained within the illustrations is in general a reasonable representation of dress etc. before the Hispanic conquest.

Huaman Poma shows only Inkas using the yauri and, in particular, shows it being associated with the Inka elite (see Huaman Poma 1978 [1613], Figure 3 showing Auca Runa, the warlike people, before the expansion of Tawantinsuyu; none of the combatants uses the yauri). The yauri seems to have been manufactured in bronze and the spread of this technology, as opposed to copper working, can be attributed archaeologically to the expansion of the Inka (Costin *et al.* 1989; Lechtman 1979).

The use images of warriors carrying yauri, to decorate kero used by Pakayoc during religious festivals for drinking *chicha* (maize beer), associates this weapon with the elite of Inka society. It was during these events that the yauri was carried in front of the Wacas/gods as they entered the sacred enclosure of Cusi Pata (Joyous Square) (Cobo 1990 [1653], 145-8; El Inca 1987 [1609], 415-7). Here is a direct association between the divine and a particular weapon, whose very name can be translated as 'sceptre' (Brundage 1967, 206).

The exclusion of other peoples from using the yauri may be the result of an ethnic preference expressed by the Inka. However, the manufacture of the yauri would be nearly impossible without the use of bronze, due to the difficulties in casting such a large volume of copper into a complex shape. The expansion of bronze use has been identified as one of the mechanisms by which the Inka controlled the previous elites in the area they conquered (Owen 2001, 285-9). The yauri, with its almost extravagant use of bronze, would reinforce this message. The dominant group, i.e. the Inka, used a weapon whose manufacture was exclusively under their control. This correlates the use of bronze, military superiority and the Inka in one elegant package that would have been widely understood throughout Tawantinsuyu.

The Sling

The sling can be described as the almost ubiquitous weapon or tool of the highland peoples. Certainly it has a long ancestry in Peru. On the arid coast, examples dating to *c.* 1100 BC have been recovered (Pozorski and Pozorski 1987, 25). Many of those dating to the Early Intermediate and Middle Horizon periods (Means 1919) were manufactured in a manner identical to those made today (Cahalder 1980). Unfortunately, due to the wetter conditions of the highlands, no slings survive from this area but their antiquity is demonstrated by the incised rock slabs at Chavin (*c.* 1200 BC) (Burger 1983, 30, Figure 19).

The sling's social context before the rise of the Inka seems to have been related to people who were not part of the elite. Its does not figure at all within the iconography of the Tiwanaku-Wari horizon. Moreover, when it is shown on Moche pottery, it is being used by individuals who lack elaborate clothing, suggesting that they were not part of the elite of that culture (Means 1919, Figures 1-2), or it is being used by individuals (in even less elaborate clothing) who appear to be defeated in combat by members of the Moche elite (Alva and Donnan 1994, Figure 138; Donnan 1976, Figure 22). Although rocks are shown 'flying' through the air on Nazca pottery, and slings represent a regular find in the burial grounds of the coast (Kroeber and Collier 1998, 78, 80, 83, 254; Means 1919), slings are rarely shown on the highly decorated pottery of this period; atl-atls predominate (Kroeber and Collier 1998, Figure 316). At the end of the Middle Horizon, some of the weapons of the elite seem to have been abandoned and, in their place, there is little archaeological evidence of weapons at all. The later Hispanic period accounts make it clear that the sling was one of the principal weapons of the highland peoples during the Late Intermediate Period (Huaman Poma 1978 [1613], Figure 3; El Inca 1987 [1609], 476), before the expansion of the Inka.

The sling, from the very first, is associated with more than just the elite of Inka society. Illapa's (the storm god of the highlands) silver clothing was thought to flash as lighting in the sky as he cast his sling, and its crack as the stone was released was interpreted as thunder (Cobo 1990 [1653], 32). The sling is quite unexpectedly placed directly in the realm of the divine. Illapa, though god of storms, was seen as a beneficial deity, as it was he who brought the rains so necessary for the non-irrigation agriculture of the highlands (Cobo 1990 [1653], Chapter 7). Illapa Inti was an idol, adopted by Pachacuti Yupanki as his *warri* or brother, following the defeat of the Chanka (Cobo 1990 [1653], 33, 54). This act associates the thunder god and his principal attribute, the sling, not only with the ruling elite but, with its mythical culture hero, who is described as individually reorganising the Inka as they expanded their empire (see Bauer 1990 for a differing interpretation).

The sling's association with the elite of Inka society is further strengthened by the myth of the founding of

Cuzco by Manco Capac. As part of the origin myth of the Inka, Manco and his three or four brothers and sisters emerge from a cave, and are directed by Inti the sun god to travel until a golden staff, he has given to them, sinks into the ground. There, they are to found the city that will become the navel of the world (Cobo 1990 [1653], 105; El Inca 1987 [1609], 42). One of the brothers was called Ayar Oche. He is set in a heroic mould and is portrayed not only as brave and strong, but as wielding the sling, with which he is able to flatten mountains and carve valleys, literally to reshape the earth (Beranzos 1996 [1557], Chapter 4).

The other three brothers are jealous of Ayar Oche's powers and, on a pretext that they have left certain golden objects in the cave of origin, they persuade him to return and bury him alive (Beranzos 1996 [1557], Chapter 3). Just before the brothers reach the valley of Cuzco, they rest on a hill and lay claim to the land by casting four sling stones to each of the cardinal points (Cobo 1990 [1653], 104). Before they enter the valley of Cuzco itself, Ayar Oche appears to them, now magnificently decked with rainbow coloured wings (Beranzos 1996 [1557], Chapter 4). At first, the brothers are terrified, believing he has come to take vengeance for their earlier actions. However, Ayar Oche has only come to tell them of the future of the Inka and to instruct them that the hill, from which they laid claim to Cuzco, should now be named Wanacuri (the rainbow), and that the initiation ceremonies of the Inka should take place there (Cobo 1990 [1653], 127; El Inca 1987 [1609], 356-60). These ceremonies included the manufacture of weapons (El Inca 1987 [1609], 369), primarily the sling, which were subsequently 'given' to the young warriors from the Inka wakas[9] (Cobo 1990 [1653], 129). The sling is being associated again with divine beings (in the form of wakas), and with the very founding myth of the Inka Empire. It would appear that the weapon of the non-elite in the other three case studies has finally found expression within a purely elite context.

It is during Manco's attack on Hurin (lower) Cuzco that the first evidence to the contrary enters the narrative. It must be remembered, that despite their divine connections, the founders of Cuzco were human. They were, to be sure, descendants of the sun god, Inti, but were not divine themselves. Therefore, they operated as the custodians of what was right and proper for the Inkas as an ethnic group. This could be adapted to changing circumstances. The fact that Manco Capac is described as marrying his sister is probably an attempt to legitimise this practice, that seems to have started with Tupac Inka in the late fifteenth century. Under these circumstances, the following is of particular importance.

During the initial attack on Cuzco, Mama Wako (one of Manco Capac's sisters) killed a local Indian with a cast from her sling, tore out his lungs, inflated them and rushed into the village brandishing this grizzly balloon. The villagers fled in terror (Beranzos 1996 [1557], 16). Suddenly, though associated with victory, the sling is being used by a woman, who in Inka society played no part in war. This motif is repeated in the narrative relating to the defeat of the Chanka by Pachacuti Inca – Chañan curi coca, defends the adobe village of Cuzco against the Chanka so valiantly that they were forced to retire (De Gamboa 1999 [1572], 92). Chañan curi coca is not described as using a particular weapon but, given the fact that the sling represented the primary weapon of the mass of the Inka forces, it seems likely that she would have been defending her home with the commonest weapon of the household. Here, the involvement of the household in defending the infant Inka Empire is brought to centre stage. Although archaeological evidence suggests that the Inka had been consolidating their power for sometime before the Chanka threat (Bauer 1996, 48), the mythology of the state projected Pachacuti Inka as the culture hero of the empire, introducing almost all the features that were necessary for its expansion. He was in legendary terms the founder of Tawantinsuyu. Yet, in the crucible of this creation a woman is seen defending her community, most probably with the sling.

The use of the sling outside the elite is not restricted to these events. It is associated with many of the Inka's enemies, in particular the Chachapoyas, who were intractable in defence of their Ecuadorian polity (El Inca 1987 [1609], 476; Salomon 1986). In Hauman Poma, they are always shown with long hair, the ultimate mark of the 'barbarian' in Inka culture, yet they use the sling (shown in Brundage 1967, 226). It would appear as though there is no clear socially defined context for the sling. There is perhaps a concession to the elite nature of the Inka within this context. The Inka's sling is described as golden (Cieza de Leon 1959 [1553], 221). Whether this is an attempt to associate the elite with the god Illapa, or an attempt to show the uniqueness of the Inka's sling (or a combination of both) is problematic. However, it would appear to be an expression of clear difference, not unlike the Roman descriptions of Gallic swords. Both cultures used the same basic weapon, yet one, the *gladius* of the Romans, is placed within the context of civilisation and innate organisation, and the other, the long slashing sword of the Gauls, in the context of the 'wild barbarian' world outside the Empire. A golden sling may be projecting a similar ideology: the Inka is using the sling, but it is different, special, and associated not only with the elite use of gold, and its connotations of divinity, but directly to the divine, due to its portrayal of Illapa's sling.

Huaman Poma also shows farmers using the sling to scare away animals that attack maize (Huaman Poma 1978 [1613], Figure 15), placing the sling within a purely peasant context, a world away from the Inka elite. Yet the sling remains central to many of the ceremonies that underpinned the Andean concept of the world. They were used to drive away *supay* (evil) from Cuzco, during the annual cleansing of the sacred city (illustration by Huaman Poma shown in Brundage 1967, 200). They

[9] Waka is a complex descriptive term. In essence, it represents 'holiness embodied'. Anything could be waka: a mountain, a single pebble. Waka, in this context probably represents the 'ancestors' of the Inkas, i.e. various mummies and sacred rocks.

were used to celebrate victory, and cracked like whips among the warriors as they danced in front of the Inka. Given its universality, not only in space and ethnicity but also in social groups, why did the Inka elite continue to use the sling at all?

The Inka were probably unable and, for that matter unwilling, to escape the ideology that the ayllu represented, the concept that leaders were no more than a redistribution mechanism, that, in a reciprocal relationship, took the goods from the ayllu and then redistributed them to all members of the group. The reality was that, even before the rise of the Inka, such relationships were unbalanced (Hastorf and D'Altroy 2001) and this difference between the local elite and the mass of the population seems to have been reduced during the Inka period. There is also significant evidence that the diet and health of those groups conquered by the Inka improved during their occupation (Hastorf 2001(b), 160; Scandefur 2002). The non-elite sections of subject populations would therefore have felt material benefits from being part of Tawantinsuyu. This, coupled with the Inka's expression of commonality, would perhaps have been one of the elements that masked the true inequality of the imperial system. The sling may have acted as one of the props to support the ideology that the Inka elite represented the appropriate world order. The sling's very ubiquitous nature would have acted as a strong indicator of the unity of all peoples of Tawantinsuyu with the Inka. The elites' continued use of this weapon may represent an attempt to express continuity with the ideology that the whole of Tawantinsuyu was one 'super-ayllu' (Conrad and Demarest 1984, 97).

The nature of state warfare would have meant that it was unlikely that the Inka elite and, much less the Inkas themselves, would have been directly engaged in the actual fighting. They would have served as the commissariat of the conscript forces drawn from across the empire. The weapons used by the Inka elite would have had a ritual part to play in any conflict, thus symbolically defeating them. Clearly, before the establishment of the empire, the Inka is likely to have been involved in combat and would have used the weapons associated with their ethnic group, which would have included the sling. The continued use of the sling in a state context serves the very real purpose of indicating expressed commonality.

The Spear

The spear in the context of pre-Hispanic Andean warfare is differentiated from the dart by the fact that it is used without an atl-atl.

It would appear that before the Late Intermediate period, the spear was non-existent or that warfare was dominated by the atl-atl. The evidence of the Hispanic authors suggests that most later spears were manufactured by fire-hardening points of wooden shafts without the use of metal or stone tips (El Inca 1987 [1609], 369). The recovery of such weapons would clearly be unusual within a highland archaeological context and, even on the coast, with its exceptional conditions of preservation, there appear to be no examples. When spears are shown in the drawings of Huaman Poma, there is clear differentiation between the elite and the commoner (1978 [1613], Figure 4). Spears associated with the elite are invariably shown as decorated with feathers, either just behind the head (which was presumably either metal or stone) or along the entire length of the shaft (Huaman Poma 1978 [1613], Figure 25(b)). Exotic feathers, available only from the Montaña, would in all probability have represented a form of wealth finance (see Earle 1990, 80 for a Hawaiian example). The Hispanic references to spears used during the Late Intermediate period, in the Mantaro Valley, describe them as sharpened sticks (Vega 1965 [1582], 169, quoted in Hastorf 1993, 98), presumably lacking any form of stone or metal head. The lack of decoration and their simple manufacture would correlate with the limited nature of Sausa elites during this period. While the differentiation between the elite and the rest of the population is clear, the spear still carries a liminal message. As a weapon it cuts across social distinctions and needs to be further enhanced with wealth finance objects, to draw it away from the mass of the population. The use of decoration to define social status is drawn into clearer relief when the spear is compared with the yauri. Only rarely does Huaman Poma show the yauri as decorated with feathers and then only at the end of the shaft (Huaman Poma 1978 [1613], Figures 5(a) and (c)). Normally the yauri is shown without any decoration (Huaman Poma 1978 [1613], Figures 5(b), 25(a-b)). It would appear that the weapon itself carried enough social distinction. Unlike the spear, it required no further enhancement to distinguish it.

Causes of Conflict in Late Intermediate and Inka Peru

Before the end of the Middle Horizon (*c.* AD 1000), warfare in Peru seems to have followed a broadly similar pattern, on both the coast and in the sierra. The evidence is limited but it would appear from the representations on Nazca pottery (AD 300-600) and their northern contemporaries, the Moche (AD 200-600), that the atl-atl was one of the most widespread weapon types. In both cultures it was associated with a crescent-shaped knife (*tumi*), which appears to have been used for the decapitation of slain enemies in Nazca culture (Verano 1995, 203-18) or, among the Moche, as part of a sacrificial system (Alva and Donnan 1994, 132). Both groups also used the sling. Actual remains have been recovered from a number of Nazca sites (Means 1919) and iconographic evidence shows its use among the Moche. However Nazca decorated pottery rarely shows individuals using the sling, and those on Moche pottery do not wear the elaborate clothing that designates a member of the elite. The purpose of warfare in general terms seems to have been either the taking of trophy heads and the ritual associated with this act (Verano 1995, 203-18) or the Moche tradition of taking prisoners for sacrifice to an iconographically restricted group of deities or their human representatives. The iconography of the much earlier Chavin (*c*. 1200-500 BC) would also suggest that the taking of prisoners (as potential sacrifices (Verano 1995,192-5) or trophy heads) represented one of the main rationales behind conflict. The use of the atl-atl would correlate well with this hypothesis. A dart could be individually identifiable, which may have been of paramount importance in a context of taking trophy heads, where an individual's prestige may have rested on his proven abilities as a warrior. Atl-atls, though relatively easy to manufacture, are also easy to control, as they have no function outside warfare or hunting, both of which are generally seen as activities of an elite. This type of warfare, assuming that the interpretation of the 'staffs' of the Tiwanaku monoliths as atl-atls and quivers of darts is correct, would also seem to extend into the Middle Horizon.

Following the collapse of the Tiwanaku-Wari Empire and the apparent pacifying influence that it generated, the social expression of warfare radically altered, at least in the Sierra. The pucaraconna, although known in the preceding eras (Pozorski and Pozorski 1987, 95) became increasingly common, although not universal as they appear to be absent from the Cuzco region (D'Altroy 2003, 209). If the evidence from the Mantaro region is in any way typical, they also became major centres of population, containing a significant proportion of the polities' population (D'Altroy 2001, 65-96). The archaeological evidence for warfare, in the shape of identifiable weapons, all but disappears. There is no evidence for lithic points that might be associated with the use of the atl-atl and the only stone 'mace' heads recovered appear to be from the preceding Middle Horizon contexts.

Evidence for warfare comes from the Hispanic accounts recorded in the first century after the conquest. The information they contain is often scant but they give the only example of warfare within the context of a weak elite, where defensive structures were used and the ideal of commonality seems to have been paramount.

The Hispanic sources make no reference to the taking, or even the assailing, of pucaraconna in pre-Inka times, and it might be reasonable to infer that such assaults did not form part of the normal expression of Late Intermediate warfare in the Sierra. This is buttressed by references to the Inka finding some pucaraconna difficult to besiege successfully, even with all the resources available from the expanding empire (Cieza de Leon 1959 [1553], 277; De Gamboa 1999 [1572], 162-3; El Inca 1987 [1609], 478-9). One pucara seemed so impregnable that Tupa Inca had to resort to deception to take it (Cobo 1990 [1653], 144-5). The relatively simple dry-stone walls and occasional ditches (Pozorski and Pozorski 1987, 95-8; Wilson 1987, 59) allowed these centres to remain inviolate until the expansion of the Inka.

Andean warfare at the time of Hispanic contact was divided into two separate types, *Ch'auxa,* real war (Hastorf 1993, 54), and *Tinku*, ritual combat (Platt 1986, 239-40). For the Late Intermediate period, all the Hispanic authors describe a context of continual and incessant warfare, a situation where no particular polity was able to gain supremacy over another, and raid and counter-raid were common. Huaman Poma (1978 [1613], 270) aptly describes this period as that of *auca runa*, the people of war. Three basic rationales behind warfare are given by the Hispanic accounts: gaining land (Cobo 1990 [1653], 34, 96; Huaman Poma 1978 [1613], 28); raiding (Betanzos 1996 [1557], 20); and taking women (Huaman Poma 1978 [1613], 28).

The first of these would appear to relate to the expansion of the Inka and an attempt by the conquered polities to align their pre-Inka activities with the dominant indigenous and European ideologies. If the evidence from the Mantaro valley is indicative, the concentration of population in pucaraconna tends to suggest that the acquisition of land by conquest was not a practical option for most Sierra ayllus. In order to undertake a productive day's work, the maximum walking distance to fields would be about 5km (a 10km round trip). Since warfare was seen as incessant and irresolvable between competing polities, it is difficult to see how land as a commodity could have played a significant role within the ideology of warfare. If expansion with its consequence of land control was successful, a smaller and smaller number of larger groups should have emerged in the Andean highlands. This does not appear to have been the case; the number of competing groups seems to have been relatively stable, indicating the inability of one polity to dominate another. The desire for land might have become one of the rationales behind warfare following the Inka conquest, when groups were moved from their traditional defended sites to the valley bottoms. This resulted in an overall increase in the health of the population (D'Altroy 2003, 330), and the telescoping of this situation back in time may have created the 'myth' that land conquest was one of the primary driving forces behind Late Intermediate warfare.

Raiding for camelids seems not to have formed a central role in pre-Inka warfare, and the single reference relates to the pre-state expansion of the Chanca in 1438, when their leader, Uscovilca, promises his captains effectively all the wealth that the Inka polity can provide (Betanzos (1996) [1557], 20). There is some evidence of the elite consuming the more meaty parts of camelids within the Mantaro valley (Sandefur 2001, 187-8). This might be expected if raiding for these animals were commonplace, as the disposal of enemies' livestock in such a redistributive manner would have served to reinforce the Zinchi's social position. Of course, the camelids could have been added to existing herds to provide wool for cloth making. Cloth was one of the great social indicators in pre-Hispanic Peru, and there is evidence that textile production was concentrated within compounds that have been identified as belonging to the elites (Sandefur 2001, 195). As with the concept of land conquest, taking animals for their fleeces, as productive elements within a mixed agricultural system, presents its own problems. Raided animals could have been used to replace the elderly or sick of a herd, or merely to add to its numbers. This would in turn have had consequences on the exploitation of the Puna, and the carrying capacity of this land. None of this, of course, excludes raiding for camelids but it does suggest that such raiding may not have been paramount in warfare.

Capture (kidnapping) of women would leave little or no archaeological evidence, but the descriptions and the sense of fear presented in some of the accounts does suggest that this was a very real threat: *They carried off women and children* (Huaman Poma 1978 [1613], 28). It would appear that kidnapping women did not result from the desire merely to have captive workers to undertake menial tasks. Traditionally, in highland Andean societies, women brew *chicha* (maize beer) and undertake most of the weaving. The use of *chicha* in local ceremonies is well attested in the Hispanic accounts, and such generosity was one of the buttresses that supported the elites. The other was gifts of cloth. Textiles held a very significant position in pre-Hispanic Andean life. The ability of one section of a community to produce more *chicha* and cloth would have been an effective way of increasing an individual's position with the ayllu, and kidnapping women would have allowed this increase in production to occur.

Tinku took, and still takes place, between rival towns and villages during local fiestas (Platt 1986). It is referred to in Hispanic accounts (Cobo 1990 [1653], 215; Rowe 1946, 308-9) and, at least in some form, pre-dates the Spanish conquest. Tinku is highly symbolic, and the majority of the images projected would seem to be at odds with a Christian interpretation of the world. This would suggest that, in essence, the primary function of tinku within a social context has remained constant, probably as a form of resistance to colonial rule. This is not to suggest that recent ethnographically observed tinku is the same as that practised before the arrival of Europeans, but that the broad basis may share a commonality. Tinku would appear to be not only a way for competitive groups to express themselves (often very violently), but seems to be part of a fertility ceremony that requires blood to be spilt on the ground to ensure next year's crops, and to express the duality that is core to the ayllu concept of the world. The whole of tinku is imbued with strong sexual overtones, including the use of maize bread being thrown from the top of a round tower, to symbolise semen being spilt on the ground of Pachamama (effectively, the earth goddess). Tinku also reaffirms social cohesion within a group, by stressing the difference between groups and the need to defeat the 'enemy' allyu (Platt 1986). It has been suggested that tinku may reflect the type of warfare undertaken in many highland polities before the expansion of the Inka (Hastorf 2001(c), 319). If this is so, it is clear that warfare had strong connotations of ritual, and the simple economic goals reported by the Hispanic authors would appear to be their cultural interpretation of the complex Andean evidence.

Potential inferences relating to the social significance of weaponry

> *A sign or symbol only acquires meaning when it is discriminated from some other contrary sign or symbol* (Leach 1976, 49)

Context is the most significant element in any archaeological interpretation (Hodder 1986; 1990). The exploration of the social context of weapons in the foregoing case studies has provided a foundation of information that allows certain inferences to be generated about why some weapons appear to be favoured and others rejected. Within the following exploration, the 'evidence of absence' will be as important as the artefactual evidence available from the archaeological record.

The evidence for warfare has been explored in Chapter Three. Within the hillfort dominated zone, during the Middle Iron Age, the only weapons for which there is any significant archaeological evidence are the sling and the spear. The mathematical modelling developed in Chapter Four strongly suggests that the latter were throwing rather than thrusting weapons. There is only limited evidence for the use of the sword and none for the use of archery (see Chapter Two). This is in sharp contrast to the preceding Late Bronze Age when swords are relatively common (Brown 1982; Colquhoun and Burgess 1988; Cowen 1967). There is some limited evidence of the continued use of archery into the early Late Bronze Age (Green 1980, 195), but evidence of sling use is absent.

In an attempt to illuminate the social context of weapons as opposed to their mechanical or functional aspects, the boat sacrifice at Hjortspring, Denmark, will be used as a counterpart to the evidence from the hillfort dominated zone. Denmark, like Britain, was on the 'periphery' of Mediterranean influences, and was not integrated into that world until the Roman conquest of Gaul in the first Century BC. However, over large parts of Denmark the societies seem radically different from those in the hillfort dominated zone. There are few massive ditched enclosures that can be interpreted as fortifications, and those that do exist are seen as primarily ritual centres, as opposed to military installations (Becker 1982, 65 (quoted in Randsborg 1995, 69); Ethelberg 1993). Settlement patterns appear to be more oriented towards aggregated villages, defined by a shallow ditch and bank (Ethelberg 1993; Madsen 1997; Jensen 1982, 207), with distinct and archaeologically identifiable social hierarchy. This social difference was also displayed in funerary ritual, where fine La Tène decorated swords and other weapons were interred with certain individuals (Ethelberg 1993; Madsen 1997). The deposition or sacrifice of weapons in watery contexts was also a feature of this period, as were the human sacrifices preserved as bog bodies (Glob 1969) dating from *c.* 750 BC to AD 1 (Jensen 1982, 280). The boat sacrifice at Hjortspring is unique, not only in its scale but also its preservation. It included a large hoard of weapons and armour (Rosenborg 1937, 40-50). The deposition is interpreted as a defeated enemy's panoply (Randsborg 1995, 38). The sacrifice took place sometime around 300 BC, a date that correlates well with the Middle Iron Age of Britain and it provides significant counter-evidence from a distinct, contemporary, geographically adjacent culture.

The potential that the Hjortspring sacrifice contains the full range of weapons carried by a particular force has significant implications in terms of its usefulness as an interpretative tool. Unlike individual inhumations, the likelihood for selective deposition that reflects the desires of the buriers' social group (e.g. to represent the dead individual as a 'warrior' by including weapons, or to increase their social standing in the afterlife by the inclusion of grave goods that may not reflect the individual's actual social standing or occupation) is reduced. It appears that the Hjortspring deposit may contain a significant quantity of the weapons from a defeated enemy, as swords, ring mail, both metal- and bone- or antler-tipped spears, wooden shields (two distinct types), pebbles (probably hand thrown stones, but just possibly sling shot), turned wooden vessels, not to mention the boat itself, have all been included in the deposition. This indicates a wide technological and, potentially, a wide social base for the weapons involved. The inferences gained from the case studies will be limited to four main weapon types, all of which were used by contemporary societies that were within geographical reach of the Middle Iron Age peoples of the hillfort dominated zone. These are swords/daggers (the Arras Culture of Yorkshire (Dent 1983; 1985; Stead 1965, 68)); archery from the Early La Tène of northern France (Mercer 1970, 201); spears; and the sling.

The Social Context for Middle Iron Age Weapons

The Sword

The 'Celtic' warrior is often seen primarily as a swordsman (Cunliffe 1991, 488; Pleiner 1993, 33-5) and yet, in the Middle Iron Age of the hillfort dominated zone, the evidence for the sword, is limited. There is extensive evidence for swords in the Late Bronze Age (Brown 1982; Colquhoun and Burgess 1988; Cowen 1967) and occasional evidence for swords and daggers in the Early Iron Age (Jope 1961). In the Late Iron Age, swords are deposited in both watery contexts (Piggott 1950, 25-8; Stead *et al.* 1981, 61) and in burials (Collis 1973). The limited use of such weapons in the Middle Iron Age of the hillfort dominated zone is in all probability a social expression, as it is unlikely to be the product of biased recovery.

The sword in three of the case studies holds a particular place within their social contexts. It is associated with either an established or an emergent elite. This social distinction is perhaps best illustrated by the way in which the sword is *removed* from the context of battle by the Archaic Greeks, when the socially levelling hoplite phalanx was developed. The sword seems to lose its special status, expressed in burials and iconography, and be relegated to very much a second class weapon. As the citizen militia gained not only social but also military strength, the sword's association with the aristocracy became untenable, and it lost its place within the social expression of the new dominant group.

Within the Hjortspring deposit, eleven single-edged swords were recovered (Rosenberg 1937, 40-2). This correlates very closely with the number of ring mail coats in the assemblage (ten to twelve, their fragmentary nature making an exact calculation impossible (Randsborg 1995, 26)). The existence of ring mail is a strong indicator of access to increased resources for certain individuals, which in turn is potential evidence for some form of elite. Ring mail not only requires large amounts of iron, but more importantly, requires the retention of specialists for its manufacture. The large number of ring mail coats from Hjortspring would suggest that they were locally made (i.e. around the Baltic basin), rather than imports from any great distance. The limited evidence of swords (and the complete lack of evidence for ring mail) within the Middle Iron Age of the hillfort dominated zone would suggest that, either these societies did not have such extensive or exclusive access to resources, or that they chose to express status through other means. Iron was used to manufacture a wide variety of agricultural tools, including sickles (e.g. Cunliffe 1984b, Figure 7.8; Cunliffe and Poole 1991, Figure 7.10; Stanford 1974: Figure 78; Williams 1951, Figure 18, 9), bills (e.g. Cunliffe and Poole 1991, Figure 7.9; Musson 1991, 144), knives (e.g. Curwen and Williamson 1931, 30-1; Gresham 1939, 123 (1.D and 2.D); Richardson 1951; Stanford 1981, Figure 58, 5) and the like. The skills needed for the manufacture of edged agricultural tools (e.g. bills and sickles) could have been turned to the production of simple swords, as the temperatures required to achieve reasonable cutting edges for such items are nearly identical (Hodges 1989, 84). However, the manufacture of ring mail is complex requiring a drawplate (*ibid,* 76). This may have been beyond the skills of the majority of Middle Iron Age blacksmiths, as none of the iron working from Middle Iron Age sites requires much sophistication (*ibid,* 86).

The weights of more utilitarian artefacts such as hook-shaped cutting tools, adzes and hammers from Danebury (weighing between 30-379g) when compared to spearheads (weighing on average 55g) certainly infers that iron as a commodity was widely available, and that its use was not confined to 'prestige' items only. This is further reinforced by the recovery of currency bars from potential Middle Iron Age contexts (Allen 1967). The most secure Middle Iron Age find of such bars occurred at Worthy Down, Winchester (Hooley 1921, 321-7), where thirteen carefully deposited sword-shaped bars were recovered from a pit sealed by a layer containing 'Early La Tène' pottery, indicating a Middle Iron Age date (the finds also included human and animal bones, pot boilers and possible sling stones). The iron bars themselves almost certainly came from the Jurassic Ridge deposits in Northamptonshire (Ehreneich 1991, 71), and therefore represented a significant investment in a commodity when they were deposited in the pit.

The deposition of apparently utilitarian agricultural artefacts has been the centre of significant debate, relating to the ritual nature of 'rubbish' in the Iron Age as a whole (Brown, Field and McOmish 1994; Hill 1995a; McOmish 1996). If the agricultural tools found on hillforts (e.g. Alcock 1972b, 153-4, Plate 59,) do represent votive deposits (and there is no reason why they should not), the absence of swords from this context only serves to buttress the inference that these weapons did not form a significant part of the social expression of Middle Iron Age peoples of the hillfort dominated zone. It is a generally recognised anthropological tenet that votive deposits contain items that are intrinsically valuable to the society that is undertaking the deposition. If iron tools and currency bars do form part of such a ritual process, the fact that they are manufactured in iron may carry specific messages relating to the availability of metal within the community. It would not be unreasonable to assume that if swords were viewed as socially valuable the they too would be deposited within an archaeologically recoverable context, as seems to have occurred in the preceding Late Bronze Age/Early Iron Age (Burgess, Coombs and Davies 1972; Fox 1946, 92, Plate XXXIII; Stead 2003, 49-56) and the succeeding Late Iron Age (Collis 1968; Stanford 1974, 167, Figure 77; Wheeler 1943, 227) within the hillfort-dominated zone, and outside this area during the Middle Iron Age (Stead 1965; Dent 1985). They were not. This is a strong indication that the use of swords was not integral to the expression of status within the Middle Iron Age societies of the hillfort dominated zone. They may have deliberately eschewed the use of the sword, as an expression that went beyond the requirements they gave to their leaders. The sword, as explored within the case studies, appears to be associated with an expression of military might and an association with the divine. The use of the sword in combat appears to have been represented as giving its wielder access to another context, an entrance into heroic time. Although more difficult to substantiate, the limited evidence for swords within the Middle Iron Age of the hillfort dominated zone may indicate that its leaders had lost this attribute, and expressed whatever power they had through other media.

Archery

By the end of the Late Bronze Age in Britain, archery equipment, and therefore archery itself, seems to disappear from the archaeological record (Green 1980). Evidence for the reintroduction of archery in the earliest Iron Age comes from Pimperne, Dorset, where two iron arrowheads were recovered from the post holes of the early house (Harding *et al.* 1993, 51, Figures 12-13). Dating evidence suggests that the site was occupied between 700-400 BC (*ibid*, Appendix 1). Evidence of bow use is then absent until the end of the Iron Age. There are iron arrowheads from Danebury (Cunliffe 1984, 366, Figure 2.104), Gussage All Saints (Wainwright 1979, 108, Figure 1114), Maiden Castle

(Wheeler 1943, 272, Figure 88, no.9) and Allards Quarry (Williams 1951, 67, Figure 23, 8).

The use of the sling has been seen as one of the reasons why archery diminished so completely, at least within southern Britain. This interpretation is based upon the premise that archery and sling use are mutually exclusive (Childe 1951). This is not the case. Although missing entirely from Early Mediaeval Ireland, both the sling and the bow were used concurrently in Israel, Greece and Peru. It is the *social* context that differentiates the two weapons. In the case studies, the bow is strongly associated with elites. This seems to relate, in part, to the complex nature of their construction. In the two Old World cases (Late Geometric and Early Archaic Greece and Iron Age Israel I and II), the bows used appear to have been of a composite manufacture. From the archaeological evidence, it would appear that this type of bow did not spread into north-western Europe, until the Roman expansion (see Rausing 1967, 52-6, for a differing interpretation of some Scandinavian Bronze Age rock art). Archery seems to be associated with only two activities, warfare and hunting, both of which relate to elite individuals within sedentary agricultural societies. It is within this social expression that inferences relating to the social significance of archery within lowland Britain must be made. The manufacture of simpler self-bows and arrows, though requiring skill and knowledge, does not appear to have been outside the capabilities of a wide spectrum of society, Theodora Kroeber (1987, 189-92) provides a description of Native American bow manufacture, using simple widely available tools.

There is no evidence for archery from the Hjortspring boat nor from contemporary occupation sites (Randsborg 1995, 42). It is only later, after *c.* AD 100, that there is evidence for archery in Iron Age Scandinavia (Rausing 196, 56-8, Figure 21).

The use of the bow and the inclusion of archery equipment in the form of arrowheads (Green 1980) and 'bracers' (Harbison 1976, 28-31) in what appear to be high status graves (Edmonds and Thomas 1987) seems to have been widespread in the Early Bronze Age. However, it is impossible that memories of this association survived into the Iron Age, to influence the social exclusion of archery. A mechanical explanation that the technology was unavailable is also weak. Ash (*Fraxinus excelsior*) which is suitable for both arrow shafts (Greenland 1994, 23) and bow staves (*ibid,* 61-2) grew both on the chalk downs of Wessex (Evans 1975, 123) and along the Welsh Marches (Musson 1991, 163, 165; Oldfield *et al.* 1986, Figure 3). Although in decline, elm (*Ulmus* sp.) continued to grow at least in the Welsh Marches (Oldfield *et al.* 1986, Figure 3), and this timber is known to have been used as early as the Neolithic for bow staves (Clark 1963). Other suitable timbers such as field maple (*Acer campestre*) (Rausing 1967, 153) would have been available in southern Britain (Robinson 1981, 261). The knowledge of toxophily seems to have been widespread in continental Europe during the Late Bronze Age/Early Iron Age. The inclusion of arrows and a possible bow within the rich grave of the Hallstatt prince at Hochdorf (Biel 1981, 216) confirms that archery, here at least, was seen as a socially acceptable pursuit for a clearly politically powerful man. This inference is buttressed by the Lashe Vessel, a Hallstatt pot that shows on its black surface an engraved scene of a deer hunt (Rausing 1967, 56). The hunters themselves are riding animals that also carry antlers. This would appear to draw the iconography away from the real world into a more spiritual expression; the hunters could represent shamans or similar, undertaking some form of spirit journey. This would appear to link archery with an expression of power, which would correspond well with Earle's connection between 'chiefs' and ideological (divine) power (Earle 1997, Chapter. 5).

The importation of Hallstatt/Early La Tène daggers into Southern Britain during the sixth/fifth century BC (Jope 1961) is a strong indication that the elites of these two areas were sharing some common ideals of social expression. Although the distribution of Hallstatt/Early La Tène daggers is concentrated in the Thames Valley (Jope 1961, Figure 10), there are some outliers in Somerset and Suffolk. This suggests that the original distribution of these elite weapons may have been wider than the archaeological evidence might indicate, as does the deposition of a sixth century BC Hallstatt sword in Lynn Cerrig Bach (Fox 1946, 96). If the material manifestation of the elite package of ideals (De Marrais *et al.* 1996), that seems to associate with the daggers also included the use archery, as a status marker, which the evidence from Pimperne (Harding *et al.* 1993, 51), suggests, then the rejection of archery by the Middle Iron Age peoples of the hillfort dominated zone may be the result of a political shift in ideology. The potential erosion of elite power during the Early Iron Age (Bradley 1991, 60; Earle 1991b, 97) may have required a shift of emphasis towards a 'group-oriented chieftainship' (Renfrew 1974). This, in turn, may have necessitated the abandonment of both archery and the dagger, as tools of exclusion, and the adoption of weapons of inclusion, such as the sling, as the requirement to express commonality with the mass of the population, became the paramount political statement.

The Spear

From the evidence provided by the case studies, the spear appears to be the most socially liminal weapon of those studied. In Late Geometric and Early Archaic Greece the use of spears seems widespread in both the *aristoi* and *domos*. Later it becomes the dynamic symbol of the social levelling created by the hoplite ideal. In Iron Age I and II Israel, the spear is associated with kingship, but here it also seems to be associated with acts that lay outside normal social relationships (e.g. kin slaying). In Early Mediaeval Ireland, there seems to be an attempt to confine the use of metal-tipped throwing spears to the elite. This very ploy indicates that the use of entirely wooden spears was possibly widespread. In Late Intermediate and Inka Peru, the spear, as opposed to the dart, does not seem to appear until after the collapse of Tiwanaku-Wari, when the allyu may have first made its appearance (Isbell 1997). This ability to cross social boundaries is clearly illustrated by the number of such weapons recovered from Hjortspring, where a total of 169 spears were recovered. Of these, 138 were in iron, and 31 in bone/antler. There was also significant variation in size and potential use. Randsborg (1995, 26) divides the iron spearheads into four distinct categories:

1. Thirty-one short broad spearheads (javelins, 50-160mm long by 50mm wide at the middle).

2. Thirty-four narrow spearheads (javelins 130-290mm long by 32mm wide at the middle).

3. Eight long narrow 'bayonet-like' spearheads (lances 250-435mm long and 27-56mm wide at the base of the blade).

4. Sixty-four broad spearheads (lances 120-365mm long by 62-75mm wide at the base of the blade).

Although some of Randsborg's interpretations are based on typological and mechanistic principles, in that the basic design of the spears must have responded to the evolution of both defensive and offensive weapons, the correlation between the number of 'bayonet-like' spearheads and the number of swords or ring mail is remarkable implying that these weapons carried social messages of power and prestige. These weapons would seem to be too large to serve as throwing weapons and were probably thrusting spears.

The range of spearhead types recovered from Hjortspring is noteworthy, and the apparent diversity of design within a single context may indicate local manufacture of these weapons. There is very little standardisation, which would perhaps be expected from a single controlled workshop. This level of variation exists within the range of spearheads from the Middle Iron Age of the hillfort dominated zone. This would suggest that each spear was made to individual requirements rather than to a set pattern, which may in turn suggest that the spear represented in some form a household weapon with associations going beyond the practice of war. The ideology of household or individual production is buttressed by the discovery of 31 spear points manufactured in bone/antler from Hjortspring. These would have been easily made by any competent individual, and would have represented an item where control of the raw material would have been impractical or impossible.

Within the context of the Middle Iron Age of the hillfort dominated zone, the lack of any large spearheads (unlike both Hjortspring and Owslebury (Collis 1973), which seem to be associated with an elite expression) would suggest that the spear is not necessarily being projected as a marker of social distinction but may represent a visual expression of social inclusiveness. The simple nature of their manufacture would not have been beyond the capabilities of any competent individual, as wrought iron working requires little specialised skill (Hodges 1989, 86). Although dating to the Early Iron Age, the conversion of an existing saw blade into a spearhead is perhaps one of the most telling examples of this (Cunliffe 1984b, 361, Figure 7.19, 2.102). The Middle Iron Age spearheads may therefore represent, at a certain level, household production, which may in turn be seen as an expression of the community, as opposed to an elite.

Despite this possible association, the case studies infer that the throwing spear is associated with an emergent elite. From the mathematical modelling undertaken in Chapter Four, it is probable that most Middle Iron Age spearheads found within the hillfort-dominated zone were throwing rather than thrusting weapons. This presents a conundrum. On the one hand, it would appear that the small, simple, and undecorated nature of Middle Iron Age spearheads points to them acting as a social expression of cohesion and, on the other, the mechanical properties indicate a weapon that is designed to exclude. Yet the case studies also indicate that spears were the most widespread and ambiguous of all the weapons studied. If other social constraints were in operation, the use of warfare as a means to transfer power to a smaller group or to expand the power base of a limited elite would have been hampered until alternative and complementary means were used. This may have been what happened as southern Britain came into contact with the expanding Mediterranean world in the late second/first century BC, and trade in exotic goods gave new opportunities for elite expansion and social differentiation (Cunliffe 1984c, 32-8; Bradley 1984,155-6).

The Sling

In the case studies, the sling, like the sword, has a consistent image. In Early Mediaeval Ireland, the sling is portrayed only as being used by those individuals who had not yet reached or could not ever, due to their divine parents reach, adulthood. In Iron Age I and II Israel, the sling appears to be placed among the mass of the population; certainly it is not associated with the emergent royal houses. In Late Geometric and Early Archaic Greece, the sling is firmly placed within the realm of the non-elite, and later non-citizens, despite the hoplites' ideal of a common citizen soldiery. Only in Late Intermediate and Inka Peru is the sling associated with a defined elite. It would appear that the allyu was so powerful, in its community ideology of social levelling, that the Inka elite either did not have time to fully escape its confines before the arrival of the Conquistadors, or were manipulating it in manner not apparent in the available evidence.

There is no evidence of slings within the Hjortspring deposit, hundreds of fist-sized stones where recovered from in and around the vessel (Rosenberg 1937, 37-9). However, the size of these would tend to suggest that they were intended to be thrown by hand rather than cast by a sling.

The sling is probably associated with non-elites for a number of complementary reasons, each of which on its own may not be significant but, in combination, make the use of the sling inherently unlikely as a marker of exclusive social differentiation.

Before the manufacture of lead sling shot, pebbles as ammunition would have been very difficult for any group to control. It is not impossible to give such objects special meaning and enable them to act as signifiers of difference, e.g. the concentric rings of pebbles found at Knowth (Eogan 1984, 33), or the unusual pebbles found at pre-Palatial Minoan sacred locations (Nowicki 1994, 43), but attempting to exclude the handling or use of all stones by an agricultural community would have been impractical and impossible. The use of the sling as a tool is also a probable contributory factor to its exclusion from elite display. The widespread availability of an artefact militates strongly against its use as an agent of exclusion, as does its association with everyday manual tasks. Finally, there is the ease of access to the materials of manufacture. Slings are made from textiles; it is impractical to attempt to control all forms of textile manufacture. Later Irish sumptuary laws relate to the colours of cloth certain classes could wear (Kelly 1998, 263), not the type of cloth. This may be because controlling the quality of cloth produced would have been difficult, as an individual's ability to manufacture fabrics from the same basic raw material would have varied considerably and could not, realistically, be confined to a particular class or caste. It is not surprising that the sling was not widely used as a metaphor of exclusion between differing social strata, but it can be inferred by correlation that it represented a tool of social inclusion.

The ayllu structure of Late Intermediate Peru provides the best evidence from the case studies for such a proposition. As discussed, the ayllu was (and still is) a self-replicating social unit, whose ideal is a socially level polity, where the leaders represent the will of the majority rather than their own. Recent (possibly apocryphal) evidence from the highlands of Bolivia serves to demonstrate this. One leader of an ayllu attempted to maintain his control and pass it directly to his son, in effect creating a hereditary control, rather than relinquishing power to another family. Some members of the ayllu prevented this by the simple expedient of killing the leader (Hastorf and D'Altroy 2001, 16). The strength of the ayllu was expressed through direct action, stressing the collective nature of this social construct. The sling is one of the primary weapons or tools associated with the ayllu. This does not appear to be a post-conquest phenomenon. The Spaniards allowed the native populations to retain the sling, as it represented little danger to their military superiority and was a useful agricultural tool. It seems to extend into the prehistoric past of Peru, even before the expansion of the Inka. The case study relating to Late Intermediate and Inka Peru strongly suggests that the sling is a materialisation of the community ideal. Can this association be seen in the archaeological evidence of the Middle Iron Age?

Sling projectiles have been recovered from both Middle Iron Age domestic sites and hillforts. There appears to be no distinction in the raw material from which a projectile is made between site types. Carved chalk, clay (both baked and unbaked) and natural pebbles occur on both domestic sites and hillforts. There is, however, a significant bias towards hillforts in the number of natural pebbles deposited, with only a limited number being recovered from domestic sites (given the early date of many of the excavations, individual pebbles may not have been seen as significant; this is certainly the case with some modern excavations on the Welsh Marches (Stanford 1974, 44, 188; 1981, 20), which may produce some limited skewing of the data). This does not appear to extend to the other forms of projectile, where roughly similar numbers have been recovered from both settlement types.

The differentiation between hillforts and domestic sites relates to the scale of storage. Simply put, hillforts have massively greater numbers of natural pebbles than any domestic sites. Given the apparently undifferentiated access to the 'raw' material for sling projectiles, it is clear that the material that is being stored is not communicating any message of exclusion. However, the level of storage may in itself be a subtle means of expressing social control (see below).

The social expression of weapons

The evidence of weaponry from the Middle Iron Age of the hillfort dominated zone is limited in its scope and nature. Only the sling and the spear appear to have a widespread distribution, the sword seems to be more constrained in its use. Both the sling and spear could have had significant use outside the context of warfare. The case studies strongly suggest that there is a certain level of cross-cultural similarity in the way in which certain types of weapon were viewed. Social exclusion was most easily expressed by materials that were easy to control (De Marrais *et al.* 1996, 19). Those weapons that require either 'exotic' materials or require specialists for their manufacture feature most strongly in emergent or established elites, as they express an ideology of wealth and social exclusion (Earle 2002, 238). By correlation, those weapons whose manufacture require materials that are readily available and/or need little specialism in their manufacture associate most strongly with the non-elites.

Clearly, the sling falls within the latter category and by inference is associated strongly with a non-elite expression of war. The spear is more problematic. This weapon crosses the boundaries of social expression with ease, and it is perhaps best interpreted within the context of the other weapons it is associated with, in the case of the Middle Iron Age, the sling. The small amounts of iron (17.5-58g (Cunliffe and Poole 1991, Microfiche 28, C11-C12)) used in their manufacture as compared with agricultural tools (47-201g (Cunliffe 1984b, 349)) would suggest that spears were not an expression of social status, at least within a material or wealth context, and may have related to social position, e.g. the assumption of adulthood or warrior status, neither of which necessarily carry elite connotations.

From the evidence of absence, it would appear that warfare in the Middle Iron Age of the hillfort dominated zone was not part of the cultural package associated with an expression of elite activity. The exact opposite would appear to the case. The weapon evidence would strongly suggest that war was the responsibility of the community, and the types of weapon involved would suggest that individual actions were not necessarily to be celebrated, as they could not easily be identified. Though an individual's spear may have been recognisable, individual sling stones would have been anonymous, unless they were marked in some way, and the large-scale storage seen at many hillforts would argue against this possibility. If ammunition was drawn from the communal store before conflict, it would suggest that an ideology of sharing the responsibility of the ensuing confrontation may have existed, which would militate against the use of individualising marks.

CHAPTER FOUR

EXPERIMENTAL AND MATHEMATICAL MODELLING RELATING TO 'SLING SHOT' AND 'SPEARHEADS' OF THE MIDDLE IRON AGE

The Middle Iron Age of the hillfort dominated zone has produced evidence of only three potential weapons. Of these, only two have a widespread distribution: large numbers of naturally rounded pebbles, clay ovoids, chalk ovoids; and a number of wrought iron spearheads. Both of these vary in shape and size and it has been assumed that this variation indicates differing use. In the following discussion, an attempt will be made to demonstrate the functional limits of these two weapons.

This chapter is divided into two sections. The first will deal with the use of the sling, and the second will attempt to correlate form with function for the spearheads, using a simple mathematical model.

The Sling

It is often assumed that the sling is practically impossible to master for anyone other than an individual who learns to use it from childhood (Griffiths 1989, 261-3). This premise coupled with a historical prejudice against the sling has led to little or no experimental work being undertaken, particularly when compared with archery (Clark 1963; Hamm 1989; Hardy 1995; *Miller et al.* 1986; Payne-Galloway 1981 [1903]; Pope 1923). This problem is compounded by the poor ethnographic record. Archery remained a recreational activity of the new leisured classes from the sixteenth century (Ascham 1985 [1545]), following its decline as a military weapon (Bradbury 1985, 155-8), until the present day when it has a widespread following. It is also due in part to that association, that it has been widely covered in ethnographic and historical literature (e.g. Bradbury 1985; Lake and Wright 1974 [containing over 5000 references to archery articles, books etc.]; Rausing 1967).

The atl-atl, likewise, has an extensive corpus of experimental work attached to it (Browne 1940; Howard 1974; Butler 1975; Raymond 1986), resulting from the discovery and interpretation of early lithic points in America as dart heads (Chapman and Chapman 1983, 38; Haynes 1966; Meggers 1972, 13). This in turn prompted a series of reconstructions of how the atl-atl was used in casting a dart, based partly on experimentation and partly on ethnographic observation.

The sling remains essentially untested, within an experimental framework. One of the primary practical problems is that, unlike using a bow or atl-atl, there is no moment of inaction when using the sling. The bow when at full draw (Helgeland 1975, 24- 5), and the atl-atl when the dart is resting on the spear thrower (Raymond 1986, Figure 2), both have moments when an instructor or experimenter can adjust the posture or the operative's method to improve their technique. This is not the case with the sling. Its motion is integral to its use.

The sling is also a liminal artefact. It does not conform to the ideological ideal of a weapon. It does not require great physical strength to use, nor interestingly, does it require any major use of tools in its production. The Inka Pakuyoc were trained to manufacture slings from local grasses with no tools whatsoever (El Inca 1987 [1609], 369). Most experimental archaeology exists in a realm of male-dominated activities or at least those that are perceived in today's society as male-dominated (Coles 1973; 1979). Much work has been carried out on the methodologies of craft production: woodworking, metalworking, lithic and pottery production have all been explored. Partly due to a lack of archaeological evidence, weaving and textile production have not been so well investigated.

The sling often requires the use of textile technology for its production (Cahlander 1980; Ochsenschlager 1993, 45-6, 54, Plate II.3). It does not require the cutting of timber, the working of stone, or the smelting of metal. The sling is perhaps seen as belonging almost to the 'feminine' world of textiles, weaving and cloth production. There is no evidence that textile production was considered a female or male task in the prehistoric past. However, it is all too easy to allow our stereotypical images to colour interpretations. In Andean society, both sexes work textiles, but women are not considered intelligent enough to undertake the complex manufacture of a braided sling (Cahlander 1980, 19). The sling is a complex piece of textile technology and, as such, can be remarkably difficult to reproduce. It was not until the 1980s that the methodology of sling braiding in the Andes was finally understood in the western world, despite having been commented on over sixty years earlier (Means 1919, 324).

The process of manufacturing a bow from basic components was often long and complex, in particular composite bows which use a variety of materials (Kroeber 1987, 190-1; Hamm 1989, 66-72; Rausing 1967, Chapter VII). Likewise, the sling has to be manufactured carefully with a considerable knowledge of the limitations and potential of the material used. The sling is not a simple weapon (*pace* Sharples 1991a, 82). The ammunition may be readily available in the shape of stones (any stone will do but this is discussed further below), clods of dried earth and even potsherds (Ochsenschlager 1993, 47). However, an artefact that allows stones to be cast over distances exceeding 80m, at speeds in excess of 100kmph has to be manufactured to high, albeit traditional, specifications.

The Method of casting a sling stone/shot[10]

The process of casting a sling stone can be broken down into a series of operations. It must be stressed that, the process of casting a sling stone is more approximately one continuous (and hopefully) smooth action. The process of slinging experimentally was achieved by a mixture of imitation and practice. The method described by Korfmann (1973, 38) was useful. News reports of the *intifada* in Palestine during the mid-1990s showed the sling being used and the method was then imitated. The addition of information from television documentaries about Andalusian and Peruvian shepherds (BBC 1996; 1998), allowed the development of a method of casting that not only works but also has a reasonable depth of at least modern authenticity to it.

Two basic ways of using the sling are often described by modern authors (Cunliffe and Poole 1991, 489; Korfmann: 1973, 38). Firstly, in the vertical plane to the side of the body, either directly to the side or at a slight angle from the true vertical. This method has been recorded throughout the areas where slings are still used today. Photographs of South American and Spanish slingers clearly demonstrate it (BBC 1996; 1998; Cahlander 1980, Cover; Hubrecht 1964, Figures 1-2; McEwan *et al.*1997, Figure 47). This allows the sling stone to be launched upwards at varying angles of release, with the resulting variation between range against initial velocity. The second method that is usually described is whirling the sling literally around the head. A trawl through the ethnographic literature produced only a single example: a colour photograph of an Ethiopian woman using a sling to scare monkeys (Leonard 1973, 153). The woman is standing on a platform constructed within the crown of a tree. The platform has a low circular band of straw at the approximate level of the waist, which would make it difficult to use the sling in the true vertical plane. The sling being used is long. It is not possible to ascertain its exact length from the image, but it appears to reach from the top of the head to the knees, which would make it impossible to use in the vertical plane on the ground. The extra length of the sling would increase its range significantly without requiring any greater strength (Monckton 1921, 39).

If the projectile is truly being cast horizontally, then there would be a significant reduction in range when compared with a vertical release. There is, however, no evidence that this method was widespread in classical Europe. The impression that this method was used in Western Europe seems to have developed from a misinterpretation of iconographic evidence mostly from the Mediterranean.

The silver rhyton recovered from Shaft Grave IV in Mycenae, dating to *c.* 1550-1500 BC (Warren 1989, 121), shows three naked slingers who appear to be advancing to the front of a group of other figures including two kneeling archers (Figure 4.1). Their posture is such that their slings are held above their heads with the left hand holding the 'pouch' end of the sling, the other end obviously gripped and held taut by the right hand. It would be easy to interpret this posture as the start of a swing horizontally around the head. However, the right arm is clearly bent in all the figures, which leads to the interpretation of this as an accurate representation of the stage when the stone (or within a Mycenaean context lead shot (Evans 1928, 344-5, Figure 196)), is being pushed into the pouch prior to casting to the side of the body. The same can be said for the representation of the ninth century BC Syrian slinger from Tell-Halaf (Buchholz 1965, Figure 3) (Figure 4.2), the slinger shown on the sixth century BC Etruscan tomb painting from Tarquinia (Strong 1968, End Plates) (Figure 4.3), and the sixth century BC vase painting of Herakles attacking the Stymphalian birds (Carpenter 1998, Plate 198) (Figure 4.4). The auxiliary or mercenary slingers carved on Trajan's Column seem to be preparing to use their slings to the side of their bodies (Lepper and Frere 1988, Plate XLVII). The sling thongs may be particularly short as the figure is shown casting a large (fist-sized) stone, though they may have been shortened to allow them to be included in the composition (Figure 4.5).

Another rarely mentioned ethnographically attested method of slinging comes from the Marquesas Islands, where a French sailor (Jean Baptist Cabri) became acculturated in Polynesian society during the closing years of the eighteenth century. As part of his integration, he learned to use the sling in what appears to be an indigenous manner. He is shown in a lithograph dating to 1812 (Feest 1980, 71). The sling was passed behind the back, with the left hand holding the pouch, in which there appears to be a large stone about half the size of the figure's clenched fist. The use of large stones is not unknown historically, and Xenophon records Persian slingers using stones within this size range (Anabasis 3.16 and 18). The right hand holds the other end of the sling, stretching the cords across the shoulders. The sling is cast by a quick 'flick' of the right hand and lower arm. This method would require considerable amounts of space, as the sling would travel in a horizontal curve. The sling is long, requiring a reach right behind the caster.

There are then at least two and possibly three defined methods of using the sling. Each of these has a measurable effect on the sling's performance. In experiment, a choice of method had to be made. It cannot be known for certain how Middle Iron Age peoples used slings. There are no iconographic representations or historical references that could have assisted in the choice of method to be used to generate the trajectories that were be applied to Middle Iron Age hillfort profiles. In the end it was decided to use those trajectories created by using the sling in a semi-vertical plane. The rationale for this choice is as follows:

1. The only iconographic evidence from Western Europe seems strongly to indicate that this method was known, at least within the Mediterranean world. There is no hint in the later Classical authors that the sling

[10] The term cast has been adopted for the process of projecting a sling shot. The term cast is used in archery and is the closest analogy.

Figure 4.1 The Silver Rhyton from Shaft Grave IV, Mycenae *c*. 1550-1500 BC

Figure 4.2 Ninth century BC Syrian slinger from Tell-Halaf

Figure 4.3 Sixth-century BC Etruscan tomb painting from Tarquinia

Figure 4.4 Sixth-century BC vase painting of Herakles attacking the Stymphalian birds

Figure 4.5 Slinger from Trajan's Column

was used in a different manner by Later Iron Age peoples of north-western Europe or the British Isles. In fact, there are few references to sling use at all outside Gaul (Strabo 4.4.3), and the use of stones as weapons is referred to only within a general context (Caesar *de Bello Gallico*, II.6, IV.32). This paucity in the record is, of course, no guarantee that North Europeans used the sling in a similar manner to Mediterranean peoples. However, the classical writers were fond of highlighting the peculiar habits of 'barbarians', and the differing use of a common weapon may have been noteworthy, as with the contrasting methods of spear or sword use by the Gauls (Dionysius of Halicarnassus 14.10.17; Polybius 2.30.8; Strabo 4.4.3).

2. The description by Vegetius II, 23 (Milner 1993) of how to use the sling and, in particular, of reducing the number of turns before the stone is cast from four to one is not compatible with the Marquesan method, although it would have been compatible with the horizontal method.

3. The weight range of pebbles (40-70g) assumed to be sling stones (Clay 1924, 86; Cunliffe 1984b, 425; Stevens 1934, 641; Wheeler 1943, 49) accords well with historically attested lead sling shot (30-60g), (Greep 1987, Figure 5; Robinson 1941, 418-43, Plates CXXIV-CXXX). The weight of the pebbles does not correlate well with the extra weight that appears to have been required for efficient use of the Marquesan method.

4. The use of the sling semi-vertically to the body achieves greater distances, and represents the most onerous method in terms of both velocity and distance, to 'test' any defensive capability of Middle Iron Age hillfort defences.

The use of the sling, like throwing a stone is instinctive. There is no method of aiming, unlike instinctive archery, where either the gap between the hand of the archer and the point of the arrow, or the distance between the point of the arrow and the intended target can be used to estimate the range (Helgeland 1975, 24; Hill n.d; Stamp 1979, 18). The development of lead sling shot with its standard size, and, more importantly, constant weight, would have increased the likelihood that the 'muscle memory' required to constantly strike a particular target would be more easily 'recalled' each time the sling was used. Some pre-metallurgical peoples would go to great lengths to shape soft stone into identical shot to improve accuracy (Monckton 1921, 38). Although the majority of sling projectiles from Britain are natural waterworn pebbles, significant numbers of clay shot have also been recovered. These would seem to indicate that, at times, a level of accuracy beyond that required when using variable weight stones was needed for either warfare or hunting. The recovery of clay shot from both domestic settlement sites and from military contexts of hillforts would indicate the interchangeability of such artefacts (see Appendix 1.1 for full references).

Unfortunately, there are no surviving archaeological remains of slings from the Middle Iron Age of Britain, or for any earlier or later prehistoric periods. Therefore, a reasoned decision on the type of sling that was to be used in the experiments had to be made, in an attempt to replicate the probable trajectory of a Middle Iron Age sling. Throughout the world, three basic materials have been used for sling manufacture: textiles, wool, cotton or similar (Cahlander 1980, 29; Means 1919, 324-7; Ochsenschlager 1993, 43); plant fibres, grasses, rushes, palm leaves (Hubrecht 1964, 92; Lindbolm 1940, 26, Figure 4; Monckton 1921, 39); or leather (Hubrecht 1964, 92; Strutt 1969 [1801], 61). Leather appears to have been the least widely used material, and then often in combination with some form of textile. All of these materials would have been available to Middle Iron Age peoples but there is no evidence to determine which would have been the most likely to be used, it is entirely possible that all three may have been used either simultaneously by differing ethnic groups in Britain, or in chronological sequence. Outside the Andes (Cahlander 1980), there are no comprehensive descriptions of sling manufacture. The lack of detailed information on production makes the replication of ethnographic slings problematical and of historical ones nearly impossible. Other than in the vaguest terms (i.e. the iconography), there is no evidence of what an ancient western European sling may have looked like, although the classical authors do give some references to the materials from which slings in the Mediterranean were manufactured (Strabo 5.1.1-2; Iliad 13.827), and some incomplete references to their method of production (Livy 38, 29.8).

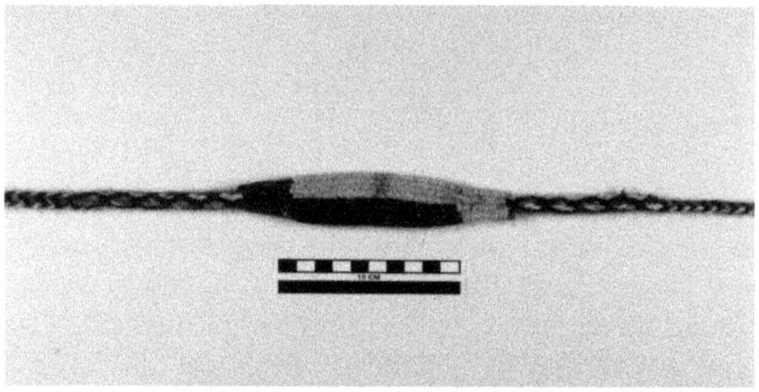

Figure 4.6 The Sling Pouch

The sling used for the experimental work was Bolivian. The rationale for this choice was that it was manufactured in wool. This commodity would have been readily available to the majority of Middle Iron Age peoples, if the distribution of spindle whorls is any indication of yarn production. The production of slings from plucked or sheared fleeces does not demand any further form of technology. An equally important element is that the sling used was manufactured expressly as a sling, in a mixed agricultural economy. The experiment had no input into its production. It was manufactured in the Altiplano using camelid wool (either llama or alpaca). The main cords are braided (Cahlander 1980, 18-20), doubling to a quadruple braid, as the double split pouch is reached. The sling is 1780mm long in total. The main cords are approximately 6mm in diameter. The quadruple braided section is, in total, 160mm long (80mm either side of the pouch) and approximately 12mm in diameter. The pouch is 155mm in total length and has a maximum width of 40mm (i.e. 20mm for each side of the split) (Figure 4.6).

The yarn used in the production of the sling was hand produced, with all the inherent irregularities that this entails. Due to the extremely arid conditions of the coastal regions of Peru, both fragmentary and complete slings have been preserved, made in both cotton and camelid hair, dating to at least 1200 BC (Kroeber and Collier 1998, 78, 83, 254; Pozorski and Pozorski 1987, 25). They were manufactured using methods identical to contemporary slings (Means 1919, 324). The Andean sling therefore has a long traditional history of production. This provides the artefact with a chronological pedigree that could not be matched had the experimental sling been manufactured following observations of extant ethnographic evidence.

The use of an Andean sling, though separated from the Middle Iron Age temporally, spatially, and culturally, does allow the demarcation of certain bounds of inference relating to its capability to be drawn, bounds that in all probability would have operated in the Middle Iron Age as much as today.

Casting from a sling

The following is a detailed description of how a sling projectile is cast using the semi-horizontal method. Without a detailed knowledge of how the sling operates, its limitations cannot be fully appreciated, and any interpretations of how it was used in warfare cannot be developed, with any degree of depth.

Figure 4.7 Method of holding sling cords

The stages of casting a sling stone or shot for a right-handed individual can be broken down as follows:

1. The index finger of the right hand is passed though the sling cord loop at the end of the shorter cord (Figure 4.7).

2. The longer cord is gripped between the index finger and the thumb of the right hand. The cord is gripped firmly but not too tightly.

3. An approximately ovoid, water worn pebble, about 50mm long with a circumference of about 100mm, weighing approximately 50g is placed in the pouch of the sling, which is held in the left hand. The use of the right hand is not sufficiently encumbered with the two cords to prevent assisting in the loading of the sling. The pebble is laid so that its longest axis is at right angles to the pouch (Figure 4.8). Any stone within the weight range of a 'normal' sling stone (40-75g) can be cast using a sling. However, the more irregular the shape the more erratic the flight path of the stone, making aiming practically impossible. This may have had little impact in warfare, if the enemy were closely ranked or presented a mass target, as often seems to have occurred in state level warfare, at least in Western Europe. In pre-state warfare, such levels of organisation are comparatively rare, and the use of more regular ovoid stones may have been required to improve accuracy when casting against a more dispersed enemy. If the stone is particularly long in its axis, there is often an audible 'hum' or 'buzz' when the stone is in flight. This is also the case with replica Iron Age clay sling shot (Mytum *pers. comm.* 2004).

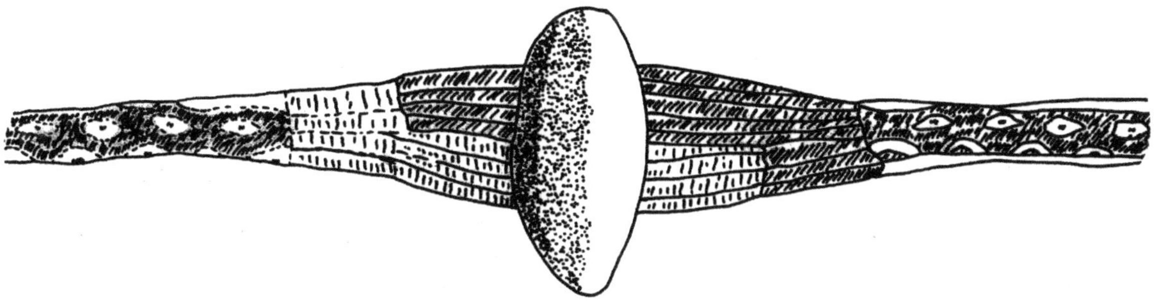

Figure 4.8 Position of stone in pouch.

4. The stone is now pushed into the sling pouch by stretching the sling upwards and pressing the stone with the thumb (of the left hand) against the index finger with the pouch between them. This allows the stone to form the flexible pouch to its surface and prevents it from flying out when the sling is revolved (Figure 4.9). It is during this phase that it is possible to 'aim' at a target using the orientation of the body as the sight. The head is turned to face along the shoulder line and the body is positioned so that the feet are at near right angles to the torso. It is this posture, recorded on the silver rhyton from Mycenae, and on other iconographic representations that seems to have led to the interpretation of the sling being swung around the head horizontally.

5. The sling is carefully allowed to drop so that it hangs to the caster's side (Figure 4.10). The left arm is extended directly forward and slightly bent to act as a counter-balance when the turning of the sling commences. The right arm is also extended slightly away from the body, so that the sling does not catch when revolved. The legs have already been separated to create a stance where the load of the body is evenly distributed, and good balance can be maintained. It is important to note that it is not possible to bend the body backwards to increase the height of a sling cast, as it is with archery. Even on a steep slope, the slinger must maintain an 'upright posture' by bending the leading leg to accommodate the gradient and thus maintain a vertical position for the torso.

6. The sling is now turned.[11] In line with the recommendations of Vegetius 11,23 (Milner 1993), the sling is revolved four times before the final cast. This motion can be generated using the wrist (Korffmann 1974, 38). However, a more steady and even circle can be described by also using the lower arm (Figure 4.11). During the initial revolution of the sling, the caster can focus on the proposed target. Four turns of the sling also gives the caster time to respond to the varying weights of the stone. From experience, it is remarkable how quickly it is possible to recognize relatively small variations in weight.

7. The release of the sling is the hardest moment to describe. Unlike the above, it does not easily separate into a series of identifiable actions. It is a single flowing, complex motion. The release of the projectile involves the major part of the upper body, as the upper torso is twisted to 'face' the proposed target while the right leg also twists to accommodate this movement, while balancing. The arm that has been turning the sling now describes the widest arc that is possible without the sling touching the ground. The projectile is released merely by opening the thumb and forefinger that held the longer cord. As the projectile has a considerable momentum, the release must occur before the optimum cast position has been reached (45°), otherwise the projectile will leave the sling pouch too early or too late and will be projected at too low or too high an angle, which results in a poor flight path (Figure 4.12).

The average initial velocity of the stone is 25.38ms^{-1}, although the range varies between 36.96ms^{-1} and 21.6ms^{-1}. The stone is visible, at least in the later stages of its flight, when it is generally falling to the ground and is thus seen against a non-illuminated background. When projected against the sky or similar bright surface (the sand of a beach for instance) there is insufficient contrast to allow the stone to been seen from the caster's position. Third party observations seem to indicate that the flight path is not a true parabola and therefore drag may play a part in the description of the trajectory. However, this may be exaggerated by a foreshortened viewpoint. Given the relative danger of observing the flight of a sling stone, it was not considered safe to stand too far forward of the caster's position. Whether a projectile would be visible from the point of view of a target is not something that is possible to discover safely. However, given the relatively high speed and small size of a stone, if not actually invisible, the projectile would be very difficult to follow by the intended target.

Three observations that are particularly germane to the use of the sling relative to hillfort defences are apparent from the above description.

1. The sling requires a clear area to the immediate front and rear of the caster. Even tussocks of grass can easily interfere with the sling causing it to snag.

[11] Turn has been adopted to include the motion of the sling when revolving through 360^0 of arc.

Figure 4.9 The sling is loaded and the projectile is pressed into the pouch while sighting at the target

Figure 4.10 The sling is allowed to 'rest' to ensure that the projectile is correctly loaded and to judge its weight

Figure 4.11 The sling is turned using the wrist and lower arm

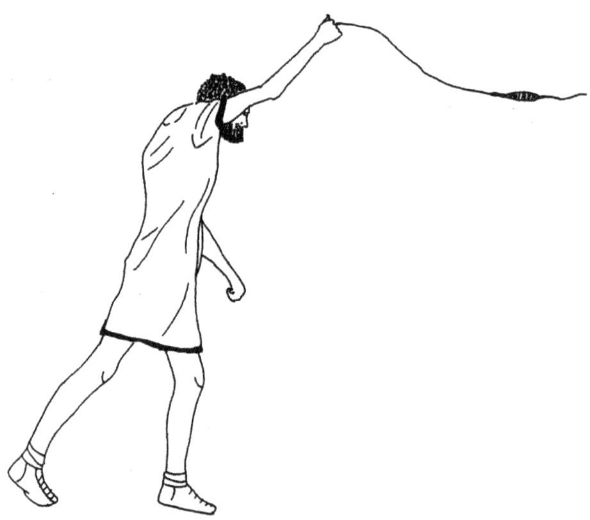

Figure 4.12 The projectile is released

The problem of undergrowth can be alleviated in some measure by angling the sling slightly off the vertical and describing an arc that is projected towards the caster's head. This cannot be accentuated too much, otherwise the final release becomes more difficult, as the angle of the body has to be so significantly altered.

2. Due to the length of the sling, the point of release of the projectile is quite low to the ground, often less than 100mm, demanding that any obstruction in front of the caster either be very low or at reasonable distance. Parapets to walls of North Coastal Peruvian pucaraconna, where the old structure has been encased in a new rebuilding are no higher than 0.8m. This has been interpreted as a direct result of use of the sling in defence (Topic and Topic 1987, 48). Such a low wall would of course leave the majority of the slinger's body unprotected from enemy fire, so a different solution must have been devised to protect the defender from a disruptive or deadly barrage.

3. The caster's stance demands that any wall or bank is wide enough to accommodate the requirements of balance and forward movement when releasing the projectile.

If the sling represented the primary distance weapon in the Middle Iron Age, this may have had a direct bearing on the construction of the linear defensive features associated with hillforts. The walls, apparently substantially over built, at least in terms of width relative to height, and the lack of identifiable parapets on the majority of earthen ramparts may be a result of the space required when using the sling as detailed above.

The Experimental method

The main difficulty with any experiment in archaeology is knowing whether the results achieved have any realistic bearing upon the problem that the experiment has been developed to solve. It is unlikely that it will ever be possible to ascertain how the Middle Iron Age peoples of Britain used the sling. Also, there is no way of knowing if the method used in the experimental work is representative of sling use in the past. All that can be said with any sense of justification is that the experiments carried out prove how far the experimenter can cast a sling stone, of a particular weight, on a particular day, at a particular initial velocity and angle of release. None of these observations can be held as proof of how Middle Iron Age peoples used the sling, or how far they could cast a stone.

In light of the above, what use or purpose does experimentation with the sling serve? Like any experimental work the results provide potential limits, potential bounds within which an exploration of artefact use can be made. The sling has a particular mythology associated with it, and specifically the distance over which a sling projectile could be cast. Ethnographic and historical ranges vary from a minimum of 30m (Enriquez De Guzman 1862 [1543], 99), to a maximum of 400m (Garcia 1972,104), at which point the sling would have been able to out-range even the mechanical siege weapons of the early imperial Roman army. The range of a reconstructed Roman catapult, casting a 3.6kg ball was between 320m and 457m (Payne-Galloway 1981 [1903], Appendix p.10).

Clearly, the upper limits of these ranges cannot be an accurate representation of reality. If this was the case, slingers would have been able to destroy the crew of any siege weapon before it was able to have any effect. The fact that slingers were unable to inflict wounds at anything like these ranges is corroborated by the statement attributed to the Spartan king, Archidamus. When seeing the use of mechanical weapons for the first time, he reputedly cried out in anguish, '*Man's valour is no more*' (Plutarch *Mor.*191E, 219A), as the enemy could kill at such ranges, thus removing the requirement for direct conflict between hoplite forces. The sling was well known to the Greeks as a weapon, and if it had the capability to out-range mechanical weapons, why would Archidamus have felt so aggrieved?

Further anecdotal evidence against such extreme ranges may be gleaned from the fragmentary poems of the Spartan Tyrtaeus, composed *c.* 640-600 BC. Tyrtaeus (F11.28) tells the Spartans, *'do not stand with your shield beyond the range of missiles'* (trans. van Wees 2000). This would mean that, at the extreme range suggested by some authors, the troops standing beyond the range of the sling would have to be some 400m away from the battle line, as the gymnetes (light or naked troops who certainly used the sling) are described as mixing with the ranks of the hoplites (Tyrtaeus F11.35-8). Recent research has shown that even a fit modern athlete is practically exhausted having run 1.6km in hoplite armour (Donlan and Thompson 1976, 340). This implies that the Spartans, who were outside missile range, would have their effectiveness as combat troops significantly reduced if they had to traverse some 400m before entering battle. Of course, it could be argued that these ranges represent the maximum achievable by the sling and not its effective range. However, unlike the arrow or atl-atl dart, the sling stone generates little drag. It therefore its terminal velocity is similar to its initial velocity. This in effect means that the sling's maximum range is also its effective range. As the sling mechanically lengthens the forearm of the caster (in a similar fashion to the atl-atl), it requires a certain weight to operate at its maximum efficiency. Unlike the bow, where the energy is stored in an elastic medium and a lighter object can be cast further, almost the exact opposite is true of the sling. Very light projectiles are unable to generate the required momentum to travel any significant distance. It is therefore impossible for a slinger to cast a lightweight object a great distance.

Given that the ranges quoted for the sling vary to such a great extent, what can be considered to be a 'realistic' range? In general, classical writers seem little interested in everyday life. This is, of course, understandable, as it was commonplace and thus common knowledge. Therefore, little detail relating to sling has survived. Garcia (1972, 104) states that the Sicilian historian Timaeus claimed that the slingers of the Balearic Islands could cast a sling stone some 600 paces. However, Timaeus' work does not, itself survive, references to it are possibly to be found in the work of Diodorus Siculus (Grant 1995, 140, Footnote 29), and almost certainly in Strabo. Neither of these authors makes any mention of a measured range, although they do comment on the differing slings used by the Balearic Islanders to achieve variable distances.

The only ancient source that makes any direct reference to the ranges associated with distance weapons is Vegetius I.16 (Milner 1993). He mentions the sling and refers to the range, at which archers and slingers are to train striking a target. But he only includes staff slingers and does not cover the use of the ribbon or hand sling. The range given is 600 feet[12] and this appears to be a maximum, as Vegetius clearly states that the troops should practice *even* at these ranges. The distance of 600 feet/paces seems to have been confused not only between two distinct forms of the sling, but also between classical authors. This measurement has been used by some modern authors to calculate that a sling in the hands of a professional can cast a projectile approximately 200-350m (Connolly 1991, 49; Griffiths 1989, 261-3).

12. The Roman foot is 10 imperial inches or 254mm so the range given by Vegetius is about 152m.

Only one other author records any information relating to the potential range of the sling in the classical world. Xenophon in his Anabasis (III, 3, 16, and 18) records that, during the retreat of the Ten Thousand, the Rhodian contingent was persuaded to act as a rearguard using their slings. Xenophon records that they were able to out-distance *most* of the Persian archers and *all* of the Persian slingers. The latter used stones the size of a man's fist, and the Rhodian slingers were using lead slingshot presumably weighing about 50g. Some modern authors have used this relative information relating to the different ranges between the Persian archers and Rhodian slingers to arrive at an absolute range for the sling. The information for the maximum range achievable by ancient Persian archers rests on a commemorative stele from the Greek Black Sea settlements. This records that Anaxagoras of Olbia cast an arrow some 1640ft or approximately 500m (Rausing 1967, 142).

Such ranges are not mere fantasy. They have been achieved in modern times using specialised powerful bows, similar to Scythian ones (Credland 1985, 40) and specially made arrows (Payne-Galloway 1981, 7; Yucel 1997, 72), that are so extremely light (approximately 19g (Payne-Galloway 1981, 7)) that they would simply have insufficient momentum to have any serious effect, which excludes them from the arena of war. It is, of course, possible that Xenophon was referring to this specialist type of flight archery. However, given the overt military nature of the Anabasis, it is more probable that he referred to the distances achievable when the Persians were using war arrows.

Bow Type	Length (inches)	Draw Weight (pounds)	Draw length (inches)	Distance of flight arrow (yards)
Igorot	61.5	26	28	100
Mohave	67	28	40	110
Paraguay	71	60	25	170
Athabascan	68	60	25	125
Luiseño	55.5	48	26	120
Navaho	44	45	26	150
Yurok	54	30	28	140
Alaskan	56	80	26	180
Yaqui	59.5	70	28	210
Yana	55	42	26	205
Blackfoot	47.5	45	25	145
Apache	41	28	22	120
Cheyenne	45	65	20	165
Hupa	47	40	22	148
Osage	47.5	40	20	92
Blackfoot	40	40	20	153
Andaman Island	62	45	20	145
South American	74	50	28	98
Solomon Island	74	56	26	148
Tartar (1)	74	48	36	112
Tartar (2)	74	100	30	100
Japanese	88	57	32	156
Negrito	76	56	28	176
Polynesian	79	48	28	163
Average		49.8		144

Table 4.1 *Archery Statistics*

Korfmann (1973, 39), using the same passage from Xenophon, also attempted to interpret the range at which a slinger can cast a projectile. By analogy with modern American hunting bows, which can cast arrows up to 250m, he concluded that the Rhodians, and therefore other ancient world slingers, could cast their shot at least 400m. Unfortunately, the logic of such assumptions is severely flawed. There is no evidence concerning the type or strength (draw weight) of the bows used by the Persians. It is evident, from the extensive work carried out by Pope (1923) on bows collected from a wide spectrum of cultures across the world, that there is considerable variation in draw weights and the distance to which they could cast an arrow (Table 4.1).

It is also clear from Pope's work (1923) that there is not necessarily a direct correlation between the bow's draw weight and the distance which it can cast an arrow. The extreme example is a Tartar bow with a draw weight exceeding 50kg and yet it would not cast the arrow provided with it further than 91m (Pope 1923, 75-6). The danger of Korfmann's approach is further exemplified by early European colonial ethnographic observations. The Native Americans of Virginia were first contacted by the English in the late sixteenth/early seventeenth centuries, by which time the bow had all but fallen from use among the English. The Powhatans could shoot an arrow accurately over some 36.5m and cast randomly up to 110m (Rountree 1989, 42). Clearly, without a detailed knowledge of the type, construction and use of late Achaemenid Persian bows, Xenophon's statement is of little use in determining the range of the sling.

By the start of the nineteenth century, with the European expansion into empire, the use of firearms was rapidly adopted wherever possible by indigenous peoples. As a result, more traditional distance weapons such as the bow and sling were only maintained for either sport or as toys. Like the classical authors before them, the early ethnographers tended to concentrate on elements in other cultures that were unusual or different, and the sling fell within that category. The number of references to the sling increased in the late nineteenth and early twentieth century, culminating in the descriptions by Lindblom (1927, 1940)

Location	Date	Reported Range	Approximate Equivalent in metres	Reference
Peru[1]	1543	30 'paces'	30	Enriquez De Guzman 1862 [1543], 99
Peru	Recent	200 feet	60	Burns (*pers. comm.*)
Tibet	1895	300 yards	274	Rockhill 1895, 714
New Britain[2]	1884	100 yards	91.4	Powell 1884, 162
Madagascar[3]	1913	75 yards	69	Linton 1913, 242
New Guinea	1921	200 yards	182.8	Monckton 1921, 38
Nigeria	1925	100 yards	91.4	Meek 1925, 116
North Africa	1927	200 metres	200	Langlet 1927, 146
Fiji/Hawaii[4]	*c.* 1930	50 paces	50	Wheeler 1943, 49
Arabia[5]	1950s	30-50 yards	27-45	Peddie 1997, Chapter 5 footnote 10
Majorca	1963	200 metres	180	Hubrecht 1964, 93
Turkey	1970s	200 metres	200	Korfmann 1973, 37
Britain[6]	1994	40 - 90m	40 - 90	Griffiths and Carrick 1994, 7
Britain[7]	2001	80m	80	Time Team, 2002

Table 4.2 *Recorded ranges of sling casts*

[1] Records the distance at which a sling stone broke a hand-held steel sword.
[2] Records the distance at which the local population could hit birds in trees, or as seldom missing a mark by more than 2-3 yards (1.8-2.7m). This is stated as a maximum range.
[3] The local slingers are reported by their neighbours as being 'poor shots'. There is no indication as to whether this refers to range, accuracy or both.
[4] The range at which the slingers could hit sticks placed in the ground; according to the report, they could cast further but with little accuracy (unfortunately Wheeler does not reference his source).
[5] Records the distance at which Arabs in Oman were able to knock small game over.
[6] Recorded the distance observed in experiment with 'Roman' slings.
[7] Recorded as a maximum range using beach-washed pebbles.

In Table 4.2 is a tabulated description of modern experimentation and early modern ethnographic references to the ranges of the sling.

It is clear that there is huge variability in the observed ranges recorded for sling projectiles (arithmetic mean = 123.6m; median = 91.4m). It is difficult to ascertain if any of these ranges relate to a measured demonstration or to an estimated observation. Unfortunately a vivid memory is not necessarily a guarantee of an accurate one. The problem with many of the above references is that it is very difficult to estimate the range of a sling cast from observation alone. Non-empirical experimentation has shown that a majority of individuals with limited knowledge of the sling's capabilities consistently estimate the range to be between 100-150m. This is certainly in excess of what experimentation indicates is achievable, and shows the difficulty in observing ranges rather than making any form of measurement. Another major difficulty is that sling stones in particular tend not to embed themselves in the ground, unlike arrows or atl-atl darts. They usually bounce, depending on the hardness of the ground; thus a reasonable estimation of the point of impact can be difficult.

A number of problems presented themselves when attempting to construct a method of measuring the distance to which a sling stone could be experimentally cast. Each of these had to be overcome or their effect limited. Firstly, sling projectiles tend to bounce. An obvious answer to this difficulty was to choose a location were the exact point of impact could be recorded. The sling stone also needs to be cast over a level expanse of ground, so that the influence of any slope was minimised. Clearly, using a weapon once used in war to kill or maim is a dangerous operation, and a location where people were either absent, rare or easily observed needed to be chosen.

In consideration of the above, the North Norfolk coast was chosen. Although certain areas are heavily used by tourists, beyond these the beaches tend to be quite empty. Anybody approaching could easily be seen long before they came into range of the sling, so work could be stopped until they passed out of danger. A vitally important factor is that at low tide, a wide expanse of hard sand was exposed.

The advantage of saturated sand is that it is relatively hard and when compressed does not reform into its original shape, but leaves an imprint of any impact. The beaches are also flat. The only disadvantage is the often strong winds. These can have a major impact on the distance that a projectile will travel. In order to prevent significant bias entering the results, only on days when the wind speed was less than 8kmph (5mph) was the experimental work carried out.

Unlike arrows, atl-atl darts, or lead sling shot, the pebbles used as sling projectiles varied in shape/volume (7.5-11cc or 29.3% variation) and weight (26.8-94.3g or 28.4% variation). The percentage difference between these naturally occurring stones is significant. Therefore, in order to construct an 'average' trajectory for an 'average' sling stone, ninety stones were cast, and their individual flight characteristics calculated. The shape of the stone undoubtedly plays a role in the flight of the projectile. However, by casting a wide range of stones, each basically ovoid allowed the formulation of the 'average' trajectory to be as genuine as possible.

The stones used were sea worn pebbles collected along the East Anglian coast. Most were flint. No measurements were taken at the time of collection. Decisions concerning the suitability of weight and shape for each pebble were made instinctively. Only after this entirely conscious choice had been made, were any details of the pebbles recorded. Clearly, there was a conscious choice in the collection of sling stones by Middle Iron Age people and, whereas there is no accurate way of reproducing that choice, the above method is probably the closest.

Each stone was sprayed with yellow aerosol road paint, so it could be easily seen once cast. The paint had one other advantage. It is not designed to adhere for long periods and will wear off in a matter of weeks. The smooth surface of the pebbles makes the paint's adherence relatively poor. Although under ordinary handling conditions no paint was removed, at the point of impact the abrasive sand removed some of the paint and a deposit of paint was left, making the identification of this location easier. Each stone was also numbered.

Each stone was cast. At the moment of release, a stopwatch, accurate to 100th of a second, was started and this was stopped once the stone was observed to strike the ground. The initial impact was usually quite deep, about 10mm, and, as stated above, a smear of yellow paint normally accompanied this impact 'crater'. Once the initial point of impact had been identified, a ranging pole was held at this point, and the distance over which the stone had travelled was measured using an optical rangefinder with an accuracy of ±0.9m at 75m or 98.8%.

The results of the experimental work are detailed in Appendix 2.1-2.3.

Effectiveness and accuracy

Since the experimental data seem to indicate that the velocity of a sling projectile does not significantly lessen over distance, it is reasonable to conclude that the final impact speed of the projectile is close to the initial velocity. Unlike an arrow or atl-atl dart, the maximum range of the sling stone is also its effective range. The relatively high drag of an arrow or atl-atl dart significantly reduces its velocity. This in turn leads to a significant reduction in impact. A light flight arrow is capable of tremendous distances but would have little capacity to cause damage at the termination of its flight. The same is not true of the sling. The force required to penetrate a human skull, one of the best defended areas of the body, is 0.23kg per square millimetre (340lbs per square inch, Pare (1969 [1569]). A sling stone with an 'average' mass of 56.7g and an 'average' velocity of 25.38 ms^{-1} could achieve this level of impact.

Effectiveness should not be understood to equate directly with accuracy. The rhetorical language of ancient western sources makes it clear that it is possible to be very accurate with the sling. The Benjamites are recorded as being able to cast a stone at a hair's breadth and not miss (Judges 20:16). Likewise Livy (38:29:7-8) records, not only the ability of the slingers of Aegium to cast through rings at a 'long distance', but also their ability to strike any part of the face at which they aimed. Strabo (3.5.1-2) records the famous story of Balearic children not being allowed to eat until they had struck a piece of bread fastened to a pole. Of course, accuracy varies between individuals and such sources may be either referring to exceptional instances/individuals or have become exaggerated in the telling. Unfortunately, the sources do not give any indication of what distance such feats could be achieved. The ethnographic record also holds few clues. As stated above, the Spaniards, during the siege of Cuzco in 1536, record that a hand-held sword was broken at 30 paces (Enriquez De Guzman 1862, 99). During the late nineteenth century, slingers in New Britain were recorded as being able to hit birds in trees or rarely miss a target by more than two or three yards (1.8-2.7m) at a range of 100 yards (91m) (Powell 1884, 162). Such a range is recorded as a maximum. Whereas a tolerance of 1.8-2.7m would be acceptable when casting at a massed body of troops, where the requirement for accuracy is not high, it would be of little use in hunting or in more dispersed combat.

Wheeler in his Maiden Castle report included a reference to Polynesian slingers (Wheeler 1943, 49). At 50 paces, they could hit sticks placed in the ground and, according to the report, could cast further but with little accuracy. Peddie (1997, Chapter 5, Footnote 10) records a similar range for Omani Arabs hitting small game. These appear to be the only clear references to the ability of slingers to hit a specific target at a given range. This evidence would tend to indicate that, at ranges in excess of 50m, slingers could not be considered accurate. This accords well with modern use of archery equipment without complex sights. Under these circumstances ranges in excess of 40m are considered to be too great to consistently kill relatively large animals such as white tailed deer (Helgeland 1975, 88).

Assuming that the above does represent the accurate range of the sling, then a further assumption may be justified. When casting from an elevated position, a slinger will automatically outrange any opponent below. From the experimental observations it is clear that the greater the elevation, the greater the difference in range. If an elevated slinger is able to cast 50% further than an opponent, it has been assumed that this represents outranging. If, however, that figure doubles to 100%, it has been assumed that this represents *significant or total* outranging. Under the latter circumstances, it may have been impossible for slingers down slope to manoeuvre themselves into a position to return fire before suffering sufficient casualties to force a withdrawal. The above observation has particular relevance when attempting to understand the use of both artificial and natural slopes in hillfort defence.

The Spear

Middle Iron Age wrought iron spearheads vary considerably in size, shape, and degree of manufacturing skill, from the simple flat leaf-shaped blade from Thundersbarrow (Curwen 1933, fig. 5.6), to the elegantly named 'flamboyant' type recovered from Bredon Hill (Hencken 1938, 13).

There has been a tendency in the past to classify a spearhead's potential use in relation to its size (Hencken 1938, 76; Stanford 1974, 165). Simply put, the existing interpretations equate large spearheads with thrusting weapons, and small ones with 'javelins' or throwing weapons. As Osgood *et al.* (2000, 16) rightly observed, such a classification is too simplistic, and may ignore subtle differences in spear design. The following is an attempt to apply a more rigorous approach, as a step towards resolving this typological issue. This is problematic due to the inadequate information often presented by the material remains. This in itself presents other problems of data, which have not have been entirely satisfactorily answered by this methodology. The following, hopefully, will present a framework from which more concrete interpretations may be drawn.

It is often stated that spears have a dual function of war and hunting (Xenophon, *Cynegeticus*). However, within settled agrarian communities, it is not always easy to separate hunting from 'warlike' intent. Hunting was seen as an extension to, and training for, warfare (Plato *Laws* 7.823b-824b). Classical authors also routinely included the hunting of humans as slaves within this context (Plato *Laws* 7.823d). This form of activity would almost certainly meet some level of violent resistance, which would be difficult to differentiate from warfare.

For a throwing spear to be as effective as possible, there are certain constraints relating to its weight. Tasmanian spears have an average weight of 600g (Noetling 1911, 78).[13] The weight of a modern Olympic javelin is 800g. It is clearly possible to throw spears that are both heavier and lighter than these figures but, within the constraints of average biomechanics, such a weight range represents the optimum (Cotterell and Kanninga 1990, 165-6). For a spear to achieve reasonable flight, its centre of gravity should be within the first 40% of the shaft (Cotterell and Kanninga 1990, 174). For ease of calculation a point approximately one third of its length away from the head has been used in this study. It should therefore be

13. The aborigines of Tasmania were the only ethnographically recorded people to retain the hand-thrown spear (as opposed to casting using a spear thrower) as a primary hunting weapon, (Noetling 1911). They never developed or adopted the spear thrower presumably as it was of little use for their hunting technique.

possible to test a sample of Middle Iron Age spearheads against this optimum.

Clearly it will not be possible to calculate the original weight of a spear from its head alone, as any length of shaft will balance at the required point, assuming that the counter-weight extension is achieved. However, it should be possible to calculate the length of a spear if a weight of 600g is assumed. This mass, as it is at the lighter end of the range of optimum weights, will provide a minimum potential length for a throwing spear. The only information required is the weight of the spearhead, the diameter of the spear shaft, and the density of the timber used for the shaft itself.

One of the main difficulties in attempting this correlation is the poor state of preservation of many Middle Iron Age spearheads from the hillfort dominated zone. They are often fragmentary or so heavily corroded that only an estimate of their original shape can be given (Curwen 1933, Figure 5; Williams 1951, Figure 18). The information included in many reports, although sufficient to identify the size and shape of spearhead, does not provide adequate information for reconstructing it, and therefore any attempt to calculate volume and mass is impossible (Cunnington and Cunnington 1913, Plate III; Gresham 1939, Plate IV; Stanford 1974, Figure 76).

Fortunately, a series of spearheads from Danebury was recovered not only in reasonable condition, due to the chalk geology, but also nearly complete (the slight loss of the presumably sharp edges should have no significant effect on the mass of the spearhead). The weights of the spearheads have also been published (Cunliffe 1984b; Cunliffe and Poole 1991).

Spear shafts also present an interpretative problem. None of the spearheads so far recovered from the hillfort dominated zone have sufficient remains of their shafts remaining to allow identification of the timber used to species level. Outside this area ash appears to have been favoured for the manufacture of spear shafts (Fox 1946, 12; Stead 2003, 56-9), and it has long been recognised as one of the best timbers for such use (Tabor 1994, 25; Hodges 1989, 121). There is clear evidence for its use in the early historic periods of Europe (Iliad 19.390; Iliad 4.47, TBC 1:3797), though a wider variety of timbers appear to have been used in the Bronze Age of Britain (Coles *et al.* 1978, 10-5). Ash grows throughout the British Isles and is tolerant of the lime-rich soils that cover much of the hillfort dominated zone (Edlin 1970, 209). It also coppices well (Tabor 1994, 25), and could have represented a perennial crop for pre-industrial peoples.

The results of the experimental work (see Appendix 2.4) indicate that the majority of spears recovered from Danebury have inherently unstable flight characteristics. This is based on three main assumptions: two are based on the limits of human mechanics, which are unlikely to have altered significantly; the third to ethnographic observations relating to spear weights. It is the last that is open to the most criticism, as the choice of the mass of a spear is at present archaeologically unrecoverable.

There is a limited number of burials dating to the Late Iron Age that contain spearheads as part of the assemblage of grave goods (four of the seventeen listed by Collis 1968). Clearly, this evidence cannot be directly compared with that recovered from the Middle Iron Age but may provide some limited comparative data. If the spears were placed in the grave unbroken, it would be reasonable to assume that they could not be longer than the grave itself. This provides a starting point for a brief discussion correlating spearhead size to potential use, in conjunction with the mathematical method described in Appendix 2.4. There is clear evidence for the spear shaft at Owslebury being broken (Collis 1973, 126), and potential evidence from Wetwang (Dent 1985, 88).

At Owslebury (Collis 1973, 126), the iron head of the spear was thrust into the solid chalk edge of the grave cut near the skull. The iron ferrule was recovered parallel to the pelvis, suggesting that the spear had been too long to fit in the grave and had been broken in two at the time of deposition.

Assuming that the full length of the grave was occupied by the broken spear with the head still attached, and that the remaining section of the spear with the ferrule also extended to the southern end of the grave, a maximum length of 3.1m can be inferred. The length of the spear, therefore, falls between a maximum of 3.1m and a minimum of 2m, assuming that it only occupied the total length of the grave. The standard length of a hoplite spear appears to have been in the range of 2.2-2.3m (Anderson 1991, 22), indicating a potential correlation of the spear as a thrusting weapon. The spearhead from this grave was long and slender, measuring approximately 320 by 40mm.

Unfortunately, the spearheads recovered from Wetwang, though combining both large and small spearheads, cannot be used to approximate shaft length, as angles at which the heads rested clearly indicated that they could not have been accommodated within the grave (Dent 1985, 88). The spears were either broken as part of the funerary ritual or were left projecting from the grave mound.

Among the graves excavated at Grimthorpe, one contained a panoply of weapons including the remains of shield fittings, a sword and the fragmentary remains of a small spearhead approximately 50mm in length (Stead 1968, Figure 11). Assuming that the spear was placed with its shaft unbroken in the grave, its overall length would have been approximately 1.25m. Its short length and small head would indicate that this spear was intended for throwing, as the limited extra reach provided by the spear would not appear to give any significant benefit to the user, unlike the example from Owslebury.

Another potentially short spear was recovered from a burial at Whitcombe (Collis 1973, 125-6). The position of the small spearhead (among the lower ribs of the crouched burial) would certainly imply that the spear could have been placed in the grave unbroken. If this was the case, the maximum length of the shaft would have been approximately 800mm, which is more in keeping with a dart than a spear.

The limited evidence from later Iron Age burials would suggest that there is a correlation between spearhead size and the length of the shaft. Although the sample is far from statistically viable, the conclusion that small spearheads were associated with short (and therefore light) spears and large spearheads were associated with long (and therefore heavy) spears appears to be justified. It may also be reasonable to telescope this analogy back into the Middle Iron Age, and suggest that the mathematical modelling of the Danebury spearheads is not solely a manufactured result.

The nature of Middle Iron Age spearhead manufacture would also imply that they were designed for throwing rather than for individualised close combat. This interpretation is based on the observation that most of the recovered spearheads from the Middle Iron Age are quite poorly manufactured (Brailsford 1962, 6). The widths of the blades of Middle Iron Age spearheads also imply that they were intended for throwing. The broader the blade in comparison to its length, the greater the likelihood of causing significant haemorrhaging, which was the main cause of death with such slowing moving projectiles. Of course this does not preclude their use as thrusting weapons, but these tend to have narrower blades making withdrawal from the victim easier. Most Middle Iron Age spearheads are hammered blades with the tang formed from the blade itself by simply hammering around a cylindrical object (Brailsford 1962, 6; Cunliffe 1984b: Figure 7.19; Cunliffe and Poole 1991 Figure 7.18; Cunnington and Cunnington 1913, Plate III; Curwen 1933, Figure 5; Williams 1951, Figure 18.8). This infers that the finished article was intended not for display, but for a more 'utilitarian' function, where recovery was not necessarily paramount. This would correlate well with their possible use as throwing weapons.

Conclusions

The Sling.

For effective use, the sling needs space. Unlike the bow, dart or javelin, it is simply not possible to sling in a massed formation. This would be true of whichever ethnographically observed method was used. Not only does the sling need space, it requires open and clear space. It is useless in areas of scrub or trees. It also necessitates the use of low breastworks or none on ramparts and walls. This, in turn, requires another solution in terms of defence.

Although any experimental work is inherently problematic, as there is no direct observation of the archaeological record, and the experimenter's own preconceptions cannot be totally excluded, the results of the work undertaken with the sling do appear to fit within the mean range of the observed ethnographic record. Assuming that the results of the work undertaken using the sling would have been within the range of experience of Middle Iron Age peoples, what does this evidence say about the design and possible defensive properties of hillforts in Britain? This will pursued this in more detail in Chapter Five.

The Spear.

The conclusions relating to the use of the spear in the Middle Iron Age are based on a series of theoretical observations relating to the average biomechanical potential of human beings and the assumed mass of throwing spears. Clearly, these observations can be challenged. However, without some attempt to structure the discussion of spear use in the Middle Iron Age, the interpretations are thrown back onto typological descriptions that categorise, but do not place the artefacts into any social context.

From the empirical evidence generated from the Danebury finds, and subjective observation of other spearheads recovered from Middle Iron Age contexts, it would appear that it would not be an unwarranted interpretation that the majority of spears in use during the Middle Iron Age were throwing weapons.

Chapter Five

Survey of Hillforts

A survey of twenty hillforts, representing approximately a 10% sample of those constructed within the hillfort dominated zone, was undertaken to test the hypothesis that the arrangement of banks, ditches and dry-stone walling and the location of hillforts had direct relevance to the use of the sling in warfare. The hypothesis that hillfort design and sling use were linked gained wide acceptance following Wheeler's interpretation of the various earthworks in relation to sling use, subsequent to his excavations at Maiden Castle (Wheeler 1943, 48). Work at Danebury (Cunliffe 1986; Cunliffe and Poole 1991) appears to support Wheeler's observation. However, no method of testing this hypothesis has been developed.

In order to test this, a series of profiles were constructed to show the existing eroded 'defensive' features. Clearly, it would not be possible to produce a detailed contour survey of each hillfort, as the amount of work involved would have been prohibitive. No attempt has been made to reconstruct the existing earthworks to their original condition. Any evidence from excavation reports relating to this will, however, be considered as part of the overall interpretation of the site. If the hypothesis is found to be valid when applied to the earthworks as they stand at present, after a considerable period of erosion, then their layout should have been at least as efficient, if not more so, when the ramparts etc. were in use. If the hillfort was constructed using dry-stone walling alone (as at Conway Mountain and The Breddin), the walls are shown at the minimum height that the excavator considered acceptable.

At each fort, where possible, four profiles were constructed. At the locations where this proved to be impossible, either due to steep cliffs forming one of the defensive obstructions or if the fort had been damaged by modern or ancient human activity, data from the 1:2500 series Ordnance Survey maps were used to reconstruct the general natural slope of the terrain.

In order to reduce the subjective bias of choosing locations that feel as if they would work, and ignoring those which felt that they would not, profiles were taken to the right of the main entrance as the hillfort is approached, and then at 90°, 180°, and 270° from that point.

A defensive structure is useless unless it has some way of being entered with relative ease by those wishing to use it. The most common and simple solution to this problem is the construction of a break in the dry-stone walling, or banks and ditches. True walls constructed of stone and standing within their own foundations are rare in Middle Iron Age Britain, but they do exist, and the term linear defences has been adopted to record them in the wider field of earthwork constructions.

The main problem faced by any 'designer' of a defensive structure is that an entrance must simultaneously maintain legitimate access and deny access to any group whose presence is undesirable. This may extend to groups representing a direct or indirect threat, or members of one's own group that, for symbolic or political reasons, are to be denied access at particular time.

The concentration of sling stones at the entrances of Danebury (Cunliffe 1984b, 425), Maiden Castle (Wheeler 1943, 115) and Conway Mountain (Griffiths and Hogg 1956, 80) are indicators of the importance that the original occupants ascribed to these areas as locations requiring defence. The concentrations of such large number of sling stones may be an artificial product of recovery, as only limited excavations of the ramparts at Maiden Castle and Conway Mountain have been carried out. However, at Danebury, Cunliffe undertook significant excavations to the rear of the northern section of the inner rampart and, although considerable quantities of slings stones were recovered (Cunliffe and Poole 1991, 404), these were not as great as those excavated at the entrance (Cunliffe 1984b, 425).

Entrances have to be more than a mere break in the linear defences. They have to be situated for easy access when required, so they are often on the gentlest gradient available. A gentle rise on one approach to a hill may have been one of the deciding factors for the fortification's original placement. The construction of the paved roads, at Danebury (Cunliffe 1984a, 128), and Midsummer Hill (Stanford 1981, 26) indicate that the traffic was considerable. This level of movement requires that the entrance is relatively free of obstructions and, although many hillforts display a convoluted entrance system, there are no examples where constrained or looping turns have been applied, despite their obvious military advantage. At Caer Caradoc (Church Stretton, Shropshire), the original entrance is approached by a long roadway that follows the natural contours of the hill, providing a gentle approach. However, it is overlooked along its entire length by the ramparts that rise almost cliff-like along its eastern side.

The majority of hillforts in the hillfort dominated zone have only one entrance (Forde-Johnston 1962), although through time a number of subsequent breaks have been created in the banks as a result of later erosion, either deliberately in order to gain more access to the interior for agriculture, or as a result of continual damage by cattle. However, Danebury (Cunliffe 1984a), Bury

Ditches (Clun, Shropshire), Old Oswestry (Varley 1948), and The Wrekin (Kenyon 1942; Stanford 1973), were provided with two entrances opposed to each other, roughly aligned east and west. At Danebury, the eastern entrance was blocked by constructing a significantly thicker bank along the line of the existing course of the linear defences. The western entrance was retained and it was at this entrance that the defences were elaborated with further hornworks and outer banks (Cunliffe 1984a, 25). The reduction in the number of entrances, and the elaboration of the remaining one, is an indication of the importance attached to the entrance as a potential weak point within the circuit of the defences. The central hornworks that formed part of the reconstruction is where one of the main concentrations of sling stones has been recovered. Cunliffe, used the layout of these earthen banks to postulate that the central hornwork was a 'command post', from where a slinger would be able to strike an opponent at a distance of 60m (1971, 64, Figure 17).

The later elaboration of hillfort entrances indicates the importance of these locations for the original builders. It is, therefore, not unreasonable to undertake the initial profile survey at this location. If the hillfort had two or more entrances, an intuitive decision relating to the elaboration of the entrance defences was made, and the survey was started at the apparently more complex of the two. It is assumed that elaborate construction directly relates to the importance that the original constructors gave to this location. Normally, if two entrances exist, they are directly opposed, and thus the survey locations detailed above (at 90°, 180°, and 270° from the main entrance) enabled the construction of profiles at both positions, (a detailed description of the survey method is given in Appendix 3.1)

The Study Area

Statistically, it would have been preferable to have selected the sites to be surveyed at random but, in reality, this proved to be impractical. Many hillforts are in the hands of private landowners, who are often reluctant to allow access to them, or are difficult to track down. Instead, hillforts that are, for the most part, either in public ownership, (e.g. the National Trust, Shropshire Wildlife Trust, Local Authorities or those under the protection of English Heritage) were surveyed. This may have introduced a certain bias into the sample, as it is mostly the more 'spectacular' monuments that are acquired for the nation or are under the protection of national bodies. However, a number of smaller earthworks fall within the large areas of land under the control of the National Trust and, although these had not been excavated and therefore remained undated, they were surveyed to provide some form of control of the overall process.

The sample included hillforts that were both excavated and unexcavated, and those that had direct evidence of sling use in the form of sling projectiles, and those that did not. The distribution of identified sling shot throughout the hillfort dominated zone strongly suggests that the sling was well known. The lack of finds in many of the hillforts may well represent the 'keyhole' nature of many of the early excavations, often restricted to single trenches cut through banks and ditches in an attempt to gain dating evidence. Taking the above into consideration, and given the widespread, if sporadic, recovery of sling stones from hillforts across the entire hillfort dominated zone, it would be reasonable to assume that the sling was known or used throughout the area.

To test the hypothesis that that the linear defences of hillforts were designed within a context of sling warfare, the average sling stone trajectories, which have been calculated as part of the experimental work were applied to the reconstructed profiles. From the defending ramparts, only the trajectory for a release at 45° was applied to the profile. Any other angle of release would not have contributed any meaningful data within this context: the maximum range of any projectile is achieved when it is released at 45° from the horizontal. The 'attacking' sling casts, however, were applied using the angles of release of 45°, 55°, 65°, and 75°. The latter represents the maximum that appears to be possible as a conscious release. In the experimental work, a release angle of 86° was achieved, but this was accidental and, although reaching a great height, did not gain any significant distance.

In each test case, a brief description of the hillfort will be given. For each profile, the application of the sling stone trajectories will be discussed and analysed in consideration of the possibility that the preserved layout of the earthworks had some relevance to defence, using the sling as the primary distance weapon. The detailed results of the hillfort survey can be found in Appendix 3.2.

The above statement is not free from any preconceptions concerning the use of hillforts in the landscape. This is not to deny or lessen any other use of these structures that clearly were (and for that matter still are) multi-functional. Nor is it to claim that warfare or the defence against aggressive or expansionist polities was the primary reason for hillfort construction. The potential difficulties of seeing what is not there because of the desire for it to be evident, and the hope that the foreknowledge of the pitfalls of such an approach will, if not prevent preconceived ideas, at least reduce the impact of any implicit assumptions.

Of the twenty forts surveyed, eleven had evidence of sling use. At one other, Croft Ambrey, the excavator found many pebbles that were alien to the site, but as these were not concentrated in pits or the like, he did not interpret them as sling stones and no record of their characteristics was made (Stanford 1974, 44). This is a good example of the problems associated with the recovery of 'natural artefacts', such as sling stones. There is a tacit belief that, at a macro-level, most if not all features and artefacts are recovered from archaeological contexts, but little research has been undertaken to

Hillfort	Gird Reference	Excavator	Sling shot recovered
Battlesbury Camp	ST 894 456	Cunnington 1923b	Yes
Bodbury Ring	SJ 292144	None known	No
Breiddin Hill	SJ 292144	Musson 1991	No
Bury Ditches	SO 3274 8372	None known	No
Bredon Hill	SO 958 400	Hencken 1938	Yes
Castle Hill	SJ373 012	None known	No
Conway Mountain	SH 760 778	Griffiths and Hogg1956	Yes
Crickley Hill	SO 927 161	Dixon 1994	Yes
Croft Ambrey	SO 445 668	Stanford1974	No
Danebury	SU 323 376	Cunliffe 1984a,Cunliffe and Poole1991	Yes
Earls Hill	SJ 409 049	None known (Forde-Johnston 1962)	No
Eggardon	SY 541 947	Marsh 1901	Yes
Hambeldon Hill	ST 845 126	Cunnington 1895	Yes
Hod Hill	ST 857 106	Richmond 1968	Yes
Ivinghoe Beacon	SP 960 169	Cotton and Frere 1968	No
Maiden Castle	SY 669 885	Wheeler 1943	Yes
Norbury Bank	SO 577 847	None known	No
Old Oswestry	SJ 296 310	Varley 1948	No
Pilsdon Pen	ST 412 013	Gelling 1977	Yes
Uffington Castle	SU 299 863	Miles *et al.* 2003	Yes

Table 5.1 *The surveyed Hillforts*

ascertain if this is in fact the case. However, the size, colour, recognisability and brightness of artefacts have a biasing effect on recovery (Sharples 1991b, 153), as does, most significantly in relation to the recovery of potential sling stones, their perceived academic value (Levitan 1982, 27).

Conclusion

The evidence from the field survey seems to strongly suggest that the use of multivallation does have a significant effect on the ability of an assailing force to use the sling, if it is accepted that the areas between the banks and ditches were allowed to regenerate with scrub. At none of the multivallated hillforts surveyed would it have been possible, from the evidence provided by the experimental data, to have struck the innermost bank without entering a space defined by the outer works.

However, the use of multivallation to prevent sling use does not explain the continued building of ditches and banks at many hillforts. In simple practical terms all that a hillfort's population needed to construct, in order to be protected from assault by sling, would be a large inner ditch and bank and a smaller outer work at 70m or so from the inner work, dependent on the topography of the site. However, many hillforts go well beyond this and have a series of ditches and banks, which often seem to add nothing to the overall defences. Clearly, there is another rationale behind multivallation. The difference between Bredon Hill and Old Oswestry draws this observation into focus. Despite the rebuilding of the ramparts at Bredon Hill, the earthworks have no major additional works, following their initial construction. In defensive terms, they present the optimum simple solution outlined above. Old Oswestry is very different. There are at least four different phases of construction (Varley 1948, 50-9, Figure 3), each significantly adding to the last. The rebuilding of the western entrances seems actually to hamper the defence of the inner banks by blocking the view from this point. Statements of status may well have been one of the key elements within a complex of rationales, but another potential reason for the drive for such additional works in terms of warfare will be explored in Chapter Six.

All of the entrances to the univallated hillforts seem to be vulnerable to attack by sling. The dating of many is purely typological. They may well date to the Late Bronze Age/Early Iron Age, before the rise of developed hillforts in the Middle Iron Age or, as in the case of Pilsdon Pen, to the Late Iron Age (Gelling 1977, 283). One feature that appears on three of the hillforts in the Welsh Marches (Bodbury Ring, Oakhill Castle Ring, Earls Hill) does suggest that the use of the sling may have played a part in their design. All three have what appears to be a terrace, an area of the hillside dug away to create a continuous platform running along a particular contour. This feature would fit well with use of the sling as a defensive weapon. The terraces appear to be relatively easy to create, as they are cut into natural ground, which will provide a stable surface. The terrace also has the potential of having no barrier to the front and a clearly defined space behind (perhaps allowing extra slingers to operate from the natural hillslope above), and also requires significantly less work than creating a series of banks or ditches. In terms of defence against other types of weapon, the terrace seems to offer little advantage. The differential in range between a defender and an assailant using the sling seems to be a factor of approximately two, or a distance in excess of 100m. This would allow the defender enough time to cast a number of shots at the assailant before they were in range to shoot back. This would not have been the case with shorter-range weapons (such as the throwing spear (Stanford 1974, 41-4)). Although the defender would still have an advantage, the distance that the assailant would have to traverse in order to return 'fire' would be considerably less, significantly reducing any advantage that the defender may have had. Though the amount of work involved in creating a terrace is less than making a bank, it would still be considerable, considering the limited tool kit available. If non-distance weapons were the primary defensive response, then a ditch and bank or a steeper bank would be a more 'logical' reaction. As other defensive measures were used often within the same hillfort, the construction of a terrace requires an explanation. Within the bounds of the inferred evidence of sling use, this method of defence appears to fit the available information.

Two of the entrances at hillforts with terraces seem vulnerable to attack with the sling: Bodbury Ring and Castle Ring. At Earls Hill, the outer works to the front of the main gate do seem to serve some defensive function. Without excavation at Bodbury Ring and Castle Ring, to explore the possibility that there were smaller now obliterated features, it is not possible to infer much about the reason behind such apparent vulnerability. There may have been sufficient natural scrub to make the construction of any further feature unnecessary (as is the case today with both forts). Alternatively, more ephemeral barriers such as a fence or hedge may have served the same role as the ditch and bank elsewhere. Or the entrance may simply have been protected by a palisade and the sling was not used in its defence (if the forts are of Late Bronze Age/Early Iron Age or Late Iron Age date, this may represent the most robust interpretation of the limited data). This would probably apply to Ivinghoe Beacon, which was certainly constructed in the Late Bronze Age/Early Iron Age transition.

Norbury Bank also seems vulnerable to attack by sling, but its shape and internal features may indicate that this is a Late Iron Age construction (as at Pilsdon Pen). This is when the evidence from Wessex suggests that the sling was losing some of its importance as a weapon (Sharples 1991a, 83). Crickley Hill also appears to be vulnerable, but this is not surprising as it was unfinished before its destruction (Dixon 1994, 194) and there is no way of knowing what, if any, outer works may have been planned.

That Uffington Castle is also vulnerable to the sling is of particular interest. The recent excavations (Miles *et al.* 2003) have shown that the fort underwent two phases of construction (*ibid,* Chapter 6). The initial box rampart dates to the Earliest Iron Age. Only at the very end of the Early Iron Age or early Middle Iron Age was a dump rampart constructed. It is during this phase that a deposit of sling pebbles was made (*ibid,* 185). Following the reconstruction of the ramparts, the interior showed little occupation and may even have been abandoned during most of the Middle Iron Age. The early abandonment of Uffington may have occurred when the shift towards the more widespread use of the sling was in progress. The apparent lack of external defensive features may be a response to this early stage of social change between the Early Iron Age and the Middle Iron Age, before the sling became the most common social indicator of warfare in the hillfort dominated zone.

Hod Hill stands out as the clear exception to the rule. Evidence of sling use was recovered from both Middle Iron Age and Late Iron Age contexts (Richmond 1968, 9, 20, 21, 22, 24). Yet there is no evidence of an external feature, even one as small as the potential outer ditch and bank at Battlesbury Camp. There is evidence of unfinished works, seen as a response to the threat posed by the Roman invasion. These may have been replacing earlier ephemeral features but, on the evidence available, Hod Hill was vulnerable to the sling during most of its occupation.

CHAPTER SIX

A NEW MODEL FOR MIDDLE IRON AGE WARFARE OF THE HILLFORT DOMINATED ZONE

The ideal 'Iron Age/Celtic Warrior' (often the terms are seen as interchangeable) is normally portrayed as a flamboyant hero. The model that this ideal generates is one of combat and warfare being led by a fully hierarchical elite, whose position is for the most part defined by their warrior status. The use of weapons of exclusion is inherently part of this model. The 'Celtic' warrior must be a swordsman, his secondary weapon is normally the spear, both thrusting and casting. The higher elites are either cavalry or ride in lightweight chariots. They indulge in 'ritualised' single combat, for their own personal glory or as a result of boasting and an easily ignited temper. The mass of the population is excluded from warfare or is seen as acting almost as a 'supporters club to the hero'.

The archaeological and ethnographic evidence presented above calls into question this existing model, with the lack of clearly defined social differentiation between the limited range of weapons used and the apparently contradictory nature of the settlement evidence. A new model for the hillfort dominated zone, reflecting the regional variation that characterised the Iron Age, is overdue.

Within the existing overarching model for Iron Age warfare, raiding is one of the main elements and, in fact, is often used as almost a simile for warfare itself throughout British prehistory (Bradley 1971, 79; Cunliffe 1995, 94,101; Harding 1999, 92; Osgood 1998, 89; Mercer 1989, 9; 1999, 144; Osgood, Monks and Toms 2000, 148). Yet, despite the common use of the term, there has been little or no attempt to define what raid warfare was, why it should apply to certain periods of prehistory, and the archaeological evidence that it may produce. That raiding formed part of warfare in Middle Iron Age,[14] is based on direct analogy from Early Mediaeval Irish literature, where cattle raiding in particular seems to have been endemic (though the case study suggests that women and other goods may have been considered as suitable targets of raids as well). In fact, it may be difficult to separate the concept of war and raiding within an Early Mediaeval Irish context (Lucas 1989, Chapter 5). Raiding also seems to have been a key part of war in Late Geometric and Early Archaic Greece (Finley 1979, 63-4; Jackson 1995, 71-5). The taking of goods, and in particular ovicaprids and cattle, figures prominently within the conflicts described in the Nevi'im, though the main aim seems to have been to reduce the position and power of rivals (Samuel I:27), which would imply that power within this context was more achieved than ascribed (Wason 1994, 45-7).

[14] For ease of reference Middle Iron Age will be used to refer to the Middle Iron Age of the hillfort dominated zone, unless otherwise stated.

Within an Early Mediaeval Irish context, raiding would appear to be a method whereby the elites expressed power and where able to maintain their position by redistributing the captured goods. This involved giving 'gifts' of cattle to their existing clients or by creating new clients and thus strengthening their social position. The same appears to be true in Archaic Greece, where cattle were redistributed as part of an unbalanced reciprocal relationship in favour of the elite.

In Mediaeval Ireland raiding appears to have been a significant social institution during the period *c.* AD 800-1600. This has been studied in some depth by Lucas (1989) and the following is a summary of this work. While these data may not necessarily be directly transferable to the Middle Iron Age, they do provide some insights into the motivation behind and potential outcomes of raids. Raiding appears to have been undertaken using substantial numbers of followers under the control of the elite (*ibid*, 169-71). The majority of cattle raids seem to have been unsuccessful, involving either no capture of animals or the death of the leader of the raid (*ibid*, 175). Cattle raiding seems to have operated as a test of kingship for a new king, or a method of extracting revenge within an escalating cycle of raid and counter raid (*ibid*, 159). The normal response to a raid was not to move cattle into defended man-made fortifications (this is often suggested as one of the uses for hillforts), but to move them to open, but relatively inaccessible, locations, in particular wooded areas where felled trees could provide a sufficient barrier to hamper movement within the territory of the raided party (ibid, 179-88). Even though this may have resulted in large numbers of cattle being concentrated in a single location, it had the advantage that, if the raiders found them, it would be difficult for them to move off quickly. The defenders would have ample time to respond at a location well known to themselves. It is under these circumstances that most raid leaders were killed (*ibid*, 174). The holding of cattle within confined man-made enclosures for any length of time would have had the potential to increase the spread of disease, and it was for this reason that, in Classical Greece, cattle were not driven into the city, but were kept outside the fortifications in the surrounding countryside (Thuc. 2.14.1, Xen. *Hell.* 4.6.4).

However, it is not only elite-led societies that produce ethnographic evidence for raid warfare. Many groups who lack permanent leadership undertake this form of conflict. The main objectives of such raids seem to have been slaves (for a market outside the tribal area, e.g. an adjacent state (Gibson 1990, 126-7)), wife stealing, limited gain of rare or desirable goods, and the general destruction of the enemy's resources (Haas 1990, 177). However, Hanson (1998, Chapter 6) has shown how difficult the latter was to achieve, even for well-organised armies, and so this probably represents symbolic expression of power rather than any real harm.

From the ethnographic evidence, raid warfare might be expected to produce a series of identifiable markers within the archaeological record (Steponaitis 1997, 207; Whitehead 1990, 157). These may include space as one of the main ways of avoiding or at least reducing the effectiveness of raids. In Late Geometric and Early Archaic Greece and Late Intermediate Peru, settlements are separated by difficult terrain, in both cases consisting of mountain ranges with limited passes. The early Germans, are described by Tacitus as having a large devastated area between each polity and its neighbour (*Germania* 16), which may represent a similar response where there was no difficult terrain to act as a natural barrier. The other primary response seems to be the construction of heavily occupied fortified villages or even small towns (Creamer 2001, 51-5; Haas 1990, 184-7). In the Mantaro valley of Peru, this can be seen particularly clearly as the population was concentrated in a series of large pucaraconna with populations estimated in their thousands (D'Altroy 2001, 67). The creation of walled towns within Iron Age I and II Israel may also represent such a reaction, but here the evidence is more sketchy. Though there is no evidence of heavily defended large scale settlements, Early Mediaeval Ireland does produce the ringfort, and recent work has shown that many of these features are visually interconnected, thus providing a network of defence against potential raids (Black 1994, quoted in Stout 1997, Figure 1). Though the ringfort has been seen as an undefended enclosure with only a ditch and bank (Mallory and McNeill 1991, 196-9), the archaeological evidence suggests that some at least had palisades atop their banks (Davies 1939, 40; Edwards 1990, 20; Monk 1995, 107).

The archaeological evidence from the Middle Iron Age suggests that the model of raid warfare is, if not untenable, then at least unlikely. The open villages offer no evidence of any defence and on occasion appear to have developed from earlier enclosed settlements, where the original ditch has been allowed to silt up naturally and was then built on (e.g. Winnall Down, Fasham 1995). Significantly, there seem to be no major breaks in occupation during this process. Certainly, the open nature of such settlements would indicate that their involvement in warfare was minimal.

Enclosed settlements provide more intriguing evidence. Many have elaborate entranceways formed by outstretched antennae created by excavation of a ditch. The ditch was impressive at the front, with depths of between 1.7m and 3m and steep sides. However, the backs of these settlements tell a very different story. The ditches are often shallow, 0.5-1m, and very wide, allowing relative ease of access. They may even have a series of causeways cutting across the ditch, with no evidence of these having been protected by anything stronger than a wicker or light timber gate. It is also at these locations that the banks were constructed outside the ditch or the ditch was backfilled within a short time of its original excavation. Where this is not the case, there appears to have been little maintenance of the ditch, and it silted up naturally during the often long-term occupation of the settlement. There is also little or no evidence of palisades along the summit of the bank, though there is perhaps some indication that a hedge may have been planted.

The use of ditches and banks to define the entrance to enclosed settlements suggests that the ideology of defence played a significant part in their construction. It is possible that their construction was intended to reflect the construction of hillforts, by providing a status-enhancing monument within the landscape. It would have been possible, for example, to produce a dramatic statement of power by the use of free-standing posts, creating an avenue to the entrance of the enclosure. The fact that the earthworks seem to imitate those of hillforts may be an expression of social cohesion.

However, the lack of defensive capability at the rear of these enclosures and, in particular, the rapid backfilling of many of the ditches soon after they had been excavated, suggests that the inhabitants had little to fear from random raids or attacks (seemingly even from wild animals such as wolves, which appear to have survived in this area until the Mediaeval period (Yalden 1999, 168)). The lack of clear defences would have removed these settlements from the arena of war and expressly associated them with non-violent statements of social standing. The length of occupation (often hundreds of years) would also suggest that the ideology of 'raid warfare' does not sit well with the Middle Iron Age. There is little or no evidence of destruction by fire, that perhaps would be expected if devastation of property were one of the aims of a raid (Haas 1990, 177-8). The lack of defence implies that the taking of goods or people, either as slaves or as 'wives', did not form a part of Middle Iron Age warfare either, as it may have done in Early Mediaeval Ireland, where the use of the ringfort was widespread.

In direct contrast to the single farmstead enclosures the ditches and banks of hillforts are massive, all encompassing and were continually repaired and maintained. This opposition of construction and maintenance implies an opposition of use. If farmstead enclosures seem to deny force and warfare, then hillforts seem to express both concepts fully. The density of population occupying a hillfort was not high, and certainly did not include the majority of the people within its area of influence. The very nature of the defences is convincing evidence that large numbers of people were needed for their construction, numbers well beyond the estimated total population of an individual hillfort. The banks and ditches at Hod Hill, for example, are nearly 2.4km in length. The total amount of excavated material is over 60,000m^3. Webb (1987, 162) calculated that marae (ritual platforms) in both Hawaii and Tahiti have volumes of between 14,000m^3 and 15,000m^3. Both of these societies were complex and hierarchical chiefdoms.

The figures indicate the level of motivation that would have been required to construct Hod Hill and other hillforts, and potentially the level of social control that would have been required. The evidence from Ladle Hill (Piggott 1931, 478) and Crickley Hill (Dixon 1994, 69, 181) for gang labour is persuasive and again represents a level of control stretching beyond the hillfort itself. The deep ditches and vertically-faced banks (or glacis forms) with narrow entrances are difficult to interpret as not being defensive in some form.

Multivallation also indicates an expression that goes beyond the mere practicality of warfare. In fact, at both Old Oswestry and Scratchbury Hill, the outer (and presumably later) banks either obscure the line of sight outwards to the entrance or overlook the entire circuit of the inner banks. Bowden and McOmish (1989) have argued that the ability to see many hillforts' interiors also renders them ineffective in warfare (*ibid*, 13). This may be true in their example of modern warfare (*ibid*, 13), when the effective range of personal infantry weapons is in excess of 2km and motorised transport can traverse even broken ground at speeds in excess of 50km/hr. However, it is less likely to have been a significant factor in pre-gunpowder warfare, as the example of Caernarfon Castle, quoted previously indicates. Given the potential range of long bows and crossbows (Payne-Gallwey [1903] (1981), 20-30), such over looking becomes even less important in the Middle Iron Age, when the greatest distance a projectile could be cast was less than 100m and the horse was the fastest transport available.

Hillforts have often been interpreted as refuges in the event of raids (Harding 1974, 72; 1977, 8; Avery 1986, 228; Cunliffe 1994, 73). From the case studies, it appears that it was nearly impossible for pre-state level societies to assault a fortification with any real hope of success (Earle 2001, 117) and even early states found such operations beyond their capabilities (Herodotus 7.154; Pausanias 10.37.37 (for early Greek examples); Livy 1.11.5-1.12.10 and 3.23.4 (for early Roman examples)). The experimental work and its application to surveyed hillforts would seem to buttress these observations. The description provided by Caesar, and so often quoted as the way in which an assault on a hillfort would have been undertaken, may have been, if not actually impossible, then extremely difficult and carried a high risk of failure. This does not exclude the possibility of attack during periods of extreme stress, such as famine, that would leave little evidence in the archaeological record due to their relatively brief duration. However, it does tend to suggest that the inability to attack hillforts had the potential to remove them from direct involvement in warfare; if they were, in effect, invulnerable, there would have been little or no point in attempting to assail them, as this would inevitably lead to defeat.

The lack of evidence for raids shifts the emphasis of Middle Iron Age warfare away from the unpredictable and violently random, towards a more predictable form of conflict. This could be seen as expressing the ideal of the 'Celtic hero', where only a small section of the community was actively engaged in socially delineated warfare, where individuals could achieve status, though combat, but requiring little involvement from the mass of the population. This was certainly the case in Early Mediaeval Ireland, where the elite occupied ringforts. The same situation seems to have existed in Late Geometric and Early Archaic Greece, where the formalisation of combat by the aristocracy and the geographical separation of polities was apparently sufficient, to render the use of fortifications non-existent except in areas of recent colonisation. However, in the context of Middle Iron Age warfare, a number of factors would argue against this interpretation.

Unlike any of the elites in the case studies, here is no clear Middle Iron Age evidence for distinct social differentiation expressed through the use of wealth finance objects (Hill 1995b, 49; 1996, 106-7; Stopford 1987, 70-1), though social differentiation may have been expressed by quantity of goods rather than quality (Cunliffe 1995, 92-3). The expression of social exclusion by wealth finance objects seems to appear during the Late Iron Age, when most of hillforts had been abandoned, and when those that remained acquired a different ideological association approaching the oppida of the Continent (Collis 1984b, 222-3; Cunliffe 1991, 366-70).

The apparent lack of strongly expressed social differentiation in the Middle Iron Age is supported by the evidence of the social significance of weaponry. The case studies clearly indicate that weapons carried quite specific messages. The sling appears to have been a weapon of social inclusion, expressing a desire that a polity or community be seen as a society with little differentiation. Late Intermediate and Inka Peru provides perhaps the best demonstration of this ideology, where even the Inka state seemed unwilling to abandon the inclusive ideology of the sling. For Iron Age Israel I and II, it can be argued that the sling represented the older, less centralised, form of authority that controlled Hebrew society before the rise of the kings and the state. If these interpretations can be transferred to the Middle Iron Age, it would suggest that the sling was being used as a metaphor for the community as a whole. The sling, because of its mechanical properties, can be used by a broad section of a given population; the difference in strength between individuals would not be significant in this context, enabling the whole community, which in principle at least could have included both sexes, to be involved in the defence of a hillfort, potentially widening the sense of inclusion.

The spear is more problematic but, as explored above, it is a chameleon that has a fluid social position, and is best seen in the light of the other weapons that it was associated with, rather than carrying its own strong message. The evidence from the social significance of weaponry would suggest that warfare in the Middle Iron Age was used not directly as a method of expressing social exclusion but rather as a method of inclusion, where the common ideals of a community were expressed in a known context.

The potential use of the sword seems to, at least in part, contradict some of the above assumptions. However, as stated earlier it is clear that the sword is not being used in the same way socially in the hillfort dominated zone, as outside it. Swords are not deposited in rivers, nor do they appear in any of the limited number of inhumations so far recovered. Some potential indications of what this difference may have been comes from the morphological evidence. Carr *et al.* (2005) have suggested that some of the evidence for sword use can be attributed to dismemberment, decapitation and human sacrifice. The evidence for human sacrifice within the case studies is limited. Within the mythological framework of Late Geometric and Early Archaic Greece, Agamemnon sacrifices his daughter Iphigenia (*Epitone* 3:21), using an axe or knife. Achilles sacrifices twelve Trojan princes at the funeral of Patroclus with the bronze the specific weapon is not described (*Iliad* 23, 201), given the description of Achilles hacking the princes to pieces, this could well have been an axe or sword, however, the evidence from the sacrifice of Iphigenia, suggests that the former may be the most likely and potential evidence of real sacrifice comes from the interment of a young woman in the grave of the 'Hero of Lefkandi' (Popham *et al.* 1982, 173). Within Late Intermediate and Inka Peru the most common method of human sacrifice appears to have been strangulation (Uhle 1991, 86-87) or exposure (Reinhard 1992: Schobinger 1991). There are strong indications from Early Mediaeval Ireland that human sacrifice played a part in pre-Christian religion, this also includes elements of nudity that may well tie in with expressions of divinity, again strangulation appears to be the normative sacrificial method (Stokes 1901, 314; van der Sanden 1996, 73, 81), though the recently discovered bog body from Clonycavan Co. Meath clearly shows that death was caused by three blows from an axe (Grice 2006, 20). This is paralleled in a British context, as one of the traumas suffered by Lindow Man was also an axe wound (Stead 1986, 180). The evidence from Denmark also seems to suggest that weapons of war and human sacrifice are not automatically associated, none of the Danish bog bodies recovered show any signs of weapon trauma (Glob 1969) and the lack of human bodies within the Hjortspring boat sacrifice, with its obviously military connexions, highlights this difference

The key factor is that there appears to be no direct link between the weapons of war and the undertaking of human sacrifice. The two areas of social expression seem to be deliberately separated, despite the potential connexion through the taking of prisoners for sacrifice from the field of battle. The use of the axe potentially highlights this disconnection. Axes are surprising rare with the archaeological record of Iron Age Britain, this is made more obvious due to their ubiquity in the preceding Late Bronze Age and subsequent Roman and Anglo-Saxon periods. The axe, may be rare due to its use in certain types of ceremony's in particular human sacrifice, the deposition of three axes at Fisketon (Stead 2003, 67, fig. 4.14), along with a decorated saw blade, indicates that what appear to be solely domestic artefacts may take on special meaning, dependant upon their context (*ibid*, 71).

There may be an analogy with the use of the axe as a 'weapon' of sacrifice and the limited evidence for the use of the sword in the Middle Iron Age. If it is accepted that swords carry connotations of elite status, and that at the end of the Early Iron Age/start of the Middle Iron Age elites were shifting the method of status expression then the sword may potentially have been realigned away from warfare to a religious framework, one which may have required the dismemberment of human bodies. The limited numbers of individuals recovered from Danebury is possibly more characteristic of sacrifice rather than the display of enemies (pace Carr *et al.* 2005, 174-6) when it would perhaps be expected that a greater number of bodies would be involved, within this context, swords may have been used as markers of social dominance inter-group and rather than intra-group. This may have been heightened if swords were produced from outside the bounds of the hillforts polity (Helms 1988, 49).

This potential link is buttressed by the possibility that some of the pit burials of complete human skeletons from Danebury (Cunliffe 1995, 78) and elsewhere were sacrificial victims (Alcock 1972b, 103; Wait 1985, 120; Whimster 1981, Chapter 8). There appears to be a significantly high proportion of males within the age bracket 25-35 years. (37% of the total, compared with 18.6% of males aged 30-50, 22.2% of children under 15, and 22.2% of females of all ages). If these burials do represent sacrificial victims, then the age distribution may be significant as it encapsulates the normative range for a warrior age group.

If human sacrifice did play a part in Middle Iron Age ritual, and if that ritual required the sacrifice of a healthy man, then, ethnographically and historically, that individual would most likely have been a warrior from a hostile polity (Conrad and Demarest 1984, 47-8; Earle 1997, 171; Schwimmer 1966, 64; Tacitus *Annals* 13, 57). As explored in Chapter Two, the settlement evidence from the Middle Iron Age presents a polarised society, where domestic occupation seems to skew the ideology of war by remaining essentially undefended, and yet hillforts with their relatively low occupation levels and unsuitability as places of refuge are massively fortified.

Throughout the available ethnographic literature, warfare appears to have strong ritual associations. It is easy in the mechanical world of the post-industrial era to see war as driven solely by a desire for greater and greater efficiency in killing. However, this does not appear to have been the situation in the societies explored in the case studies, where warfare played a different role and allowed individuals access not only to different times but to other realities as well (Sinos 1998, 76). It may be no coincidence that many mythological narratives seem to be inherently interwoven with warfare. It is difficult to connect directly Middle Iron Age ideologies of warfare to this concept. However, there are hints in the archaeological record, not from the hillfort dominated zone itself, but within the surrounding areas and from the case studies, that infer a potential relationship.

In the case studies, there is clear evidence of an association between nudity and the divine. This connection has been most widely explored in Late Geometric and Early Archaic Greece, where there is a widespread portrayal of nude male and female divinities (Marinatos 2000). The representation of warriors fighting nude has been considered to be an artistic convention (Boardman 1983, 27), but these objections are based on a 'logical' argument for the practicality of Greek warfare that have been recently challenged as not representing the probable nature of conflict (van Wees 2000, 132). Despite considerable Christian influence, there is strong evidence that some of the Early Mediaeval Irish may have fought naked and, there is evidence of an association between ritual and nudity in what were perceived as being pre-Christian times. There is even evidence from the Nevi'im that there was some correlation between nudity and the divine. Saul lies naked before the prophets in divine ecstasy (Samuel 1.19) and David dances naked when the Ark of the Covenant is brought into Jerusalem (Samuel II.6).

Within the overarching model for 'Celtic' warfare, the use of nudity by the transalpine Gaesatae is interpreted as having some form of magico-religious symbolism (Pleiner 1993, 25; Rankin 1996, 69). Within Classical literature, the nudity of the Gaesatae is seen from a more 'practical' viewpoint of bravery and bravado and ease of movement (Polybius II: 28). Within a British context, there are hints that nudity, warfare and the divine may have been synonymous in certain areas. None of the evidence falls within the hillfort dominated zone and some falls outside the Middle Iron Age. The evidence from Britain includes the collection of carved chalk figurines from the Yorkshire Wolds (Stead 1988), clearly showing naked men carrying swords, one of whom appears to be ithyphallic (*ibid*, Figure 1, No 38 and Plate 1). Stead, interprets these figures as gods, mythological figures or ancestors and infers that they may well have served a ritual or magical function (*ibid*, 28). The ithyphallic nature of divine representations continued into Roman Britain, where carvings showing naked 'warriors' with inferred divine associations have been recovered at Maryport (Ross 1967, Plate 37), Yardhope (Charlton and Mitcheson 1983, 148) and Wall (Frere *et al*. 1977, 394). These figures are assumed to be representative of, at least, later prehistoric religious ideologies (Green 1989; Ross 1967). All the case studies, and the limited late prehistoric evidence from Britain, would suggest that warfare had, if not divine connections, then at least a ritual context, which may have applied within a Middle Iron Age framework.

The potential nature of that framework is further illustrated by the deposition of large numbers of sling stones in pits at both Danebury and Maiden Castle and possibly by smaller numbers of pebbles, chalk ovoids and clay shot from elsewhere (Appendix 1.1). It is now widely accepted that pits in the Middle Iron Age were not solely rubbish dumps but were often filled ritually with structured deposits. It is strange, then, that the deposition of often huge numbers of pebbles, interpreted as sling stones, is still given an almost entirely practical interpretation, namely the storage of ammunition in case of attack (Avery 1986, 225; Cunliffe 1995, 94; Sharples 1991a, 83; Wheeler 1943, 115). At Maiden Castle, Pit G.6, cut into the backfill behind the stone lining of the entrance, contained 22,260 beach pebbles (Wheeler 1943, 115), and at Danebury pit No. 911 produced over 11,300 pebbles (Brown 1984, 425). However, the latter were not concentrated in a single layer or deposit but were spread throughout the fill, with some concentrations in lenses and more at the sides of the pit (Cunliffe 1984a, Figure 4.96; Poole 1995, 262). This does not seem to be a single deposit awaiting reclamation, but appears to be structured, possibly over a period of time. The recovery of 108 clay sling shot from Cadbury, which appears to form a special deposit within pit S066D, would buttress the above interpretation (Poole 2000, 247). The retrieval of large numbers of sling stones from the internal flanks of the banks at both the eastern and western entrances of Danebury would seem more symptomatic of sling stones deposited in readiness for attack, either in discrete piles or open baskets, as the 108 such pebbles recovered from Bredon Hill would suggest (Hencken 1938, 34).

Deposits from Danebury and Maiden Castle suggest that sling stones and, by extension, the sling, were associated with some form of ritual. At Danebury (Cunliffe 1995, Microfiche), the large deposit of human skeletal remains (mostly partial bodies and skulls) with animals remains and sling stones from Pit 923 (34:C3), the horse's skeleton and surrounding sling stones from pit 321 (34:B4), the single sling stone under an inverted pot in P2476 (34:F1) all infer an association between the sling and chthonic deposits, as does the deposition of 117 slingstones with the burial of an adult female at Maiden Castle (Wheeler 1943, 348). Although of significantly lower numbers, clay and chalk sling shot do not appear to have such strong associations. Although, as explored above, there is no direct correlation between the weight and the use of sling projectiles, it may be reasonable to assume, given their context, that the pebbles from Danebury, Maiden Castle and elsewhere were primarily intended for warfare.

One aspect of the overarching model for Iron Age warfare, generally seen to hold magico-religious connotations, is so-called head-hunting. Classical authors make it clear that the Iron Age peoples of Western Europe curated the heads of revered enemies and attempted to preserve them for long periods (Diodorus 5, 29). The taking of heads is clearly a significant part of warfare in both Early Mediaeval Ireland and, to a more limited extent, in Late Intermediate and Inka Peru (Huaman Poma 1978 [1613], shown in Verano 1995, Figure 2, 219 and in D'Altroy 2003, Figure 8.1). From generalised ethnographic evidence the taking of heads does not seem solely to represent the collection of trophies from enemies but carries aspects of access to divine power, the other world and social expression. Evidence from Early Mediaeval Ireland correlates with this generalisation: even within an Early Christian context, the taking of a hero's head appears to represent the most appropriate form of death and remembrance. This may represent an elite method of exclusion, as only members of the respective group would be known to each other and their killing by another would represent power and exclusivity (within this ideological context the taking a non-elite's head would not carry kudos). The practice of head-taking is also attested in Late Intermediate and Inka Peru, when the skulls of notable enemies were made into drinking vessels (Cieza de León 1959 [1553], 84; McIntyre 1975, 59). Anthropological evidence from Borneo suggests that the act of taking a head had aspects of contact with, or appeasement of, ancestors (Beavitt 1997, 211-2). Head hunting also seems to produce specific responses that should be recognisable within the archaeological record. This seems primarily to be represented by the dispersal of settlements (Gibson 1990, 133; Harner 1972, 77-8), a similar response to that engendered by chronic raid warfare. As discussed above, this seems to be an inappropriate model for Middle Iron Age warfare.

The recovery of human skulls from Middle Iron Age contexts have been interpreted as the heads of revered enemies within a 'Celtic' context (Cunliffe 1992, 76; Davidson 1988, 73-4; Hencken 1938, 23; Wait 1985, 120; Whimster 1981, 185-8). Although adult male skulls predominate, juveniles and females (Cunliffe 1995, 73-5) suggest a more complex pattern of activity. If the taking of trophy heads was part of Middle Iron Age warfare, then this is potentially another indication of the ritual and formal nature of any conflict. The settlement evidence would argue against raiding for heads, so if the male skulls are war trophies, their collection is most likely to have taken place after or during a conflict and may represent ways of accessing or appeasing divine powers.

One of the overarching characteristics of chiefdoms is the use of divine power as a cornerstone of secular control (Flannery 1972, 403; Fried 1967, 141; Service 1962, 171). Cunliffe has suggested that the evidence from Danebury indicates a chiefdom (1995, 93), earlier describing them (without actually using the term) (Cunliffe 1994, 72-3) as a staple finance chiefdom, as defined by Earle (1997, 70-3; D'Altroy and Earle 1985, 188). Although Renfrew used the term 'Group Orientated Chiefdom', this is in essence the same broad concept. Warfare was also often seen as one of the ways in which a chief could maintain power (Carneiro 1981, 63-5; 1990, 192; Drennan 1991, 280; Earle 1997, Chapter 4), not only through its undertaking, but also by threatening conflict as a cohesive force within society. The exploration of the social significance of weaponry strongly infers that within staple finance chiefdoms, weapons and, by analogy, warfare were a method of expressing commonality with the mass of the population rather than being an expression of exclusivity, as was more the case in a wealth finance chiefdom, where weapons were used to exclude certain sections of society.

Hillforts can be seen as a material expression of this concept of access to the divine. The original nature of many hillforts layouts, consisting of pits filled with ritual deposits, stake-palisades and 'ranch' boundaries (Cunliffe 1984a, 12, Figure 2.1; Hawkes 1939, 169-71; Piggott 1931, 476-8), suggests, initially, at least, many may have served as foci for ritual activity, with no overt military function or connection. The enclosure of Neolithic and Bronze Age burial monuments would also suggest that there was some attempt at contacting or protecting a physical representation of a mythological past (Barrett 1999). Certainly, the societies that form the basis of the case studies incorporated the still upstanding structures of past cultures into their relative mythologies. The use of Bronze Age and Neolithic burial monuments as homes for the Tuatha De Dannan in Early Mediaeval Ireland; the opening of tholos tombs in Late Geometric and Early Archaic Greece; the incorporation of long-abandoned cities into the migration narrative of the Hebrews; and the incorporation of the Tiwanaku monuments into Inka creation myths are all examples of this phenomena. Following the enclosure of the developed hillforts with massive earthworks or the like, the inherent difficulties in assailing such forts may have rendered assault impractical, and therefore outside the normative structure of conflict. If this was the case, why did the people of the Middle Iron Age continue to add ditches and banks to the hillforts? These would not have improved the defensive capability of the fort.

Ethnographic evidence would infer that it is unlikely that society with no overarching leadership would have produced such impressive constructions (pace Hill 1995b). Without corvée labour, neither the Hawaiians nor the Tsikwaiye of the Northern Rio Grande were able to construct features, including residential units, in excess of 12,000 m^3 (Kolb and Snead 1997, Tables 1-2), which is well below that required in a hillfort's defences. This is drawn into sharper relief when the timescale of the ethnographic examples is taken into consideration; in order to reach 12,000 m^3, about 420 years had to elapse (*ibid*, Table 1). The evidence from the construction sequence at Ladle Hill (Piggott 1931, 478-81) indicates that the construction process of this hillfort was relatively rapid and, by analogy, that most other hillforts were constructed in a relatively short period of time. This certainly infers that there was an overarching authority planning the layout and construction. In turn, this would suggest that the construction of new banks and ditches was under some 'centralised' guidance. The anthropologically generalised term, 'chief', is problematic, therefore the less emotive and less precise concept of an elite (Wason 1994, 103-4), which carries none of the connotations associated with the above, will be used. This allows social scenarios to be created, without diluting the central premise that certain individuals or small groups took control in certain circumstances, possibly with a contextual power base.

If it is accepted that material differentiation in the Middle Iron Age was a matter of quantity rather than quality and that most settlements could, in theory, be self-sufficient in almost all staples excluding iron (Avery 1976, 26; Fitzpatrick 1997, 76-7; Hill 1995b, 52; Stopford 1987, 71), this capability for self-reliance may have had the potential for fragmentation, as each individual community was able to support itself without significant economic interaction with its neighbours. This situation has the inherent danger of intra-group feuding and escalating violence. In order to maintain their position of authority, an elite would have to have had the ability to maintain group cohesion and prevent fissioning. One way of achieving this would be the creation of the ideology that a threat from another polity was greater than any internal differences (Billman 2001, 183). Circumscription is seen by anthropologists to be one of the driving forces for warfare (Carneiro 1981, 65-7; Steponaitis 2001, 215). However, it has been proposed that the elites of Late Geometric and Early Archaic Greece and Iron Age Britain may have created an ideology of circumscription, through the use of warfare, by which they were able to contain and control an essentially self-sufficient population (Earle 1991a, 12). The construction of fortifications such as hillforts may be a physical manifestation of this belief system. The building of extra and occasionally obstructive banks may also have formed part of this expression. During times of stress, both internal and external, the creation of further defences to the already invulnerable fort would not only have strengthened the social bonds of the community (Abrams 1989, 62), by co-operating in the construction process, but may have provided a growing sense of security against the perceived threat of another polity, while increasing the fear that 'necessitated' the construction of the ditches and banks. Under these circumstances, the regular maintenance or additional construction may have formed part of a process of competitive display (Kirch 1991, 131), which included warfare within its umbrella of meaning. The collection of large numbers of sling stones and their deposition in pits may have served a similar function. The requirement to collect vast quantities of ammunition, ultimately destined for concealment within a pit, would have again highlighted the need for defence and the requirement that the community should act together in times of stress, even if these were more imagined than real (Evans 2003, 258).

If it is accepted that the deposition of humans, animals, pottery etc. in the base of 'redundant' pits were acts of propitiation for chthonic deities (Cunliffe 1992, 78-9), then, by extension, the deposition of sling stones that are often associated with such deposits would potentially also represent offerings to these particular powers. The fact that many of the pebbles used as sling stones were recovered from particular deposits, or were collected from nearby beaches, could perhaps be seen as evidence of a connection between warfare and the more 'natural' chthonic or marine divinities, both of which mythologically can be shown to have held significant levels of destructive power.[15] It would perhaps not be too fanciful to extend the analogy further and see Middle Iron Age warfare of the hillfort dominated zone as part of a ritual sequence associated with the cycle of the seasons.

For Late Intermediate Peru, there is evidence that, in order to ensure that the next harvest was successful, human blood had to be spilt on the ground. In modern times (and possibly in prehistoric times as well) this has been achieved during the ceremony of tinku (Hastorf 2001 (c), 319). The ithyphallic nature of Late Iron Age figures that seem to carry ritual associations, could appear to be partly a representation of this concept. The deposition of martial objects (iron spearheads and swords) with agricultural artefacts at Fiskerton (Field and Parker-Pearson 2004, Chapter 4), Bredon Hill (Hencken 1938, 57, Figure 6, 73-4) and Llyn Cerrig Bach (Fox 1946, 54-5, 65) buttresses the connection between agricultural fertility and, warfare.

The case studies also suggest, that even in elite-led societies, the concept of the honour of a community was highly valued (Evans 1997, 37; Fisher 2000), and that any infringement of this had the potential to erupt into violence and conflict (Conrad and Demarest 1984, 44; Rountree 1988, 121). The ideal of the hero (at least in Late Geometric and Early Archaic Greece) was an individual who was supported by the community to protect them and their honour from attack; on occasion they literally paid with their lives for this position.

Under these circumstances, it would be probable that Middle Iron Age warfare was organised and undertaken at certain times of year at certain places, perhaps at locations where conflicts might be most intense, e.g. where boundaries needed to be redefined or confirmed (Tarzia 1987). Certainly the sling, with its widespread distribution, is more a weapon of open warfare than of ambush or raid, as it is difficult to use when concealed.

If so, where should hillforts be placed within this more formalised form of conflict? If hillforts were essentially invulnerable, as the experimental and ethnographic evidence suggests, and their status as defensive monuments was clear to all, as the very nature of their construction would infer, what were they defending? As explored above, hillforts do not appear to have been densely occupied and seem to be ineffective as refuges. The use of pits as places of propitiation and the construction of small square buildings, normally interpreted as shrines (Alcock 1972a, 36-7; Cunliffe 1991, 234, 236-7; Gelling 1977; Wheeler 1943, 127, 131-5; Wait 1985, 156-71; see Downes 1997, 151; Moore 2003, 54, for alternative views) may provide a possible answer. Pits and shrines are normally seen as evidence of ritual activity and as ways of accessing the divine. The defensive structure of hillforts may represent both a definition of sacred space and the need to defend that space. This would not in anyway preclude the occupation of hillforts. In fact, if anything, it would encourage the idea that hillforts were occupied by individuals of 'special' standing. Such individuals often draw much of their power by having privileged access to the divine, and the evidence of significant levels of ritual activity within

[15] It is of course more practical to see the exploitation of these resources as a response to the use of the sling. However, historical evidence would suggest that material resources were seen as a product of particular past events within an explanatory mythological framework, e.g. the creation of various useful stones from the bodies of defeated enemies (Jacobsen 1987, 256-68)

hillforts would tend to reinforce this inference. If hillforts can be seen as the definition and defence of a sacred space and potentially a sacred person(s), then warfare within the Middle Iron Age of the hillfort dominated zone would sit more comfortably as an expression of a ritual cycle, and not as part of a process for the acquisition of personal wealth (as in raid warfare) or personal power. Middle Iron Age warfare should perhaps be seen as an integrated part of the life cycle of the people that built hillforts, who collected and carried vast numbers of pebbles to 'defend' these structures, undertook formal battles using weapons that carried the social significance of a community, and, partly through these acts, defined their social constructs and their world.

APPENDIX 1.1

GAZETTEER OF SITES WITH SLING PROJECTILES FROM SOUTHERN BRITAIN (SEE FIGURE 2.1 FOR DISTRIBUTION MAP)

1. Conway Mountain

Griffiths, W. E. and Hogg, A. H. A. 1956: The hill-fort on Conway Mountain, Caernarvonshire. *Archaeologia Cambrensis* 105, 49-80.

Large numbers of sling stones recovered, oval in shape, 25-50mm in length, mostly schist or quartz. Total number 1141, provenanced as follows:

Hut one 612; Hut four 400; Hut three 59; entrance to large enclosure 36; Hut four (?) 29; occupation layer underlying the bank outside the north-east wall of the small enclosure 6; rampart summit eastern extremity of the small enclosure 2.

2. Pen-Y-Gaer

Hughes, H. and Gardner 1906: Exploration of Pen-Y-Gaer, above Llanbedr-Y-Cenin. *Archaeologia Cambrensis* 6, 241-67.

Site III Several river pebbles found within a house, dimensions 45mm x 75mm x 47mm. (p.249)

Site V, Nine river pebbles discovered, on different sites a large number of pebbles were found, it may be concluded that some, doubtless, were sling-stones. (p.50)

Site VII Thirty-four pebbles of various sizes, perfectly smooth and probably sling-stones, were also found.

Site XIV ...Several pebbles were found.

3. Pen Dinas, Aberystwyth

Forde, D. C., Griffith, W. E., Hogg, A. H. A. and Houlder, C. H. 1963: Excavations at Pen Dinas, Aberystwyth. *Archaeologia Cambrensis* 112, 125-53.

More than 100 pebbles of uniform size recovered at the base of the rampart to the eastern defences, south fort. (p.153)

4. Titterstone Clee Camp

O'Neil, B. H. St J. 1934: Excavations at Titterstone Clee Camp, Shropshire 1932. *Antiquaries Journal* 14, 13-32.

Two large pebbles, 100mm long, derived from Shirley Brook, below the fort, could have been used as pounders, potboilers or sling-stones. (p. 32)

5. Bredon Hill

Hencken, T. C. 1938: The Excavation of the Iron Age Camp on Bredon Hill, Gloucestershire, 1935–1937. *Archaeological Journal* 95, 1-111.

Site H -107 sling-stones recovered in the upper occupation layers of the fort. (p.34, Plate XVI)

Phase II - the main entrance 'the actual roadway down the entrance was composed of well-made cobbles, in which a great number of sling-stones were embedded'. (p.47)

6. Cadbury Castle

Alcock, L.1968: Excavations at South Cadbury Castle. *Antiquaries Journal* 48, 6-17.

Local clusters of sling stones, associated with pottery similar to Maiden Castle War Cemetery. (p.14)

Alcock, L. 1972b: '*By South Cadbury is that Camelot...*' *Excavations at Cadbury Castle 1966-1970* (London).

Notes eighteenth century antiquaries recovery of sling stones. (pp.14 & 28)

Clay sling shot fused to corroded 'bill hook'. (p.154, Plate 59)

Barrett, J., Woodward, A and Freeman, P. 2000: *Cadbury Castle, Somerset: the Later Prehistoric and Early Historic Archaeology* (London).

Pit D630A An unrecorded number of clay slingshot found with a hoard of iron work. The deposit had been burnt, producing sufficient heat to fuse the sling shot to the saw (pp. 6 & 83)

Site J clutch of slings stone recovered from apparent bank (J116) (pp. 69)

Deposit K668 contained a hoard of stone sling shot and the remains of a human infant (pp. 92).

The earliest deposits of the eastern guard chamber contained a cache of slingshot (pp. 93)

Total of 113 clay sling shot recovered. The majority (108) from pit S066D, which seem to be a structured deposit. The weight range is between 18-26g, the majority averaging 20g. Lengths varied between 34mm-

46mm, the majority averaging 40mm. These may have been used for hunting. (pp. 247)

Total of 2036 sling stones were recovered from the site. Generally water-worn flint pebbles. Weight range 19-172g, average 40.9g. length range 33-72mm, average 50mm. A further 477 sling stones are recorded in the site archive, and a single hoard of 302 sling stones also formed part of the finds inventory, but cannot be assigned to a particular feature or site. Certainly many more were found, but not recorded. (pp. 247)

Probable source for the sling stones is Chesil beach (pp. 266-7)

7. Maiden Castle

Wheeler, R. E. M. 1943 *Maiden Castle, Dorset* (London).

'slingstone culture'. (pp. 48-51)

weight range 14.17gm - 56.69gm with the majority gravitating to the upper figure. (p. 49)

four pits (B7, B10 and on pl.XVI) served as armouries for large dumps of slingstones. (p. 54)

Hut DB2 At one side of the entrance was a hoard of slingstones. (p. 55)

Pit B12 contained over 4,000 sling-pebbles and pit B7 produced a somewhat smaller hoard. (p.91)

Pit D14 On the side of the floor lay a hoard of about 100 sling-pebbles. (p.96)

in the pit lay 22,260 sling-pebbles (Plate CIV), obviously stored for defence of the gate - a vivid token of the deliberate and orderly preparation and of civic discipline. (p.115)

Skeleton from pit Q4, lying on the earth bottom of the pit.... The layer contained 117 sling-pebbles, with Iron Age B shards. (p.348)

Skeleton T10 Two slingstones beside the skull (Late Iron Age in date). (p.349)

Skeleton P34, two slingstones by the right humerus and a third by the right humerus (Late Iron Age in date). (p.355)

8. Danebury

Cunliffe, B. 1984b: *Danebury: an Iron Age Hillfort in Hampshire. Vol. 2: the Excavations 1969-1978: the Finds* (CBA Research Report 52 (2): London).

Slingshots: Clay. Eleven manufactured clay slingshot recovered, all from the final phases of occupation, one from cp 6 and ten from cp 7.

Two were ovoid and pointed at both ends: lengths between 40-50mm, diameter between 27-31mm, weight range 30-50 grams, all except one were baked.

Three others were spherical balls, measuring 35-40mm in diameter, two were complete and weighted 39.5 and 42.2 grams. (p.398, Figures 7.44, 7.1-7.11)

Sling stones: many thousands of water worn oval/spherical pebbles recovered. Largest single deposit approx. 11,000 from pit 911, just inside main gate. Many of the pits adjacent to the western gate contained pebbles and several thousand were recovered from the blocked eastern gate. Found in pits of all periods but increase dramatically in cp 7.

Nearest source the River Test, however, pebbles of the size collected do not predominate. Harewood Forest, 8km from Danebury, has a higher proportion of pebbles of the right size and weight. The weight range is 29.5gms to 109.5gms. The weight range when compared with that available from the natural range, infers a considerable amount of selection. (pp.425-6)

Sling stones recovered with exposed bodies in pits 935, 923 and 1073, suggesting some form of ritual association. (p.448)

Approximately 50 sling stones found with a horse skeleton in pit 321. (Figure 9.30)

The mean weight of a sample of 1,000 sling stones was 77.99 grams (MF.12: E9), with tight grouping of the distribution around the modal class, weight 69.5 grams. (MF.12: E10)

Cunliffe, B. and Poole, C. 1991: *Danebury: an Iron Age Hillfort in Hampshire, Vol. 5: the Excavations 1979-88: the Finds* (CBA Research Report 73: London).

Seven sling shots of baked clay recovered; four came from the later phases of the fort (cp 6/7), one from a cp 4 pit, one from a cp 3 pit and one was unstratified. The weights of the damaged examples were 25gms, 34gms and 35gms. Lengths 40-43mm by 26 -30mm diameter; the cp 3 example was smaller - 31m by 22mm. (p.370, Figure 7.42 (7.72-7.77))

Sling stones continued to be found in considerable numbers. One large collection found immediately behind the rampart, in a context that dates to cp 3, shows that slings were in use from the earliest phase of the forts construction. (p.404)

9. West Stow

West, S. 1990: *West Stow, Suffolk: the Prehistoric and Romano-British Occupations* (East Anglian Archaeology 48: Bury St. Edmunds).

Two clay sling shot recovered, one broken the other complete (SF2256 and SF1448 respectively). (p.60, Figure 45, 72-73)

10. Beltout

Toms, S. 1912: Excavations at the Beltout Valley Entrenchments. *Sussex Archaeological Collection* 55, 41-55.

Egg-shaped pebbles recovered from cuttings, probably sling stones. (p.48, 50)

11. Thundersbarrow

Curwen, E. C. 1933: Excavations on Thundersbarrow Hill. *Antiquaries Journal* 13, 109-33.

Quartz pebbles possibly sling stones. (p.132)

12. The Caburn

Curwen, E and Curwen, E. C. 1927: Excavations in the Caburn, near Lewes. *Sussex Archaeological Collection* 68, 1-56.

One clay sling shot recovered along with 570 beach pebbles, most from five pits: pit 44, 42 pebbles; pit 48, 169 pebbles; pit 49, 66 pebbles; pit:4A, 137 pebbles; pit 77A, 55 pebbles. The remaining 101 pebbles were collected from no less than 30 pits. The average weight was 48 grams. The closest location from which the pebbles could be derived is the seashore, 9.6 kilometres away.

13. Cissbury

Curwen, E. C. and Williamson, R. P. R. 1931: The date of Cissbury Camp. *Antiquaries Journal* 11, 14-36.

Beach pebbles recovered across the site, absent from undisturbed ground and Neolithic contexts. Pit 29 held 404 found with iron knife. From the pottery evidence the fort dates to the Middle Iron Age (Early Iron Age in report). Sling stones stratigraphically early, associated with construction of the fort.

14. Moel Trigarn, Pembrokeshire

Baring-Gould, S., Burnard, R. and Anderson, I. K. 1900: Exploration of Moel Trigarn. *Archaeologia Cambrensis* 17, 189 –211.

8 water worn pebbles in hut 1; huts 2-8, a few water worn pebbles in each; 12 number water worn pebbles in hut 9. (p.201)

15. Harding's Down West fort

Hogg, A. H. A. 1973: Excavations at Harding's Down West Fort, Gower. *Archaeologia Cambrensis* 122, 55-68. Huts 1, 2, 15 and 17 all contained small pebbles. (p.67)

16. Yarnbury

Cunnington, M. E.1933: Excavations at Yarnbury Castle Camp, 1932. *Wiltshire Archaeological Magazine* 46, 198-213.

Six chalk and seven clay sling shot recovered from Section B. (p.211, Plate XI)

17. Casterley Camp

Cunnington, M. E. and Cunnington, B. H. 1913: Casterley Camp excavations. *Wiltshire Archaeological Magazine* 38, 53-105.

Single chalk sling bullet found in pit (p.76)

18. St Catherine's Hill

Hawkes, C. F. C., Myres, J. N. L. and Stevens, C. G. 1930: *St. Catherine's Hill, Winchester* (Winchester).

Tertiary pebbles occasional finds in Rampart Section 2. Small round pebbles common. Hillfort dates to Middle Iron Age. (p.135)

19. The Trundle

Curwen, E. C. 1929: Excavations in the Trundle, Goodwood, 1928. *Sussex Archaeological Collection* 70, 33-85.

346 beach pebbles recovered from 'Early Iron Age' levels. (p.64)

20. Hengistbury Head

Cunliffe, B. 1987: *Hengistbury Head Dorset, Volume 1: the Prehistoric and Roman Settlement, 3500 BC-AD 500* (Oxford University Committee for Archaeology Monograph 13: Oxford).

Single clay sling shot, weight 19.8gms. (p.165, Plate Ill.118,168)

21. Pen-y-Corddyn

Gardner, W. 1910: Pen-y-Corddyn, near Abergele. *Archaeologia Cambrensis* 10, 79-156.

Eight water worn pebbles, one broken by heat (pot boilers?). No dimensions given. (p.147)

22. Lyneham Camp

Bayne, N. 1957: Excavation at Lyneham Camp, Lyneham, Oxon. *Oxoniensia* 22, 1-10.

Egg-shaped fragment of sandy limestone, likely to have been brought in by outside agency, sling stone? (p.7)

23. Pilsdon Pen

Gelling, P.S. 1977: Excavations at Pilsdon Pen, Dorset, 1964-71. *Proceedings of the Prehistoric Society* 43, 263-86.

Two caches of Tertiary pebbles on ramparts, not natural, although such pebbles occur across the site, neither are they a natural accumulation. (p.286)

24. Anstiebury

Thompson, F. H. 1979: Three Surrey hillforts: excavations at Anstiebury, Holmbury and Hascombe 1972-77. *Antiquaries Journal* 59, 245-318.

Many of the trenches produced smooth egg-sized pebbles, which are foreign to the natural, Lower Greensand. Their nearest province would be the Chalk Downs, 6 km from the site. (p.262)

25. Holmbury

Thompson, F. H. 1979: Three Surrey hillforts: excavations at Anstiebury, Holmbury and Hascombe 1972-77. *Antiquaries Journal* 59, 245-318.

Many of the trenches produced smooth egg-sized pebbles, which are foreign to the natural, Lower Greensand. Their nearest province would be the Chalk Downs, 6 km from the site (p.258)

26. Hascombe

Thompson, F. H. 1979: Three Surrey hillforts: excavations at Anstiebury, Holmbury and Hascombe 1972-77. *Antiquaries Journal* 59, 245-318.

Complete and fragmentary baked clay sling bullets recovered, length 38mm-44mm, average diameter of 25mm and weights of 25-35gms. (p.289 Figure 26 1-6, Plate LIV)

Sling stones were also recovered from pit 1, trench 2. (p.289, Plate LIV)

27. Glastonbury Lake Village

Bulleid, A. and Gray, H. St. G. 1917: *The Glastonbury Lake Village Vol.2* (Glastonbury).

Both fired and unfired clay sling shot recovered in their hundreds, occasional but rarer selected beach pebbles were also found. The clay sling shot measure 34.5- 39mm in length and 18.5-25mm in diameter; their average weight is 17.3gms. (pp.562-7 for a full description)

28. Boscombe Down (East and West)

Richardson, K. M. 1951: The excavation of an Iron Age village on Boscombe Down West. *Wiltshire Archaeological Magazine* 54, 123-68.

Clay sling shot from pit Q3 associated with rib bone knives, bone awl, bone gouge, bone ring, 2 bone needles, 4 chalk spindle whorls and hone stone. (p.164)

Stone, J. F. S. 1936: An enclosure on Boscombe Down East. *Wiltshire Archaeological Magazine* 47, 466-89.

Pebbles recovered from Angle ditch: Layer 2, 52 tertiary pebbles; Layer 3, 8 tertiary pebbles; Cutting I, 5 tertiary pebbles; Cutting II (enclosure ditch, 2 tertiary pebbles; linear ditch, 3 tertiary pebbles); Cutting IV, 6 tertiary pebbles; Cutting V, 5 tertiary pebbles. (p.489)

29. Fisherton

Stevens, F. 1934: The Highfield pit dwellings, Fisherton, Salisbury. *Wiltshire Archaeological Magazine* 46, 579-624.

Four oval clay sling shot (one broken), average length 42mm, average width 26mm, average weight 39gms, three from pit 57, one from pit 67.

Five 'roughly' shaped chalk sling shot (two broken), average length 44mm, average width 28mm, average weight 24gms, three from the ditch, two from pits (unnumbered in the report). (p.614)

30 Budbury Camp

Wainwright, G. 1970: An Iron Age promontory fort at Budbury, Bradford-on- Avon, Wiltshire. *Wiltshire Archaeological Magazine* 65, 108-66.

A group of nine complete or fragmentary sling missiles of baked clay, of pointed egg-shaped form, which average 50mm in length. (p.147, Figure 20, p.151-2)

31. Fifield Down

Clay, R. C. C. 1924: An Early Iron Age Site on Fifield Bavant Down. *Wiltshire Archaeological Magazine* 42, 457-96.

Tertiary pebbles and natural flints, round or egg-shaped, commonly found in pits (normally more than ten). As many as thirty found in one area. Lengths 25-37mm. Unnatural concentrations; nearest source is Blagdon Hill. 11km away '..the only conclusion to be drawn is that they were used as sling stones'. Six manufactured sling bullets also recovered, one of baked clay and five of chalk, two of which had been scorched. Their average length is 30mm, average width 27mm and their average weight 26.5gms

32. Lidbury Camp

Cunnington, M. E. and Cunnington, B. H. 1917: Lidbury Camp. *Wiltshire Archaeological Magazine* 40, 12-36.

Pit 2, flint flake with 4 whole clay sling shot and two pieces, two sling shot of chalk associated with bone implements and periwinkle shell. Pit 7, two chalk sling shot associated with bone implements and fragment of dish. Pit 10, chalk sling shot with ornamented bone, pointed bone implements, loom weight, flat pebble ground at edges. NB fragments of pottery and burnt flint found in all pits, in several there were pebbles about the size of sling shot and 'may have been used as such'. (pp.22-4)

33. Allard's Quarry

Williams, A. 1951: Excavations at Allard's Quarry, Marnhull, Dorset. *Proceedings of the Dorset Natural History and Archaeological Society* 72, 20-75.

Two baked clay sling pellets from pit 3 (uncertain date) one illustrated (Fig. 14)

Similar pellets were recovered from pit 37 (period C 'Late Iron Age') and one from an unspecified pit.

34. Swallowcliffe Down

Clay, R. C. C 1925: An inhabited site of La Tène I date on Swallowciffe Down. *Wiltshire Archaeological Magazine* 43, 59-93.

Sling bullets found in the following pits (assumed by the excavator to be 'pit dwellings'): Pits, 11, 34, 40, 48 and 66 (finds from 11 and 34 identified as baked clay, others recorded as sling bullets only). (pp.64-6)

Tertiary pebbles common. Six chalk sling bullets found (no location given). p. 86

35. Shearplace Hill, Sydling St. Nicholas

Rahtz, P. and ApSimon, A. M. 1962: Excavations at Shearplace Hill, Sydling St. Nicholas, Dorset. *Proceedings of the Prehistoric Society* 28, 289-328.

Possible sling bullet, consisting of reddish clay, composed of two pointed flattened pieces, pressed together while wet, may be a sling bullet, if made in a double mould or two single moulds, either way it was not successful. (p.323)

36. Gussage All Saints

Wainwright, G. J. and Spratling, M. 1973: The Iron Age settlement of Gussage All Saints. Part 1: the settlement. *Antiquity* 47, 109-30.

'Quantities of sling stones and clay bullets could also have a peaceful context in the hunt or for scaring predators from the fields'. (p.115)

Wainwright, G. J. 1979: *Gussage All Saints: an Iron Age Settlement in Dorset* (Department of the Environment Archaeological Report 10: London).

Phase 1. Nearly complete sling missile, pointed at both ends. (pp.100-1, Figure 4010)

Phase 2. A sling missile made of baked clay 30mm long (not illustrated). (p.101)

Phase 3. An egg-shaped baked clay missile (not illustrated). (p.102)

37. Crickley Hill

Dixon, P. 1994: *Crickley Hill: the Hillfort Defences* (Nottingham).

F1001 contained 341 pebbles (sling stones). (pp.105, 108)

Context 10733, 1 pebble; context 3351, 250 pebbles; context 3474 (part), 43 pebbles; context 3217, 32 pebbles. (p.132)

Pebbles probably derived from the Bunter beds of the Midlands, concentrations recovered from site not natural. (p.249)

38. Hod Hill, Dorset

Richmond, I. A. 1968: *Hod Hill, vol.2: Excavations carried out between 1951 and 1958* (London).

Hut 56, 117 sling stones on western side of the door. (p.20)
Hut 43, 218 sling stones; 1 clay shot. (p.21)
Hut 146, 887 sling stones. (p.9)
Hut 36a, 'contained sling stones'. (p.22)
Hut 109, sling stones mixed with charcoal. (p.24)

39 Battlesbury Camp

Cunnington, M. E. 1923b: Pits in Battlesbury Camp. *Wiltshire Archaeological Magazine* 42, 368-73.

24-25 (some fragmentary) sling bullets from pit No. 4, 'roughly made and badly baked, some partly blackened'. (p.370)

One baked clay sling bullet from pit No 88. (p.372)

40. Eggerdon Hill

Marsh, H. C. 1901: Report in the Excavation on Eggardun, 1900. *Proceedings of the Society of Antiquaries* 18, 258-62.

Pit IV, a beach pebble 68mm x 50mm long with a white quartz pebble. (p.262).

41. Chalbury Camp, Dorset

Whitley, M. 1943: Excavations at Chalbury Camp, Dorset, 1939. *Antiquaries Journal* 23, 98-121.

Site D, roundhouse outside west wall, group of about 30 sling stones associated with Iron Age 'A' pottery. (p.107, Figure 2, Section A-B)

42.Shipton Hill, Dorset

Farrar, R. A. H. 1955: An Early Iron Age fort on Shipton Hill, Shipton Gorge. *Proceedings of the Dorset Natural History and Archaeological Society* 77, 135-6.

About 1,000 sling stones recovered from trial trench across the site. (Figure B, Plate 1). Dump of sling stones recovered in second trial trench.

43. Pins Knoll, Dorset

Bailey, C. J. 1967: An Early Iron Age/Romano-British site at Pins Knoll, Litton Cheney, Final Report. *Proceedings of the Dorset Natural History and Archaeological Society* 89, 147-59.

Pit 4, below a uniform band of dark earth, pottery, animal bone, bone implements, loom weights and sling stones were recovered.

44. Dinordan, Abergele

Gardner, W. and Savory, H. N. 1964: *Dinorban, A Hillfort occupied in Early Iron Age and Roman Times* (Cardiff).

'Many small pebbles collected from the site could have been used as sling stones, but no hoards were found as would prove this use'. (p.79)

45. Grimthorpe, Yorkshire

Stead, I. M. 1968: An Iron Age hill-fort at Grimthorpe, Yorkshire, England. *Proceedings of the Prehistoric Society* 34, 148-90.

Two deliberately manufactured stones, from the local geology. Both their shape and weight would have particularly suitable for sling-stones. (p.166)

46. Stockton

Saunders, P. R. 1997: The excavations of an Iron Age settlement site at Stockton. *Wiltshire Archaeological Magazine* 90, 13-25.

Chalk sling shot illustrated (Figure 6, (8)) but not mentioned in accompanying text.

47. Uffington Castle

Miles, D., Palmer, S., Lock, G., Gosden C. and Cromarty, A. M. 2003: *Uffington White Horse and its Landscape: Investigations at White Horse Hill Uffington, 1989-95 and Tower Hill Ashbury, 1993-4* (Oxford).

Flint slingstones, average weight 63gms, with a distinctive crackled appearance, may have come from the Clay-with-Flints, but must be derived from the Tertiary beds, as are those at Danebury. (p.185)

48. Hambledon Hill

Cunnington, E. 1895: Excavations on Hambledon Hill, *Proceedings of the Dorset Natural History and Antiquarian Field Club* 16, 156-8.

Excavation apparently in the central section of the hillfort produced Iron Age A pottery, quern stone fragments and sling stones these were interpreted as being Roman by the excavator. (p.157)

49. Bishopstone, Sussex

Bell, M. 1977: Excavations at Bishopstone. *Sussex Archaeological Collections* 115, 1-299.

Two clay sling shot recovered from the top of the enclosure ditch, both manufactured from pottery fabric 3a. (p.119, Figures C6-C7)

50. All Cannings Cross

Cunnington, M. E. 1923a: *The Early Iron Age Inhabited Site at All Cannings Cross* (Devizes).

Two sling bullets, baked clay 26gm ($1^{1/16}$oz in the original, shown on Plate 26)
2 chalk bullets, between 28-26gm.
8 clay shot, weighing between 9.5gm and 29gm and one of unbaked clay, six other fragments that were too small to weigh.
'A number of small rounded pebbles about the size of sling bullets, and that may well have been used as such, were found throughout the excavation'. (p.142)

51. Thetford

Gregory, T. 1991: *Excavations in Thetford, 1980-82, Fison Way*. (East Anglia Archaeology 53: Dereham).

Five bi-conical, fired sling shot were recovered. Their similarity to the weathered flint makes their under-representation likely. (p.148, Figure 133)

52. Balksbury Camp

Wainwright, G. J. and Davies, S. M. 1995: *Balksbury Camp, Hampshire: Excavations 1973 and 1981* (English Heritage Archaeological Report 4: London).

Hollow 656 contained 27 even-sized pebbles, suggesting that they had been deliberately selected, perhaps as sling stones. (p.30)

53. Segsbury Camp

Lock, G. and Gosden, C. 1997: Hillforts of the Ridgeway Project: Excavations at Segsbury Camp, 1996. *South Midlands Archaeology* 27, 69-77.

Pit 1009, series of layers with 500+ river-rolled pebbles at its base; these pits were shallow, less than 1m deep with vertical sides, they were not of typical beehive construction.

54. Blewburton Hill

Collins, A. E. P. 1953: Excavations on Blewburton Hill, 1948 and 1949. *The Berkshire Archaeological Journal* 53, 21-64.

Some 'bunter quartzite pebbles' within ditch II (sling stones?). (p. 38)

Collins, A. E. P. 1959: Excavations on Blewburton Hill, 1953, *The Berkshire Archaeological Journal* Vol. 59, 252-73.

Thin scatter of pebbles too high to be a road surface behind the ramparts (sling stones?). (p. 263)

55. Grimsbury Castle

Wood, P. 1959: The Early Iron Age Camp called Grimsbury Castle, near Hermitage, Berks. *The Berkshire Archaeological Journal* 59, 74-82.

'Sling pebbles'. (p. 79)

56. Winnall Down

Bates, P. J. & Winham, R. P. 1985: Daub and burnt or baked clay. In Fasham, P. J. *The Prehistoric Settlement at Winnall Down, Winchester: Excavations of MARC3 Site R17 in 1976 and 1977* (Hampshire Field Club Monograph 2: Winchester), 86-90

Egg-shaped baked clay object, possibly a sling missile. Weight 40g, pit No. 8594, Middle Iron Age in date.

Appendix 1.2

Gazetteer of spearheads from the Middle Iron Age, recovered from the hillfort dominated zone.

1. Croft Ambrey

Stanford, S. C. 1974: *Croft Ambrey* (Hereford).

Socketed iron spearhead, with single bronze rivet, with possible minute ring decoration. Total length 90mm, maximum width, 21mm, shaft diameter 10mm. (p. 167, Figure 76, 1)

Socketed iron spearhead, very corroded. Total length 80mm, maximum width 20mm, shaft diameter 12mm. (p. 167, Figure 76, 2)

Both are dated to phase IV which according to the excavator corresponds with 262-102BC. (p.229)

2. Glastonbury Lake Village

Bulleid, A. and Gray, H. St. G. 1917: *The Glastonbury Lake Village Vol.2* (Glastonbury).

i. Socketed spear or javelin head, found in four fragments, approx. total length 140mm (no width given). (p.381)

ii. Socketed spearhead, short leaf-shaped, total length 86mm, maximum width, 28.5mm, shaft diameter 20mm. (p.381, Plate LXII, 16)

iii. Possible javelin head, very corroded, total length 110mm, no width given, shaft diameter 18mm. (p.381)

3. Thundersbarrow

Curwen, E. C. 1933: Excavations on Thundersbarrow Hill. *Antiquaries Journal* 13, 109-33.

Badly corroded spearhead, leaf-shaped, approx. 10mm long by 60mm wide. (p. 131, Figure 6)

4. Casterley

Cunnington, M. E. and Cunnington, B. H. 1913: Casterley Camp excavations. *Wiltshire Archaeological Magazine* 38, 53-105.

Small, leaf-shaped iron spearhead, stray find from the surface of an irregular enclosure dating to Early Iron Age or Middle Iron Age, approx. 90mm long by 30mm wide, clearly has a socket. (p.103, Plate III No.15)

5. Allard's Quarry

Williams, A. 1951: Excavations at Allard's Quarry, Marnhull, Dorset. *Proceedings of the Dorset Natural History and Archaeological Society* 72, 20-75.

Two fragmentary leaf shaped spearheads, both socketed. The illustrated example is 120mm long by 30mm wide. (p.57, Figure 18 No.8)

6. Spettisbury Rings Dorset

Gresham, C. A. 1939: Spettisbury Rings, Dorset. *Archaeological Journal* 96, 114 –31.

Nine spearheads recovered 'probably of native Iron Age origin'. No further dating is given:

i. Length 180mm, width 20mm, dia. of socket (external) 19mm;
ii. Length 135mm, width 22mm, socket 19mm (both tip and socket broken);
iii. Length 174mm, width 25mm, socket 18mm. Recorded as being found with a skeleton in October 1857;
iv. Length 165mm, width 24mm, dia. of socket 20mm;
v. Length 112mm, width 35mm, dia. of socket 17mm;
vi. Length 230mm, width 42mm, dia. of socket 20mm;
vii. Length 160mm, width 27mm, dia. of socket 20mm;
viii. Length 123mm, width 22mm, dia. of socket 20mm;
ix. Length 247mm, width 25mm, dia. of socket 21mm;

7. Danebury

Cunliffe, B. 1984b: *Danebury: an Iron Age Hillfort in Hampshire. Vol. 2: the Excavations 1969-1978: the Finds* (CBA Research Report 52 (2): London).

Four socketed spearheads were recovered; the last is from an unstratified context. (p. 362, Figure 7.19, small finds 2.100-2.103)

2.100 dates from cp 7, length 173mm, width 22mm;
2.101 dates from cp 6, length 112mm, width 20mm;
2.102 dates from cp 3, length 98mm, width 23mm (clearly fashioned from a saw blade) ;
2.103 unstratified, length 90mm (tip has broken off), width 28mm.

Cunliffe, B. and Poole, C. 1991: *Danebury: an Iron Age Hillfort in Hampshire, Vol. 5: the Excavations 1979-88: the Finds* (CBA Research Report 73: London).

Six spearheads were recovered during this phase of the Danebury excavations, small finds numbers 2.282-2.287:

2.282 dates from cp 3, length 120mm, width 28mm;
2.283 dates from cp 7, length 68mm, width 32mm;
2.284 dates from cp 7, length 265mm, width 39mm (incomplete);
2.285 dates from cp 8, length 173mm, width 64mm;
2.286 dates from cp 3, length 51mm, width 19mm;
2.287 dates from un-phased context, length 124mm, width 16.5mm.

APPENDIX 2.1

EXPERIMENTAL CASTING DATA

Number	Weight (gms)	Distance (m)	Time (sec)	Velocity (m/s)	Initial Velocity m/s^{-1}	Angle of release in degrees	Volume m/l
1	61.3	102.3	3.94	25.9	32.55	37.25	29
2	55.1	77.7	5.18	15	29.93	59.92	23
3	75.7	48.0	2.00	24	26.00	22.61	32
4	43.8	58.0	2.50	23.2	26.35	28.31	20.5
5	40.9	67.2	3.44	19.5	26.00	41.41	19
6	55.7	88.5	2.47	17.8	21.6	34.75	18.5
7	54.3	77.0	3.97	19.4	27.75	45.65	23
8	52.6	65.2	3.79	17.2	25.59	47.77	19
9	44.1	47.5	1.71	27.8	29.08	17.09	14.5
10	49.2	44.9	1.53	29.3	30.28	14.63	17
11	47.9	80.7	4.19	19.3	28.48	47.34	16.5
12	67.9	66.2	3.81	17.4	25.80	47.59	26
13	57.7	71.5	3.13	22.8	27.65	34.4	19.5
14	43.6	70.0	3.44	20.3	26.60	40.27	15
15	48.9	72.1	3.06	23.6	28.12	32.95	17
16	41.2	72.6	2.72	26.7	29.96	26.99	14.5
17	57.9	82.6	4.65	17.8	29.28	52.56	19
18	67.9	82.8	3.68	22.5	29.06	39.27	25
19	62.7	60.8	3.15	19.3	24.91	39.21	26
20	75.2	76.5	3.47	22.0	28.00	38.26	25
21	55.5	66.9	4.53	14.8	27.05	56.83	20
22	44.8	81.5	2.32	35.1	36.96	18.28	15.5
23	58.6	89.1	3.75	23.8	30.29	38.23	23.5
24	53.3	44.7	1.75	25.5	29.04	28.59	17
25	43.7	77.6	2.78	27.9	29.24	17.41	14.5
26	72.9	49.9	2.80	17.82	22.7	38.2	30.5
27	72.5	58.7	4.65	12.62	26.45	61.5	32.5
28	82.1	55.8	4.88	11.43	24.4	64.9	27
29	38.5	53.0	4.43	11.96	22.15	61.6	22.5
30	46.9	45.4	4.09	11.1	20.45	61.5	21.5
31	85.7	47.2	3.93	12	19.65	58.60	26
32	80.6	49.9	4.13	12.08	20.65	54.36	24
33	56.6	49.9	3.73	13.37	22.90	59.67	25

34	71.6	65.2	5.10	12.78	28.52	63.38	26
35	65.1	54.3	3.46	15.69	23.35	47.79	25
36	61.6	49.9	4.55	10.96	25.25	64.20	27.5
37	80.7	55.6	5.03	11.05	27.47	66.28	26.5
38	43.7	54.3	2.89	18.78	23.69	37.57	24
39	51.5	49.8	2.54	19.60	23.35	32.9	23.5
40	81.6	48.1	4.70	10.23	25.63	66.40	24.5
41	40.6	47.8	3.08	15.52	21.86	44.70	18.5
42	68.5	60.7	3.53	17.65	24.49	46.09	26
43	59.3	47.7	2.17	10.85	24.51	26.27	24.5
44	38.3	48.1	4.33	11.10	24.32	62.85	22
45	74.1	49.8	2.81	17.72	22.61	38.41	26
46	42.7	57.1	4.19	13.63	24.99	56.95	12.5
47	37.1	49.6	3.37	16.85	28.20	36.68	23
48	56.0	49.5	3.80	13.02	23.00	55.58	26.5
49	26.8	39.8	2.79	14.26	20.05	44.67	11
50	55.9	44.9	2.23	19.90	22.1	29.26	26
51	59.3	57.2	2.97	14.85	24.31	37.64	24.5
52	47.1	45.1	2.72	13.60	21.44	39.36	23.5
53	45.2	50.3	2.79	13.95	20.17	42.72	21.5
54	52.2	47.8	2.59	17.04	22.72	41.41	23.5
55	54.1	58.6	4.32	13.56	25.50	57.88	24
56	48.6	46.9	1.43	32.79	33.56	12.30	24
57	46.0	45.6	3.95	19.75	22.87	59.43	18.5
58	57.8	38.2	3.35	16.75	20.26	55.76	24.5
59	45.3	41.2	4.27	9.65	23.45	65.67	18
60	67.1	53.8	4.30	12.51	24.87	59.80	25.5
61	57.4	44.6	5.00	8.92	26.54	70.36	24.5
62	45.2	54.1	4.99	10.84	27.20	66.51	23.5
63	72.4	59.1	4.62	12.79	26.40	61.02	21
64	82.8	47.7	3.19	14.95	21.86	46.85	28
65	48.9	59.3	4.90	12.10	27.32	63.76	24.5
66	62.1	46.7	3.41	13.69	21.86	51.23	24
67	30.1	51.2	3.34	15.33	22.66	47.44	13.5
68	63.5	54.3	4.22	12.86	24.71	85.65	25.5
69	46.9	64.5	5.05	12.77	28.29	63.17	23
70	56.1	38.4	1.55	24.77	25.95	17.37	24.5
71	40.9	52.1	4.03	12.92	23.93	57.33	18
72	55.5	58.1	4.14	14.03	25.00	55.87	25
73	46.2	52.2	2.26	11.30	25.70	26.09	22
74	50.6	44.9	5.34	8.40	27.99	72.53	23.5
75	61.8	46.7	2.93	15.93	21.64	42.60	24.5
76	65.7	46.5	2.01	22.14	24.50	23.37	24.5
77	51.7	53.3	3.66	14.56	23.38	51.49	21
78	44.3	52.1	2.58	25.38	28.47	26.94	22.5
79	60.6	43.9	3.14	13.90	20.96	48.83	24.5

80	71.9	49.9	3.36	16.98	23.88	44.69	26
81	94.3	50.5	4.81	10.49	26.23	66.43	37.5
82	45.1	50.8	2.21	22.98	25.49	25.68	17.5
83	61.7	41.7	5.27	18.14	31.99	55.45	24.5
84	49.3	46.8	3.09	15.14	21.63	45.58	22
85	62.4	50.2	3.00	16.73	22.46	41.87	24
86	50.7	42.2	1.59	26.54	27.70	16.67	22
87	33.3	47.7	2.98	16.00	21.86	42.96	13.5
88	69.4	60.2	4.37	13.77	25.84	57.78	38
89	64.4	47.8	3.57	13.39	22.31	53.12	25
90	62.9	50.1	3.69	13.57	22.90	53.66	18.5
Σ	5081.2	5046.2	315.2	1535.6	2284.11	4142.15	2045
Arithmetic mean	56.5	56.07	3.50	17.06	25.38	46.02	22.72

Medium Values

Initial velocities of casts

36.96, 33.56, 32.55, 31.99, 30.29, 30.28, 29.93, 29.96, 29.28, 29.24, 29.08, 29.06, 29.04,28.52, 28.48, 28.47, 28.29, 28.20, 28.12, 28.00, 27.99, 27.75,27.70, 27.65, 27.47, 27.32, 27.20, 27.05, 26.60, 26.54, 26.45, 26.40, 26.35, 26.23, 26.00, 26.00, 25.95, 25.84, 25.80, 25.70, 25.63, 25.59, 25.50, 25.49,

25.25 m/s^{-1} Medium

25.00, 24.87 24.99, 24.91, 24.71, 24.51, 24.50, 24.49, 24.4, 24.32, 24.31, 23.93, 23.88, 23.69, 23.45, 23.38, 23.35, 23.35, 23.00, 22.90. 22.90, 22.87, 22.72, 22.7, 22.66, 22.61, 22.46, 22.31, 22.15, 22.10, 21.86, 21.86, 21.86, 21.86, 21.64, 21.63, 21.6, 21.44, 20.96, 20.65, 20.45, 20.26, 20.17, 20.05, 19.65

Mass of collected pebbles:

94.3, 85.7, 82.8, 82.1, 81.6, 80.7, 80.6, 75.7, 75.2, 74.1, 72.9, 72.4, 72.5, 71.9, 71.6, 69.4, 68.5, 67.9, 67.9, 67.1, 65.7, 65.1, 69.4, 63.5, 62.7,62.9, 62.4, 62.1, 61.8, 61.7, 61.6, 61.3, 60.6, 59.3, 59.3, 58.6, 57.9, 57.9, 57.8, 57.4, 56.6, 56.1, 56.0, 55.9

55.7g Medium

55.5, 55.1, 54.3, 54.1, 53.3, 52.2, 52.6, 51.7, 51.5, 50.7, 50.6, 49.3, 49.2, 48.9, 48.9, 48.6, 47.9, 47.1, 46.9, 46.9, 46.0, 45.3, 45.2, 45.2, 45.1, 44.3, 44.7, 44.8, 44.1, 43.7, 43.8, 43.7, 43.6, 42.7, 41.2, 40.9, 40.9, 40.6, 38.5, 38.3, 37.1, 33.3, 30.1, 26.8

APPENDIX 2.2

STATISTICAL ANALYSIS OF SLING STONE DATA

The test used was the product-moment correlation coefficient (Upton and Cook 1998:374-6)

Number	Weight (g) X_i	Initial Velocity m/s^{-1} y_i	$X_i\ y_i$	X_i^2	y_i^2
1	61.3	32.55	1995.32	3757.69	1059.50
2	55.1	29.93	1649.14	3036.01	895.80
3	75.7	26.00	1968.2	5730.49	676.0
4	43.8	26.35	1154.13	1918.44	694.32
5	40.9	26.00	1063.4	1672.81	676.0
6	55.7	21.6	1203.12	310.49	466.56
7	54.3	27.75	1506.83	2948.49	770.06
8	52.6	25.59	1346.03	2766.76	654.85
9	44.1	29.08	1282.43	1944.81	845.65
10	49.2	30.28	1489.78	2420.64	916.88
11	47.9	28.48	1364.19	2294.41	811.11
12	67.9	25.80	1751.82	4610.41	665.64
13	57.7	27.65	1595.41	3329.29	764.52
14	43.6	26.60	1159.76	1900.96	707.56
15	48.9	28.12	1375.07	2391.21	790.73
16	41.2	29.96	1234.35	1697.44	897.60
17	57.9	29.28	1695.31	3352.41	857.32
18	67.9	29.06	1973.17	4610.41	844.48
19	62.7	24.91	1561.86	3931.29	620.51
20	75.2	28.00	2105.60	5655.04	784.0
21	55.5	27.05	1501.28	3080.25	731.70
22	44.8	36.96	1655.81	2007.04	1366.04
23	58.6	30.29	1774.99	3433.96	917.48
24	53.3	29.04	1547.83	2840.89	843.32
25	43.7	29.24	1277.79	1909.69	854.98
26	72.9	22.7	1654.83	5314.41	151.29
27	72.5	26.45	1917.63	5256.25	699.60
28	82.1	24.4	2003.24	6740.41	595.36
29	38.5	22.15	852.78	1482.25	490.62
30	46.9	20.45	959.11	2199.61	418.20
31	85.7	19.65	1684.0	7344.49	386.12
32	80.6	20.65	1664.39	6496.36	426.42
33	56.6	22.90	1296.14	3203.56	524.41
34	71.6	28.52	2042.03	5126.56	813.39
35	65.1	23.35	1520.09	4238.01	545.22
36	61.6	25.25	1555.40	3794.56	637.56
37	80.7	27.47	2216.83	6512.49	754.60
38	43.7	23.69	1035.25	1909.69	561.22

39	51.5	23.35	1202.53	2652.25	545.22
40	81.6	25.63	2091.41	6658.56	709.16
41	40.6	21.86	887.52	1648.36	477.86
42	68.5	24.49	1677.56	4692.25	599.76
43	59.3	24.51	1453.44	3516.49	600.74
44	38.3	24.32	931.46	1466.89	591.46
45	74.1	22.61	1675.40	5490.81	511.21
46	42.7	24.99	1067.07	1823.29	624.50
47	37.1	28.20	1046.22	1376.41	795.24
48	56.0	23.00	1288.0	3136.0	529.0
49	26.8	20.05	537.34	718.24	402.0
50	55.9	22.1	1235.39	3124.81	488.41
51	59.3	24.31	1441.58	3516.49	590.98
52	47.1	21.44	1009.82	2218.41	459.67
53	45.2	20.17	911.68	2043.04	406.83
54	52.2	22.72	1185.98	2724.84	516.20
55	54.1	25.50	1379.55	2926.81	650.25
56	48.6	33.56	1631.02	2361.96	1126.27
57	46.0	22.87	1052.02	2116	523.04
58	57.8	20.26	1171.03	3340.84	410.47
59	45.3	23.45	1062.29	2052.09	549.90
60	67.1	24.87	1668.77	4502.41	618.52
61	57.4	26.54	1523.40	3294.76	704.37
62	45.2	27.20	1229.44	2043.04	739.84
63	72.4	26.40	1911.36	5241.76	696.96
64	82.8	21.86	1810.01	6855.84	477.86
65	48.9	27.32	1335.95	2391.21	746.38
66	62.1	21.86	1357.51	3856.41	477.86
67	30.1	22.66	682.07	906.01	513.48
68	63.5	24.71	1569.08	1332.25	610.58
69	46.9	28.29	1326.80	2199.61	800.32
70	56.1	25.95	1455.80	3147.21	673.40
71	40.9	23.93	978.74	1672.81	572.64
72	55.5	25.00	1387.5	3080.25	625.0
73	46.2	25.70	1187.34	2134.44	660.49
74	50.6	27.99	1416.29	2560.36	783.44
75	61.8	21.64	1337.35	3819.24	468.29
76	65.7	24.50	1609.65	4316.49	600.25
77	51.7	23.38	1208.74	2672.89	546.62
78	44.3	28.47	1261.22	1962.49	810.54
79	60.6	20.96	1270.18	3672.36	439.32
80	71.9	23.88	1716.97	5169.61	570.25
81	94.3	26.23	2473.49	8892.49	688.01
82	45.1	25.49	1149.60	2034.01	649.74
83	61.7	31.99	1973.78	3806.89	1023.36
84	49.3	21.63	1066.36	2430.49	467.86
85	62.4	22.46	1401.50	3893.76	504.45

86	50.7	27.70	1404.39	2570.49	767.29
87	33.3	21.86	727.94	1108.89	477.86
88	69.4	25.84	1793.30	4816.36	677.70
89	64.4	22.31	1436.76	4747.36	497.74
90	62.9	22.90	1400.41	3956.41	524.41
Σ	5081.2	2284.11	128644.35	297863	58640

$S_{xy} = 128644.35 - \frac{5081.2 \times 2284.11}{90} = -311.42$

$S_{xx} = 297863 - \frac{5081.2^2}{90} = 10989.74$

$S_{yy} = 58640 - \frac{2284.11^2}{90} = 671.57$

$r = \frac{-311.42}{.10989.74 \times 671.57} = -2.32$

The two sets of data display a strong negative correlation, therefore, the statistical analysis supports the observations relating to the scatter diagram and the two sets of data are unlikely to have any significant correlation.

Number	Weight (gms) X_i	Distance (m) y_i	$X_i\ y_i$	X_i^2	y_i^2
1	61.3	102.3	6270.99	3757.69	10465.29
2	55.1	77.7	4281.27	3036.01	6037.29
3	75.7	48.0	3633.60	5730.49	2304.0
4	43.8	58.0	2540.4	1918.44	3644.0
5	40.9	67.2	2748.48	1672.81	4515.84
6	55.7	88.5	4929.45	3102.49	7832.25
7	54.3	77.0	4187.1	2948.49	5929
8	52.6	65.2	3429.52	2766.76	4251.04
9	44.1	47.5	2094.75	1944.81	2256.25
10	49.2	44.9	2209.08	2420.64	2016.01
11	47.9	80.7	3865.53	2294.41	6512.49
12	67.9	66.2	4494.98	4610.41	4382.44
13	57.7	71.5	4125.55	3329.29	5112.25
14	43.6	70.0	3052	1900.96	4900.0
15	48.9	72.1	3525.69	2391.21	5198.41
16	41.2	72.6	2991.12	1697.44	5270.76
17	57.9	82.6	4782.54	3352.41	6822.76
18	67.9	82.8	5622.12	4610.41	6855.84
19	62.7	60.8	3812.16	3931.29	3696.64
20	75.2	76.5	5752.80	5655.04	5852.25
21	55.5	66.9	3712.95	3080.25	4475.61
22	44.8	81.5	3651.20	2007.04	6642.25
23	58.6	89.1	5221.26	3433.96	7938.81
24	53.3	44.7	2382.51	2840.89	1998.09
25	43.7	77.6	3391.12	1909.69	6021.76

26	72.9	49.9	3637.71	5314.41	2490.01
27	72.5	58.7	4255.75	5256.25	3445.69
28	82.1	55.8	4581.18	6740.41	3113.64
29	38.5	53.0	2040.50	1482.25	2809.0
30	46.9	45.4	2129.26	2199.61	2061.16
31	85.7	47.2	4045.04	7344.49	2227.84
32	80.6	49.9	4021.94	6496.36	2490.01
33	56.6	49.9	2824.34	3203.56	2490.01
34	71.6	65.2	4668.32	5126.56	4251.04
35	65.1	54.3	3534.93	4238.01	2948.49
36	61.6	49.9	3073.84	3794.56	2490.01
37	80.7	55.6	4486.92	6512.49	3091.36
38	43.7	54.3	2372.91	1909.69	2948.49
39	51.5	49.8	2564.7	2652.25	2480.04
40	81.6	48.1	3924.96	6658.56	2313.61
41	40.6	47.8	1940.68	1648.36	2284.84
42	68.5	60.7	4157.95	4692.25	3684.49
43	59.3	47.7	2828.61	3516.49	2275.29
44	38.3	48.1	1842.23	1466.89	2313.61
45	74.1	49.8	3690.18	5490.81	2480.04
46	42.7	57.1	2438.17	1823.29	3260.41
47	37.1	49.6	1840.16	1376.41	2460.16
48	56.0	49.5	2772.0	3136.0	2450.25
49	26.8	39.8	1066.64	718.24	1584.04
50	55.9	44.9	2509.91	3124.81	2016.01
51	59.3	57.2	3391.96	3516.49	3271.84
52	47.1	45.1	2124.21	2218.41	2034.01
53	45.2	50.3	2273.56	2043.04	2530.09
54	52.2	47.8	2495.16	2724.84	2284.84
55	54.1	58.6	3170.26	2926.81	3433.96
56	48.6	46.9	2279.34	2361.96	219961
57	46.0	45.6	2097.6	2116.0	2079.36
58	57.8	38.2	2207.96	3340.84	1459.24
59	45.3	41.2	1866.36	2052.09	1697.44
60	67.1	53.8	3609.98	4502.41	2894.44
61	57.4	44.6	2560.04	3294.76	1989.16
62	45.2	54.1	2445.32	2043.04	2926.81
63	72.4	59.1	4278.84	5241.76	3492.81
64	82.8	47.7	3949.56	6855.84	2275.29
65	48.9	59.3	2899.77	2391.21	3516.49
66	62.1	46.7	2900.07	3856.41	2180.89
67	30.1	51.2	1541.12	906.01	2621.44
68	63.5	54.3	3448.05	4032.25	2948.49
69	46.9	64.5	3043.81	2199.61	4160.25
70	56.1	38.4	2154.24	3147.21	1474.56
71	40.9	52.1	2130.89	1672.81	2714.41
72	55.5	58.1	3224.55	3080.25	3375.61

73	46.2	52.2	2411.64	2134.44	2724.84
74	50.6	44.9	2271.94	2560.36	2016.01
75	61.8	46.7	2886.06	3819.24	2180.89
76	65.7	46.5	3055.05	4316.49	2162.25
77	51.7	53.3	2755.61	2672.89	2840.89
78	44.3	52.1	2308.03	1962.49	2714.41
79	60.6	43.9	2660.34	3672.36	1927.21
80	71.9	49.9	3587.81	5169.61	2490.01
81	94.3	50.5	4762.15	8892.49	2550.25
82	45.1	50.8	2291.08	2034.01	2580.64
83	61.7	41.7	2572.89	3806.89	1738.89
84	49.3	46.8	2455.14	2430.49	2190.24
85	62.4	50.2	3132.48	3893.76	2520.04
86	50.7	42.2	2139.54	2570.49	1780.84
87	33.3	47.7	1588.41	1108.89	2275.29
88	69.4	60.2	4177.88	4816.36	3624.04
89	64.4	47.8	3078.32	4147.36	2284.84
90	62.9	50.1	3151.29	3956.41	2510.01
Σ	5081.2	5046.2	285335.3	302755.4	515824.9
	X_i	y_i	$X_i y_i$	X_i^2	y_i^2

$$S_{xy} = 285335.3 - \frac{5081.2 \times 5046.2}{90} = 438.06$$

$$S_{xx} = 302755.4 - \frac{5081.2^2}{90} = 15882.14$$

$$S_{yy} = 515824.9 - \frac{5046.2^2}{90} = 232890.07$$

$$r = \frac{438.06}{15882.14 \times 232890.07} = 0.00072$$

The two sets of data display a very weak positive correlation, therefore, the statistical analysis supports the observations relating to the scatter diagram and the two sets of data are unlikely to have any significant correlation.

Graphical representation of the experimental data

1. Scatter diagram showing Initial velocity opposed to mass of projectile (Figure 4.13). It was assumed that the mass of the projectile would have a direct and observable effect on the initial velocities, as the variation between the masses of the stones used was high (29%). It was expected that an inverted U-shaped curve would result from this graph, showing a reduction of initial velocity at the extreme ends of the mass of the projectiles. This was not the case: a random distribution of initial velocities versus mass occurred and a product-moment correlation coefficient (r) = -2.32 was statistically derived from these data (see Appendix 2.2). These data strongly indicate that, within the weight range of stones used for the experimental work, there is no measurable difference relating to their effectiveness as projectiles. As the weight range is comparable to that of sling projectiles recovered from Middle Iron Age sites (see Appendix 1 for references), the same results should apply

2. Mass opposed to distance (Figure 4.14*)*. It was assumed that the greater the mass of the stone, the shorter the distance it would travel. This again proved to be a false assumption. The distribution of mass versus distance appears to have a slight positive slope, but this is not clear within the scatter diagram. Although there is a general cluster at distances of 65m to 80m, the mass of these projectiles is within the weight range of 40g to 75g. A product-moment correlation coefficient (r) = 0.00072 was statistically derived from these data (see Appendix 2.2), indicating a very weak correlation between mass and distance. These data strongly indicates that, within the weight range of stones used for the experimental work, there is no measurable difference relating to their effectiveness as projectiles. As the weight range is comparable to that of sling projectiles recovered from Middle Iron Age sites (see appendix 1 for references), the same results should apply.

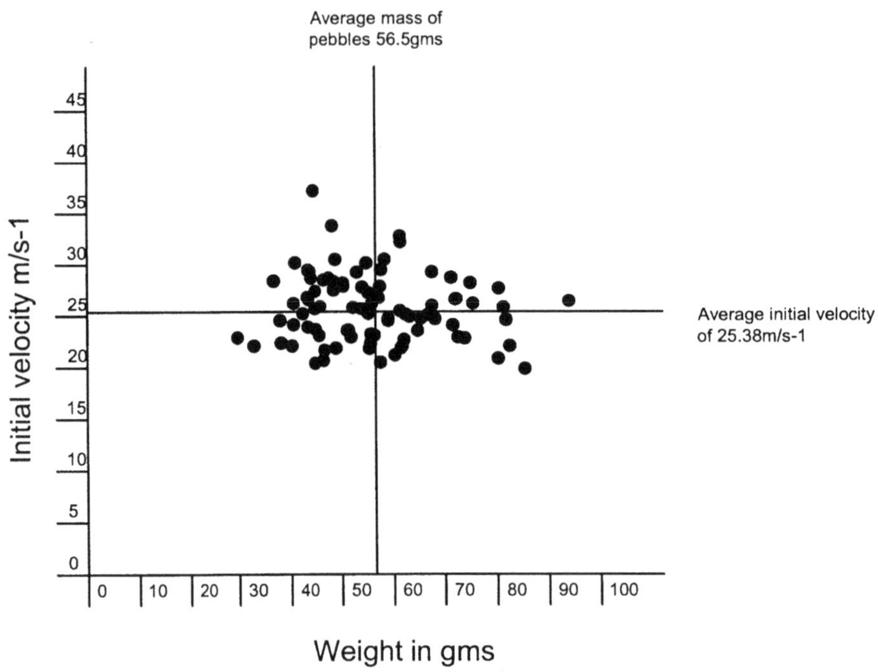

Figure 4.13 Scatter Diagram: initial velocity against mass

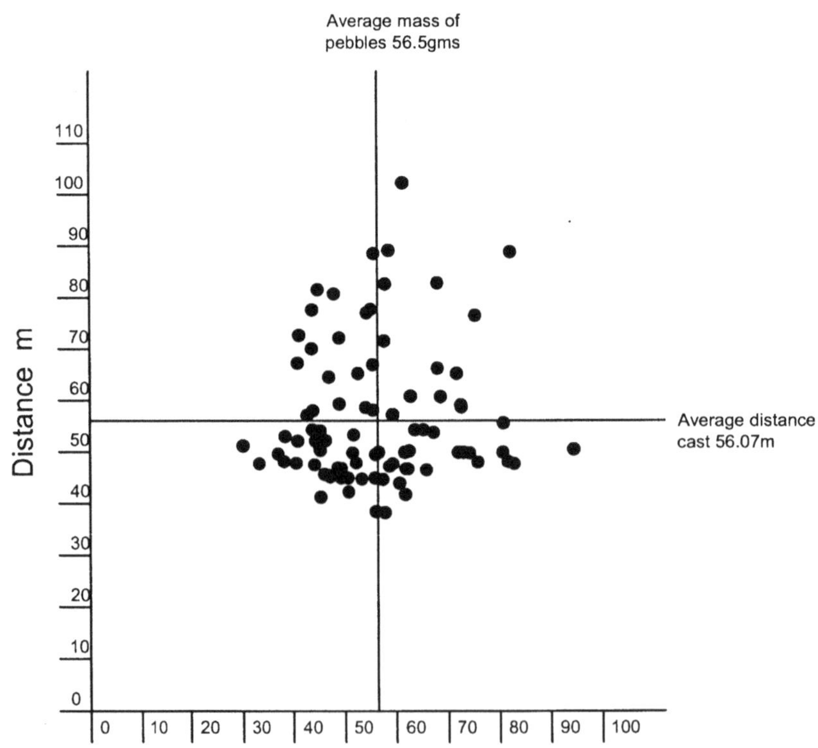

Figure 4.14 Scatter Diagram: distance against mass

APPENDIX 2.3

THE RESULTS OF THE EXPERIMENTAL WORK (THE SLING)

The data relating to the distance and time taken to travel that distance can be used to create an initial velocity for each stone and, from that, an average trajectory at the most effective angle of release, i.e. 45° from the horizontal, can be calculated. The mathematics relating to the construction of a trajectory are not complex (Sadler and Thorning 1996). When the motion of an object from point A to point B is considered, only two pieces of information are required, the distance travelled and the time taken to traverse that distance. Both these were recorded for each sling stone and the average then calculated. The mean and median were also derived to confirm that these were within acceptable limits of the mathematical average (a variance of 5% or less was accepted). Throughout these calculations, the effect of gravity is taken to be 9.81 ms^{-2}.

The initial velocity of each stone was calculated using the following:

s = Distance travelled by the sling stone, measured to the nearest 100mm.

t = the time taken from the release of the stone until it struck the ground. This was measured to the nearest 100th of a second. There is a certain latitude relating to when the stone actually left the sling pouch. By undertaking such a large number of measured casts, this problem should have been averaged out.

U = the horizontal motion of the stone

Once the distance travelled and the time taken for the object to travel that distance are known, U can be calculated by transposition.

$s = U \times t$

therefore $U = s \div t$

V = the vertical motion of the sling stone

The equation ($s = ut + \frac{1}{2} at^2$) gives the vertical motion of the object. This simplifies as $0 = (V \times t) - \frac{1}{2} (9.81)t^2$; the constant 9.81 is gravity. Therefore by transposition this equation is simplified to:

$$V = \frac{4.9 \times t^2}{t}$$ (the constant 4.9 is half of gravity)

The speed of projection =. $(U^2 + V^2)$

The angle of projection □ above the horizontal is calculated as follows:

$$\tan \frac{V}{U}$$

Example

Sling stone number seven travelled 77 metres in 19.4 seconds.

$$U = \frac{77}{3.97} = 19.4\ ms^{-1}$$

$$V = \frac{4.9 \times 3.97^2}{3.97} = 19.45\ ms^{-1}$$

speed of projection =. $(19.4^2 + 19.45^2)$ = .(754.66) =27.47 ms^{-1}. The speed of projection is the initial velocity at which the projectile leaves the sling pouch.

The angle of projection $= \frac{19.45}{19.4} = 45°$ from the horizontal

Once the initial velocities had been calculated, the arithmetic mean (average) was derived. This proved to be 25.38 m s^{-1}. This was then used to calculate a maximum trajectory for this averaged initial velocity. As with any projection, an angle of 45° from the horizontal allows the missile to travel the greatest distance for any given input of energy. The shape of the trajectory was calculated using a number of points at 0.5 second intervals. The resulting equation is set out below.

Equation of trajectory

where

u = Initial velocity (average at 25.38 m s^{-1})

∞ = Angle of release

t = time in fractions of a second (0.5 second)

g = gravity (9.81ms^{-2})

Vertical distance (y) Horizontal distance (x)

$u_v = u \sin\infty$ $u_h = u \cos \infty$

$t = t$ $t = t$

$y = (u \sin\infty)t - ½gt^2$ $x = (u \cos\infty)\, t$

at 45° of release

$\therefore y = (25.38 \text{ x } 0.707)t - ½gt^2 x = (25.38 \text{ x } 0.707) = 17.94$

$y = (17.94)t - 4.905t^2$ $x = (17.94)t$

Time (seconds)	Vertical distance (metres)	Horizontal distance (metres)
0.5	7.74	8.97
1.0	13.03	17.94
1.5	15.87	26.91
2.0	16.26	35.88
2.5	19.26	44.85
3.0	9.675	53.82
3.5	2.70	62.79
3.65 (*interpolated time)*	0.00	65.66
4.0	-6.72	71.76
4.5	-18.59	80.73
5.0	-32.93	89.7
5.5	-49.70	98.67
6.0	-68.94	107.64
6.5	-90.62	116.61
7.0	-114.77	125.58
7.5	-137.76	134.55
8.0	-170.4	143.52

Table 5.2 *Trajectory calculation for average initial velocity*

The calculations were continued to below the horizontal, so that any sling shot cast from an elevated position (e.g. a hillside or rampart) would also be described. The distance of 143.52m is the point at which the trajectory began to return and curve under itself. In this theoretical model, it would be safe to assume that at this point the sling shot would drop vertically, as gravity alone begins to act as the stone has lost all forward momentum (Figure 4.15).

Although a useful guide to the flight of a sling shot, the above calculation, in common with most theoretical models, assumes that the object has no mass and is therefore immune to the effects of drag. From field observations outlined above, this appears not to be the case for a sling stone. Drag is a complex mathematical problem, so much that many authors simply ignore its effects.

It cannot be calculated by a simple linear equation (Kibble 1985, 46) and it can only be successfully determined through experimentation. However, an approximate value for the maximum range of a projectile can be achieved, if certain assumptions are made (*ibid,* 46-7). A number of factors influence the effect of atmospheric drag on flight path: the shape of the projectile; the moisture content of the air; variation of gravity at differing heights; and the varying density of the atmosphere. The variability of these factors is likely to be insignificant in relation to a small projectile like a sling stone.

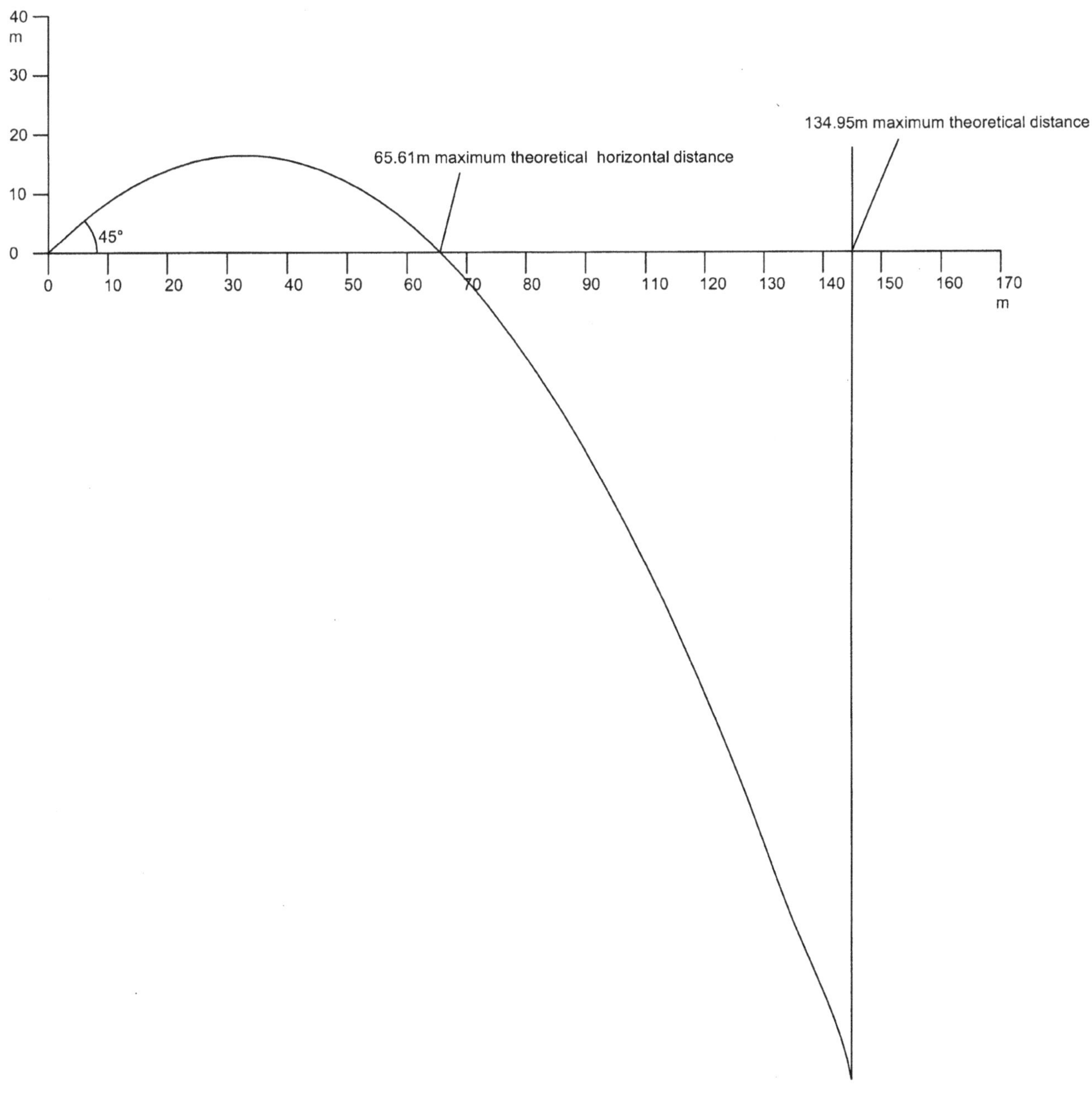

Figure 4.15 Trajectory described at 25.38ms^{-1}

The resistance that an object encounters when travelling through any medium can be expressed as a Reynolds Number. A Reynolds Number is a non-dimensional number that allows a mathematical expression for the skin resistance of the object to be made. In essence, the Reynolds Number is inversely proportional to the coefficient of drag; the higher the Reynolds Number, the lower the effects of drag. Considerable experimentation relating to the flight paths and the effects of drag on spherical objects has been undertaken. A theoretical spherical sling stone was tested to ascertain its probable Reynolds Number. Although not representing absolute reality, such a calculation should indicate whether drag has an effect, and whether it was worth pursuing this matter further.

From the data gathered, the average volume of the sling stones used was calculated as 227.2mm^3, which gives a diameter of 35.2mm. The Reynolds Number is deduced by the following:

$$Re = \frac{Vd}{\mu}$$

Where V = the initial velocity of the object

d = the diameter of the object
μ = a constant of 0.000015

Therefore, the probable Reynolds Number of a spherical sling stone would be:

$$Re = \frac{25.38 \times 35.2}{0.000015} = 59558440 \text{ or } 59^6$$

Any Reynolds Number above 10^6 provides a drag coefficient (C_D) of less than 0.1. At this point, the drag has decreased to a supercritical coefficient and the corresponding flight path is relatively close to potential theory (Hoerner 1965, 3-9). The probable drag of the spherical sling stone is so low that its flight path would correspond to that predicted for a massless particle.

It has therefore, been assumed that, although variation is bound to occur between the theoretical spherical object and the approximately ovoid sling stones used in the experimentation, drag has no significant impact on the predicted path of a sling stone. So the trajectories will be calculated using classical mechanical theory. The following statements then can be made.

The maximum distance that a massless particle projected at a given initial velocity can travel in a vacuum is calculated as $R_{max} = \frac{V_o^2}{G}$

Where: R_{max} = Maximum theoretical range

V_0 = Initial velocity

G = Gravity at 9.81ms $^{-2}$

Therefore $R_{max} = \frac{25.38 \text{ m s}^{-1}}{9.81} = 65.66\text{m}$

Calculation of Impact

Impact equals the mass of the object x its velocity x the duration of impact. The last element is difficult, if not impossible, to calculate; however, assuming a duration of only 1/500 of a second (Cotterell and Kamminga 1990, 100), the following impact in newtons is achieved: 0.056kg x 25.38 x 500 = 710.64N or 72.44kg (taking gravity to be 9.81m s $^{-2}$). The exact area of impact of an approximately spherical object is difficult to determine, as it will increase with depth of penetration. However, assuming the same dimensions used for the exploration of drag, 0.2% of the surface area of the sling stone would be sufficient to fracture a human skull. The damage that can be caused by a stone sling shot can be seen in the fractures on a number of skulls recovered from Balboa Park, Peru (Wells 1964, 49).

APPENDIX 2.4

THE RESULTS OF THE EXPERIMENTAL WORK (THE SPEAR)

The density of ash like any timber is directly related to its moisture content. It has been assumed that this would be 15%. This can be achieved without modern kiln drying and gives the timber strength and resistance to decay. At this moisture content, ash has a mass of 704.8kg per cubic metre, or 0.0000704g per cubic millimetre (James 1989, 181). To calculate the mass of the shaft, it has been assumed that the overall diameter of the spear socket measured to its outside edge would have represented the constant shaft diameter. This is the case for those rare examples where the shaft has been partly preserved (Fox 1946, 74, 98; Stead 2003, 59).

Take for example, small find number 2.286 from Danebury (Cunliffe and Poole 1991, Figure 7.18).

The weight of this spearhead is 27.2 grams and the socket has a cross-sectional area of $284mm^2$.

Assuming an 'ideal weight' of 600g, a mass of timber of $0.0000704g/mm^3$, and a balance point at one third of the length of the spear, the following can be deduced:

The mass of the shaft would be 284 x 0.2g per 1mm of length.

The first third of the spear would have a mass of 200g - 27.2g = 172.8g

Therefore the first third of the timber shaft can be calculated by dividing the remaining mass by the mass per unit of length:

$$\frac{172.8g}{0.2g \text{ per 1mm unit of length}} = 863.3mm.$$

The remaining two thirds of the shaft would have a theoretical mass of 400g and therefore a theoretical length of:

$$\frac{400g}{0.2g \text{ per 1mm unit of length}} = 2000mm$$

The total theoretical length of the spear would be composed of 100mm (length of the spearhead) plus 863.3mm (first third of the spear shaft) plus 2000mm (remaining two thirds of the spear shaft) = 2963.3mm, 2.93m.

A direct measure of a shafted projectile's ability to fly with any degree of precision can be gauged by calculating its spline or stiffness. This measures the ability of a shaft to regain its stability in flight once it has been released. If it is not stiff enough, the vibrations created in casting will cause it to oscillate too much and produce an unstable flight pattern. It is possible to cast such a spear, and at short range strike a target accurately, but, as the optimisation of spear mechanics is easy to achieve, it is unlikely that any culture would deliberately manufacture inherently unstable throwing spears.

The period of free vibration of a spear with a uniform diameter can be calculated by:

$$T = 0.25 \left[\frac{mL^3}{EI}\right]^{0.5}$$

Where m = the mass of the spear
L = the length of the spear
E = the Young's Modulus of the shaft
I = Second moment of area ($\pi d^4/64$)

Therefore, the period of free vibration of the above spear could equal:

$$T = 0.25 \left(\frac{0.6 \times 4.28^3}{17.3 \times 10^9 \times 6.397 \times 10^{-9}}\right)^{0.5}$$

This gives a period of free vibration of 0.515sec.

For a spear with a spline matched to the strength of the cast, an approximate value of 0.14secs can be calculated, assuming that the average velocity of a cast is about $22m/s^{-1}$, and the length of the forearm of the caster is approximately 0.38m (Cotterell and Kanninga 1992, Figure 7.3). This is derived from the following calculation:

$2l/v \times 0.25$
where l = the length of the caster's forearm and v = the initial velocity of the spear.

Thus the calculated period of free vibration is below that required for the proposed spear, making it inherently unstable in flight.

Below are the tabulated results for the calculations on all the complete spearheads recovered from Danebury, encompassing all ceramic phases:

Small finds No.	cp	Spearhead weight (g)	Shaft diameter (mm)	Calculated length (m) at 600gms mass	Calculated period of free vibration (sec)
2.100	7	55.7	15	3.54	0.12
2.101	6	30.5	16	4.85	0.88
2.102	3	17.5	17.5	4.14	0.58
2.282	3	58	15	5.2	1.12
2.283	7	114.8	21	1.63	0.099
2.285	8	94	20	2.68	0.23
2.286	3	27.2	17	4.28	0.515
2.287	uns	45.9	19	3.38	0.36

Table 5.3 *Calculated spearhead statistics*

Only one of the spearheads (2.100) recovered from Danebury has a stable flight pattern at a mass of 600g, and an initial velocity of $22m/s^{-1}$. It may therefore be inferred that spearheads from other sites that fall within the range of sizes provided by the Danebury examples would also have unstable flight characteristics at approximately this mass. If this is correct, it may be inferred that spears with small heads had a mass of less than 600g. This could mean that they were either designed as light throwing spears or used as short-range thrusting spears, similar to those recorded by classical authors being used by the iron-poor Germans of the 1st Century AD (Tacitus *Germanica* 6; Todd 1992, 37). Taking into consideration the evidence from the Late Iron Age burials, it is most likely that the small spearheads at Danebury where also associated with short spears.

Appendix 3.1

Hillfort Survey

Method

The first requirement of the survey was to produce a measured transect that will represent a profile cartographically. For this level of information, indicative hatched sections are of insufficient quality. In order to construct a profile accurately, a single level projection must be created. From this, a series of theoretical measurements to the ground surface is then taken, creating a two-dimensional co-ordinate for each point that has been plotted. The drawn profile represents, as far as possible, one continuous running section. The projected line was at a nominal right angle to the first bank of the hillfort's construction. This, given the curvilinear nature of many hillfort defences, is not always easy to determine, but deviation from the required section was kept to a minimum.

The survey method had to conform to the following criteria:

1. To allow the maintenance of a continuous running transect over the hillfort's surface.

2. The level of acceptable variation was to be ± 5mm in the vertical plane and ± 25mm in the horizontal plane for each measurement taken (giving a maximum horizontal deviation of 2.5m per 100m traversed.

3. To measure linear distance over the undulating ground of the hillfort whilst measuring the vertical differences at the same time.

4. Due to time limitations survey work had to be achievable within realistic parameters, a major consideration influencing the choice of survey method.

Initially, the use of optical survey equipment was considered, using the stadia method of tachometric surveying (Clark 1969, 379-82). However, this method was not a practical option, for the following reasons:

1. The surveying staff was often out of view as it only extends to four metres; the combined height of the ditches and banks was often in excess of this.

2. The distance at which accurate readings on the staff could be made is limited to approximately 50m. Many of the hillfort 'defences' extend significantly beyond this range. Although it is possible to measure linear distance by tachometric survey, the greater the distance that the chainman moves from the first station, the greater the inaccuracy that enters the recording.

3. In order to retain a straight transect, the level would have to be moved, which proved to be difficult due to the hillfort's topography, and resulted in unacceptable inaccuracy.

The Tape and Baton Survey

Tape and baton is at first sight simple but it does encompass all the requirements detailed above. The technique is as follows:

1. A transect line is laid out using ranging poles and a military sighting compass. The transect line is delineated with a 30m tape. This represented the distance over which the linear transect could be laid out with confidence, given the often vertiginous topography, while representing a reasonable distance over which to work. By locking the compass at a given angle relative to Magnetic North, the transect could be laid out in various stages, obviating the need to set the line out in one continuous operation. The tape was used for measuring distances in areas where the fall was less than 50mm in a metre. Otherwise it was primarily used as the guiding line for the transect.

2. The linear measurement and vertical height were measured using a bubble level, baton and surveying staff. The bubble level and baton consists of a 1.5m length of 25mm hollow square section aluminium bar divided into 10mm blocks for measuring distances less than 1.5m on a slope. 1.5m was chosen as a convenient length that could be managed down the often steep slopes of the earthworks while representing a reasonable length over which to measure the fall or rise of the ditches and banks. A bubble level was attached to the baton (see

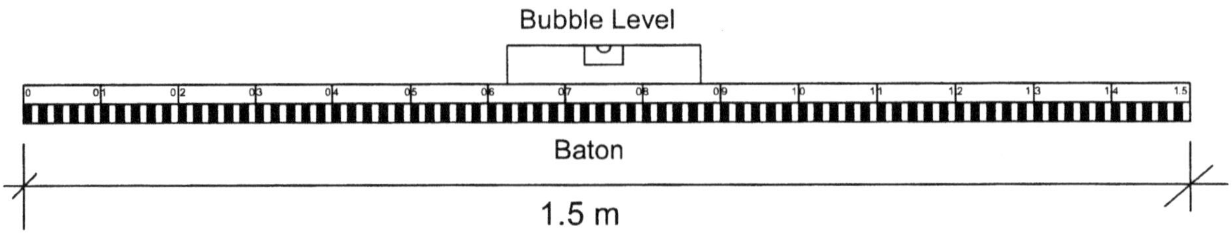

Figure 5.1 The Baton and Bubble Level

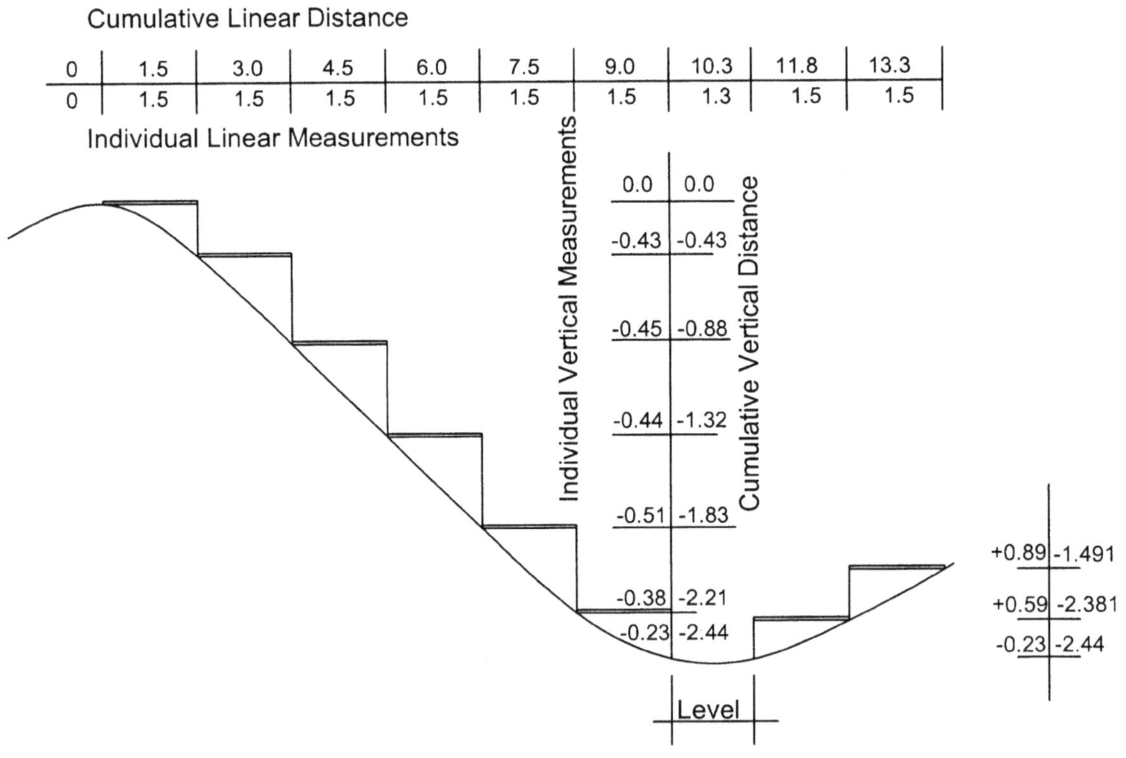

Figure 5.2 Construction of earthwork profiles

Figure 5.1). Once the baton was level, a reading to within ±5mm was taken from the staff. The baton allowed a constant linear measurement to be recorded over the varied topography of the earthworks. A vertical measurement was taken for each section of linear distance. The staff was moved to the end of the baton each time a measurement was taken, and then the baton was placed at the base of the staff before the process was repeated, until the whole profile had been recorded (see Figure 5.2). All measurements were recorded in metres or decimals thereof. Each profile was recorded on a table, an example of which is shown below:

Hillfort Name: Bury Ditches
Date: 30/07/99 Section location: Eastern Entrance
Sheet No. 1

Measured length	Cumulative length	Measured height	Cumulative height
1.5		-0.450	
1.5		-0.550	
1.5		-0.555	

The measured height was recorded as a minus measurement, when traversing down slope, and as a positive measurement, when measuring up slope. No distinction was made for the linear distance.

The linear distances were added together, giving a continual reference for any point along the total level distance traversed. The vertical distance was calculated by adding all the distances represented by a minus measurement, and those represented by a positive record were then subtracted from this running total, as these represent an upward and, thus, reducing measurement. This gives a two dimensional reference for each point of the profile. For an example see below:

Hillfort Name: Bury Ditches
Date: 30/07/99 Section location: Eastern Entrance sheet 1

Measured length	Cumulative length	Measured height	Cumulative height
1.5	1.5	-0.450	-0.450
1.5	3.0	-0.550	-1.00
1.5	4.5	-0.555	-1.555

Once the coordinates for each location had been calculated, they were drawn by plotting them, using AutoCAD release 2004 for Windows. The profile was drawn using an application know as a 'polyline'. This, in effect, creates a single continuous line, which is not divided into a series of individual segments, each corresponding to a given locale. The advantage of this is that, normally, such a drawing is shown as a series of straight lines. Clearly, this is not the case for eroded features made of a natural material that has achieved an equilibrium or angle of repose. Once all the points have been plotted, a separate feature of AutoCAD can be applied to the polyline; it is 'splined'. This creates a continuous curvilinear line that more accurately mirrors the actual ground surface. A certain level of detail relating to the exact position of each coordinate is lost, as the programme calculates the new shape of the line relative to all the points plotted. However, this loss amounts to no more than approximately 0.1 of a drawing unit or 100mm in real life. Considering the scale at which the plans are plotted, i.e. no larger than 1:200, such a loss is negligible.

APPENDIX 3.2

RESULTS OF THE HILLFORT SURVEY

Site: Battlesbury Camp		Grid Ref. ST 894 456	
Reference: Cunnington 1923b			
Sling projectiles recovered from excavation: Yes			
Drawing No. 5.1			
Section	Would any assailant have to enter an area defined by external earthworks before being able to strike the innermost bank?		Distance factor in favour of the defenders
A - A	No		1.4
B - B	No		2.0
C - C		Yes	1.5
D- D	No		1.9

Additional notes:

The survey commenced at the southern entrance, as this appeared to be the most elaborate. Protected by a large double bank and single ditch, the outer bank seemed too large to be the result of cleaning the ditch and would appear to have been purpose built as part of the defences. Despite this level of defence, the entrance seems to have been vulnerable to attack from the sling.

The same appears to be true of Section B - B. The approach to these banks is steep, perhaps making it a difficult location to assault.

Section C - C faces Scratchbury Hill. The area in front of the earthworks seems to have remained as pasture, unlike the eastern side, which has been ploughed. Here, outside the main large earthworks, a smaller bank, approximately 300mm high and 1000mm wide, follows the line of the outer bank for the entire length of the pasture (about 150m), but may have been destroyed within the ploughed area. It is not possible to be certain that this feature was constructed as part of the original fortification and it may represent a later hedge bank or similar feature (the bank lies about 10m from the toe of the outer bank of this hillfort and, as such, seems to make little sense as a later agricultural feature, since the main bank would have easily served any requirements of definition). However, when the experimental trajectories were applied to this section, it was found that any assailant would have had to cross this bank into the area defined by it and the larger earthworks. The slope of the hill on which Battlesbury Camp has been constructed is at its shallowest along this side, and may have been the original designers of the fort felt that a further defensive feature was required.

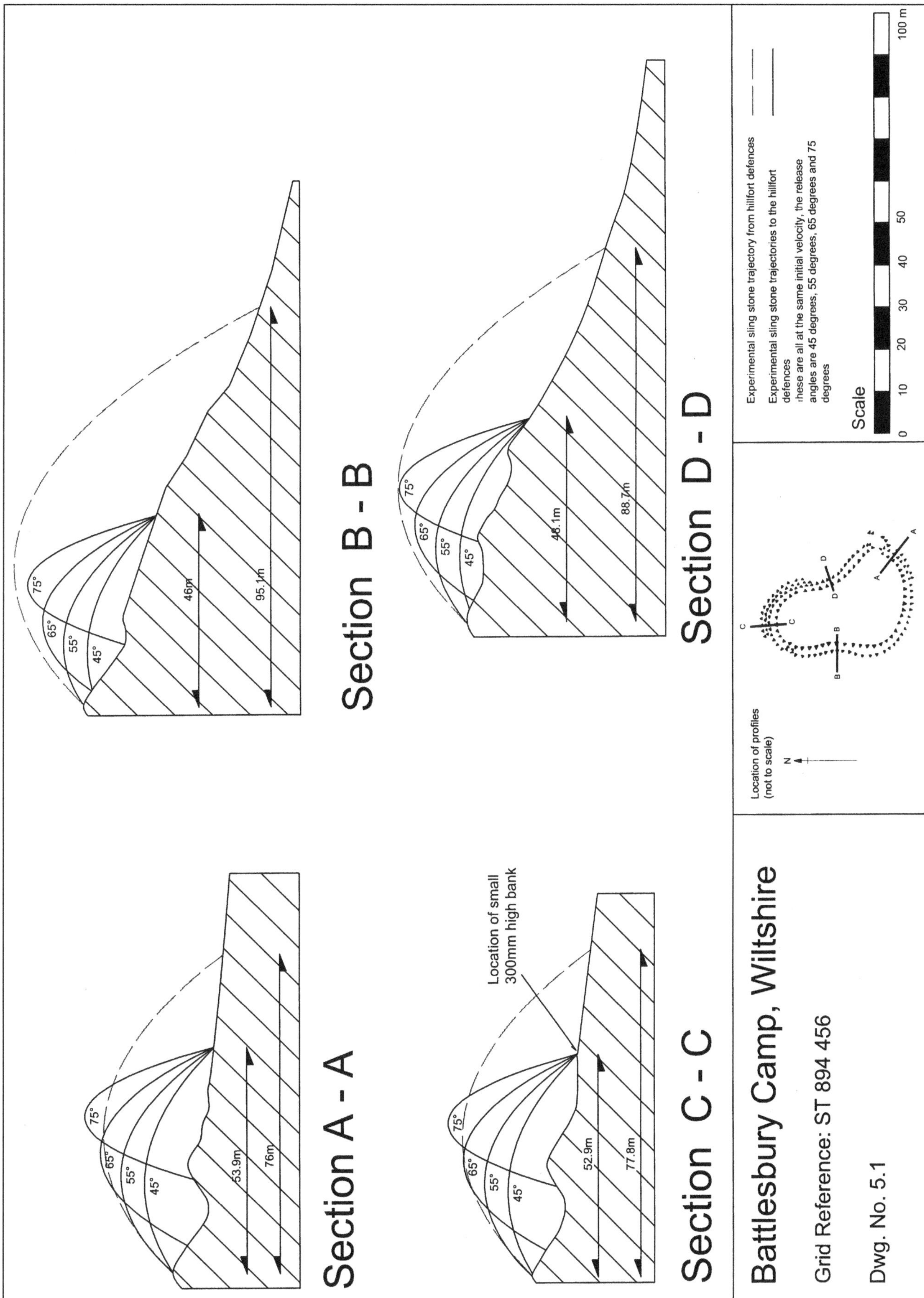
75°
65°
55°
45°
53.9m
76m
Section A - A
75°
65°
55°
45°
46m
95.1m
Section B - B
Location of small
300mm high bank
75°
65°
55°
45°
52.9m
77.8m
Section C - C
75°
65°
55°
45°
48.1m
88.7m
Section D - D
Battlesbury Camp, Wiltshire
Grid Reference: ST 894 456
Dwg. No. 5.1
Location of profiles
(not to scale)
N
Experimental sling stone trajectory from hillfort defences
Experimental sling stone trajectories to the hillfort defences
rhese are all at the same initial velocity, the release angles are 45 degrees, 55 degrees, 65 degrees and 75 degrees
Scale
0
10
20
30
40
50
100 m

<table>
<tr><td colspan="2">Site: Bodbury Ring</td><td colspan="2">Grid Ref. SJ 292144</td></tr>
<tr><td colspan="4">Reference: No excavation has been recorded</td></tr>
<tr><td colspan="4">Sling projectiles recovered from excavation: No</td></tr>
<tr><td colspan="4">Drawing No. 5.2</td></tr>
<tr><td>Section</td><td colspan="2">Would any assailant have to enter an area defined by external earthworks before being able to strike the innermost bank?</td><td>Distance factor in favour of the defenders</td></tr>
<tr><td>A - A</td><td>No</td><td></td><td>1.04</td></tr>
<tr><td>B - B</td><td>No</td><td></td><td>2.4</td></tr>
<tr><td>C - C</td><td>No</td><td></td><td>2.6</td></tr>
<tr><td>D- D</td><td>No</td><td></td><td>2.8</td></tr>
<tr><td colspan="4">Additional notes:
Bodbury Ring is more of an inland promontory fort than a 'typical' hillfort. The northern side of the spur has a ditch and bank cut across it, enclosing an area of approximately 0.7ha. The surviving depth of the ditch is about 1.85m, suggesting that this was never a particularly deep or impressive structure.

On the western, eastern, and southern flanks of the fort, the hill naturally falls away steeply. Here, no ditch or bank was constructed. Instead, it appears that a section of the hill itself has been dug away to create a level 'terrace' about 1.5m wide that encompasses the entire circuit of the promontory. The steepness of the slope (an average of 1:4) surrounding the spur of Bodbury Ring appears to have been sufficient to act as a deterrent. The 'terrace' would allow effective use of the sling, as there are no obstructions to the front or rear.

The northern bank and ditch presents more of an enigma. Despite extensive examination of the ground to the front of this area, no other features were seen. Therefore, within the constraints of this study, it has to be assumed that the flat area to the north was devoid of manmade obstructions.</td></tr>
</table>

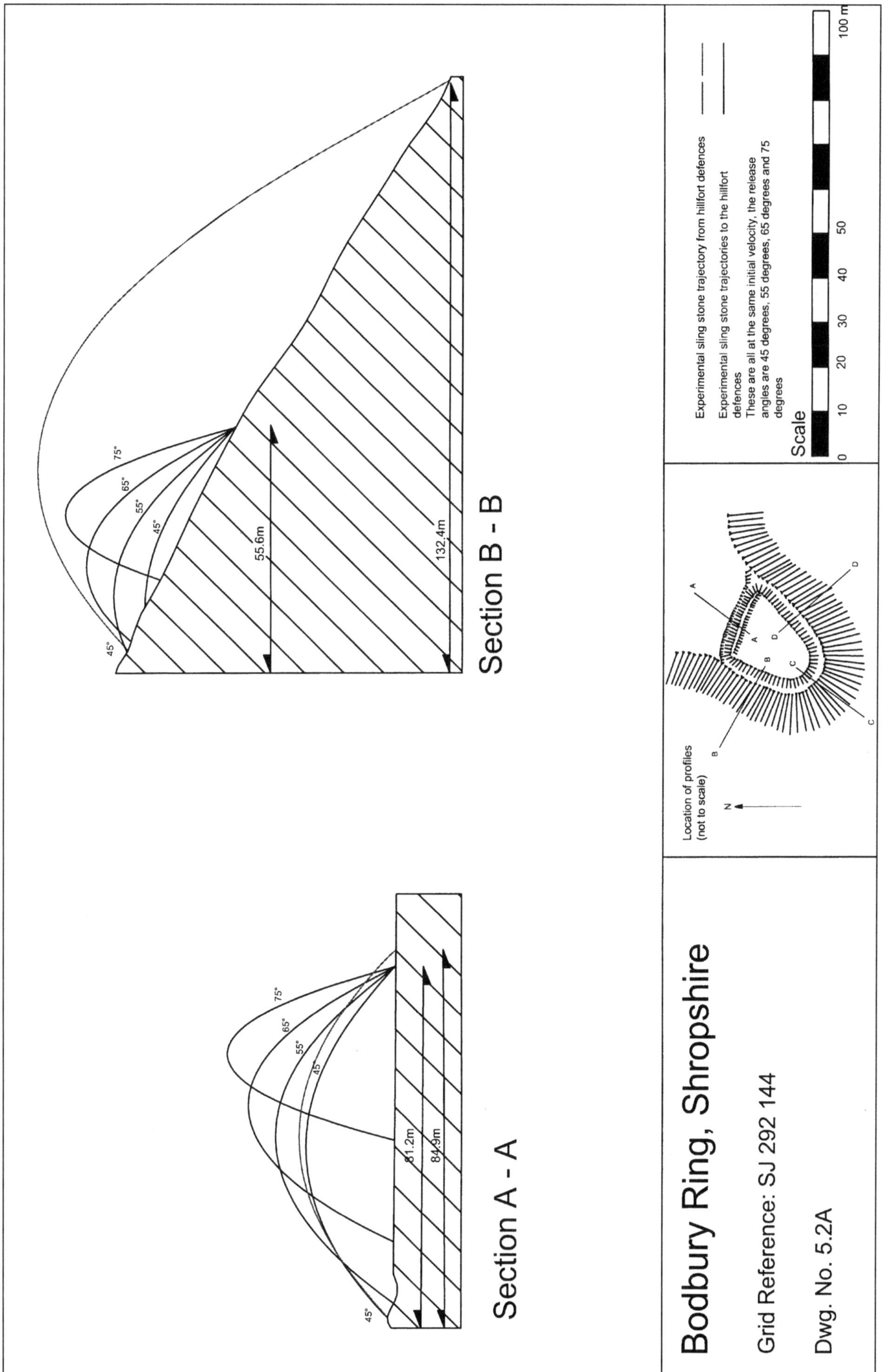
Section A - A
45°
75°
65°
55°
45°
81.2m
84.9m
Section B - B
45°
75°
65°
55°
45°
55.6m
132.4m
Experimental sling stone trajectory from hillfort defences
Experimental sling stone trajectories to the hillfort defences
These are all at the same initial velocity, the release angles are 45 degrees, 55 degrees, 65 degrees and 75 degrees
Scale
0
10
20
30
40
50
100 m
Location of profiles (not to scale)
N
A
B
C
D
Bodbury Ring, Shropshire
Grid Reference: SJ 292 144
Dwg. No. 5.2A

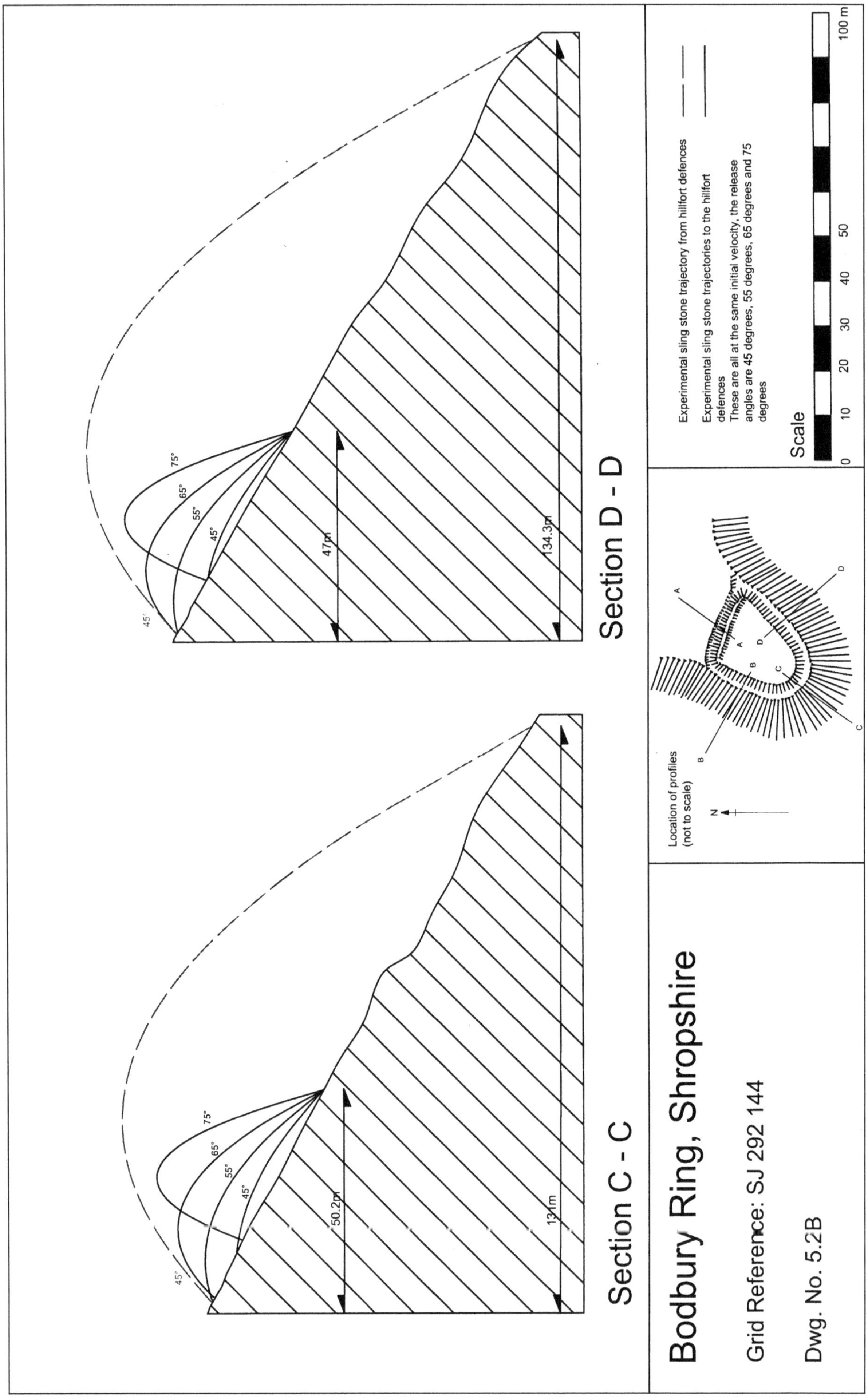
Section C - C
75°
65°
55°
45°
45°
50.2m
131m
Section D - D
75°
65°
55°
45°
45°
47m
134.3m
Experimental sling stone trajectory from hillfort defences
Experimental sling stone trajectories to the hillfort defences
These are all at the same initial velocity, the release angles are 45 degrees, 55 degrees, 65 degrees and 75 degrees
Scale
0 10 20 30 40 50 100 m
Location of profiles (not to scale)
N
A
B
C
D
Bodbury Ring, Shropshire
Grid Reference: SJ 292 144
Dwg. No. 5.2B

Site: The Breiddin		Grid Ref. SJ 292144	
Reference: Musson 1991			
Sling projectiles recovered from excavation: No			
Drawing No. 5.3 (a and b)			
Section	Would any assailant have to enter an area defined by external earthworks before being able to strike the innermost bank?		Distance factor in favour of the defenders
A - A	No		3.5
B - B	No		2.3
C - C		Yes	1.7
D- D	No		2.0

Additional notes:
The hillfort is a multivallate stone walled construction enclosing approximately 28 ha. Today, the walls are spreads of collapsed rubble extending 15-20m down slope. The discovery of the wall foundation trenches and the calculation of the volume of the rubble have allowed a reasonable assumption to be made relating to the walls' original height (Musson 1991, 33).

Limited experimental rebuilding of the walls during the excavation, using the stones available from the collapse, indicated that there was insufficient stone for the formation of a breastwork (*ibid*, 33). This feature is potentially linked with sling warfare.

The walls are also thick, approximately twice their width to height (5m by 2.2m). The earlier Late Bronze Age rampart is significantly narrower (2.5-3m) (*ibid*, 25-36). Unfortunately no estimate of their original height was possible, due to the Iron Age construction.

Although strongly protected on the south-eastern side by man-made defences, to the west by a natural cliff, and to the north-east by a rocky gully, the hillfort lacks any constructed defences on its north-western flank. Instead, it appears as though the constructors of the hillfort relied entirely upon the natural slope of the hill. Therefore, whatever means of defending The Breiddin were employed, it must have been deemed sufficient to cope with this apparent accessibility.

The main entrance of the fort faces almost due west and is approximately 70m in length. The funnelled nature of the entrance with its associated ditches (O'Neil 1937) makes this a particularly strongly defended area of the hillfort (Musson 1991, 8).

The walls have been reconstructed assuming a general height of 2.1m from ground level to the front of the wall (*ibid*, 180). No attempt has been made to reconstruct the silted ditch to the front of the wall. At section A-A, where no wall had been constructed, the defenders significantly out-range any assailants. The assailant slingers had to be within 40m of the wall before they could strike it. The defenders had a range advantage of a factor of three. At sections B-B and D-D, only a single wall exists: here, the defenders have a range advantage of a factor of two. At section C-C, an assailant would have to enter the defined space between the walls before being able to cast at the inner wall (see Drawing No. 5.3).

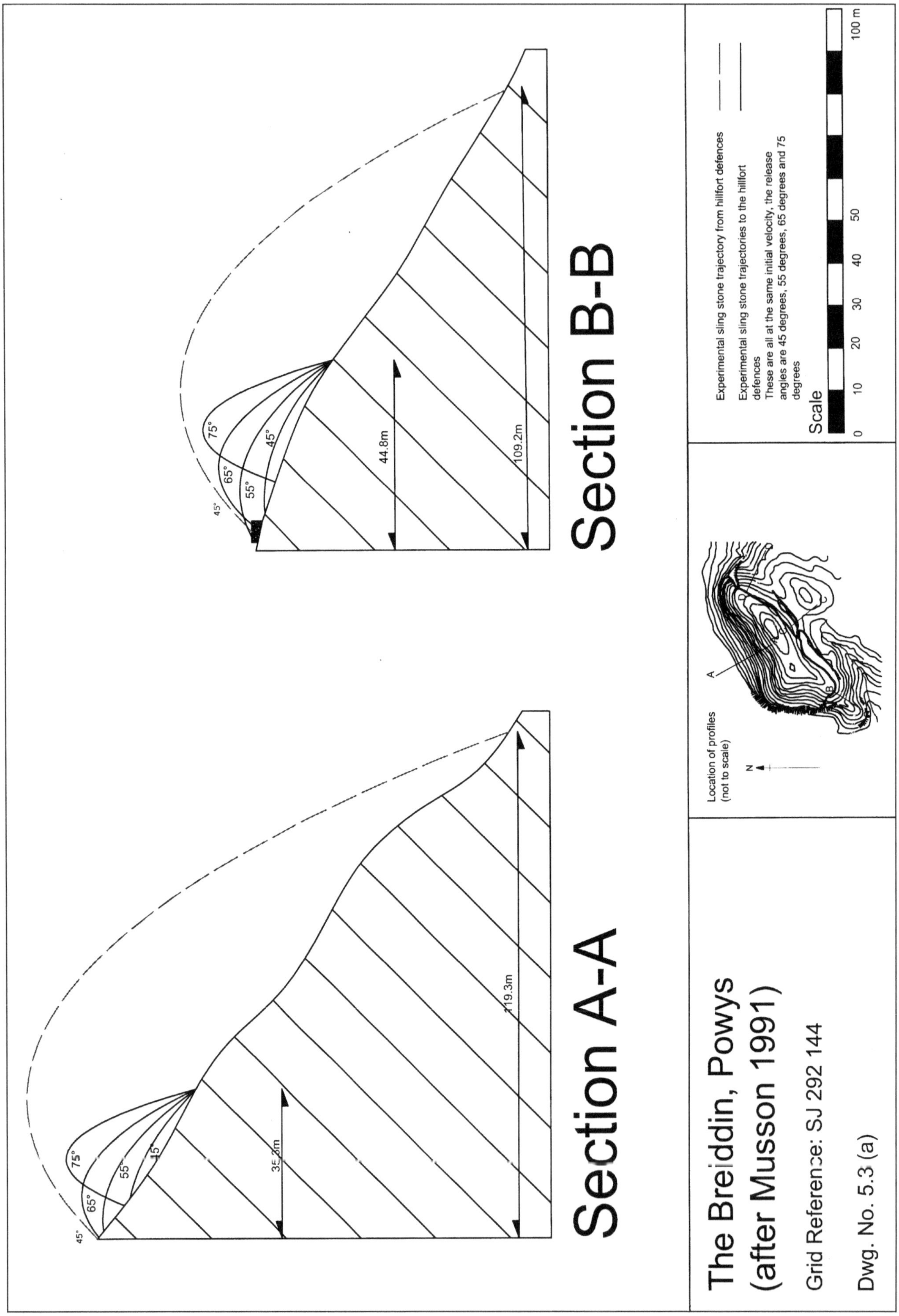

The Breiddin, Powys (after Musson 1991)

Grid Reference: SJ 292 144

Dwg. No. 5.3 (a)

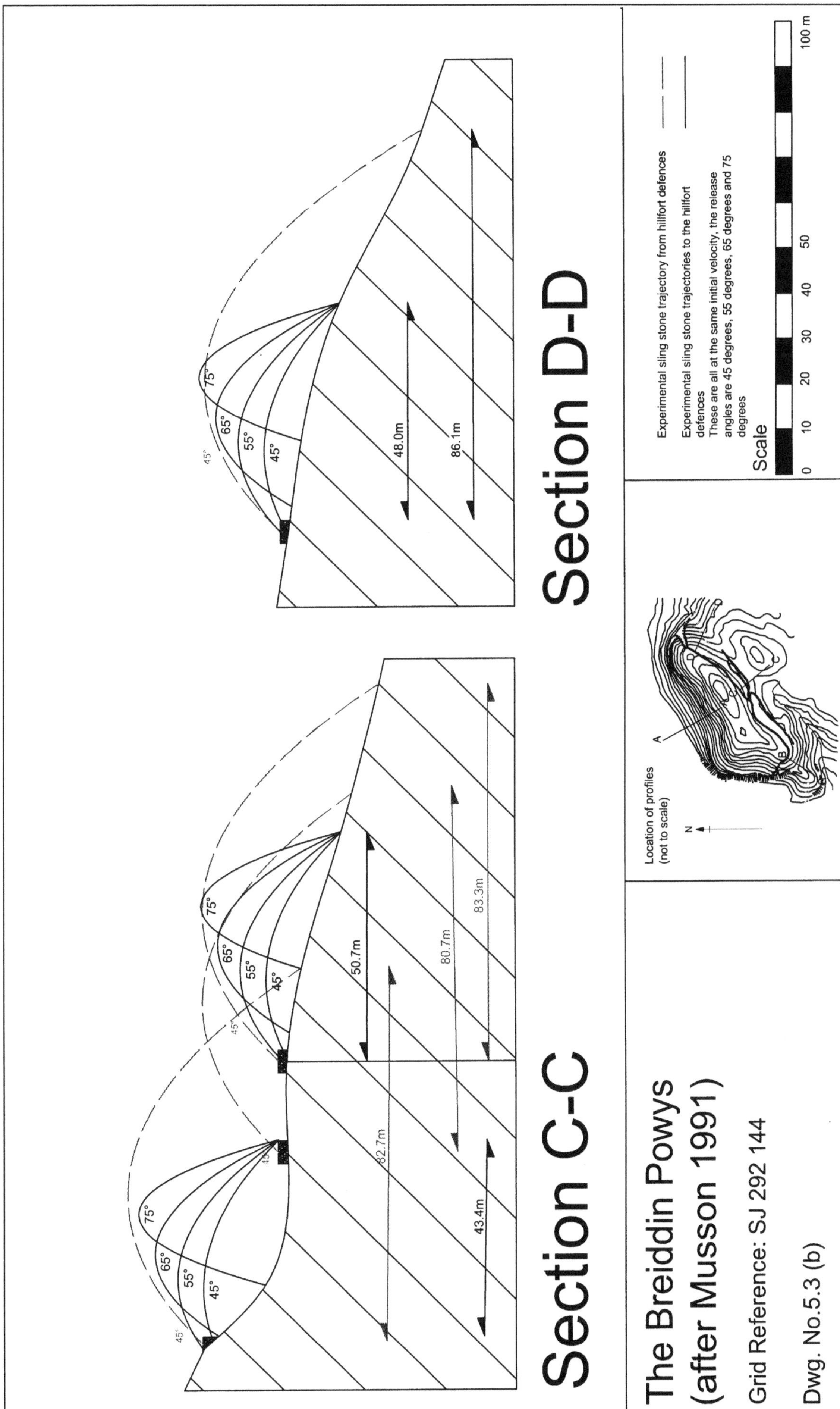
Section C-C
Section D-D
75°
65°
55°
45°
50.7m
80.7m
83.3m
82.7m
43.4m
48.0m
86.1m
The Breiddin Powys
(after Musson 1991)
Grid Reference: SJ 292 144
Dwg. No.5.3 (b)
Location of profiles
(not to scale)
N
A
B
Experimental sling stone trajectory from hillfort defences
Experimental sling stone trajectories to the hillfort defences
These are all at the same initial velocity, the release angles are 45 degrees, 55 degrees, 65 degrees and 75 degrees
Scale
0 10 20 30 40 50 100 m

Site: Bredon Hill		Grid Ref. SO 958 400	
Reference: Hencken 1938			
Sling projectiles recovered from excavation: Yes			
Drawing No. 5.4			
Section	Would any assailant have to enter an area defined by external earthworks before being able to strike the innermost bank?		Distance factor in favour of the defenders
A - A	Yes		1.16
B - B	Yes		1.14
C - C	Yes		1.1
D- D			

Additional notes:

Though Hencken (1938, 7) states that the inner and outer earthworks form two distinct phases of construction, there is, in fact, no stratigraphic evidence to support this assumption. The fact that multivallation is now known to be a feature of early hillforts supports the interpretation that both the inner and outer bank were built at roughly the same time.

Hencken dates the fort to the 1st Century BC on the evidence of duck-stamped pottery (*ibid*, 97, Figure 14). However, typologically these vessel forms can be shown to date to the fifth-first century BC (Cunliffe 1991, Figure A:18; Gibson 2002, 63), suggesting that, like many other hillforts, Bredon Hill was constructed at some time in the early Middle Iron Age.

Site: Bury Ditches		Grid Ref. SO 327 837	
Reference: None			
Sling projectiles recovered from excavation: No			
Drawing No. 5.5			
Section	Would any assailant have to enter an area defined by external earthworks before being able to strike the innermost bank?		Distance factor in favour of the defenders
A - A	Yes		1.1
B - B	Yes		1.3
C - C		No	2.2
D- D	Yes		1.5

Additional notes:
The multivallate hillfort of Bury Ditches was constructed on the summit of a hill that rises some 70m above the surrounding area, and is approximately 500m from the nearest modern water source, which lies to the north-west.

The fort has two entrances, opposed at the north-eastern and south-western ends. Although, from the plans available, it appears that the eastern entrance is the most elaborate, with an extra bank and ditch to the north side, on the ground it is the western approach that appears to provide the easiest access and seems to retain the most obvious roadway. This ditch forces traffic entering the hillfort from this direction to present their right (presumably unshielded) side to the main earthworks of the enclosure. This is perhaps more of a symbolic gesture, as it would have been perfectly possible to assail the hillfort without traversing the road, as the ditch that defines it is deep but not impassable.

Both the eastern and western entrances have in-turned banks leading into the hillfort. On the western approach, the second bank also turns into the interior on its southern side; on the northern side it turns outwards, creating a longer narrow passage. This would suggest, in the absence of any published excavation, that the first two banks, if not actually constructed at the same time, are, at least, broadly contemporary.

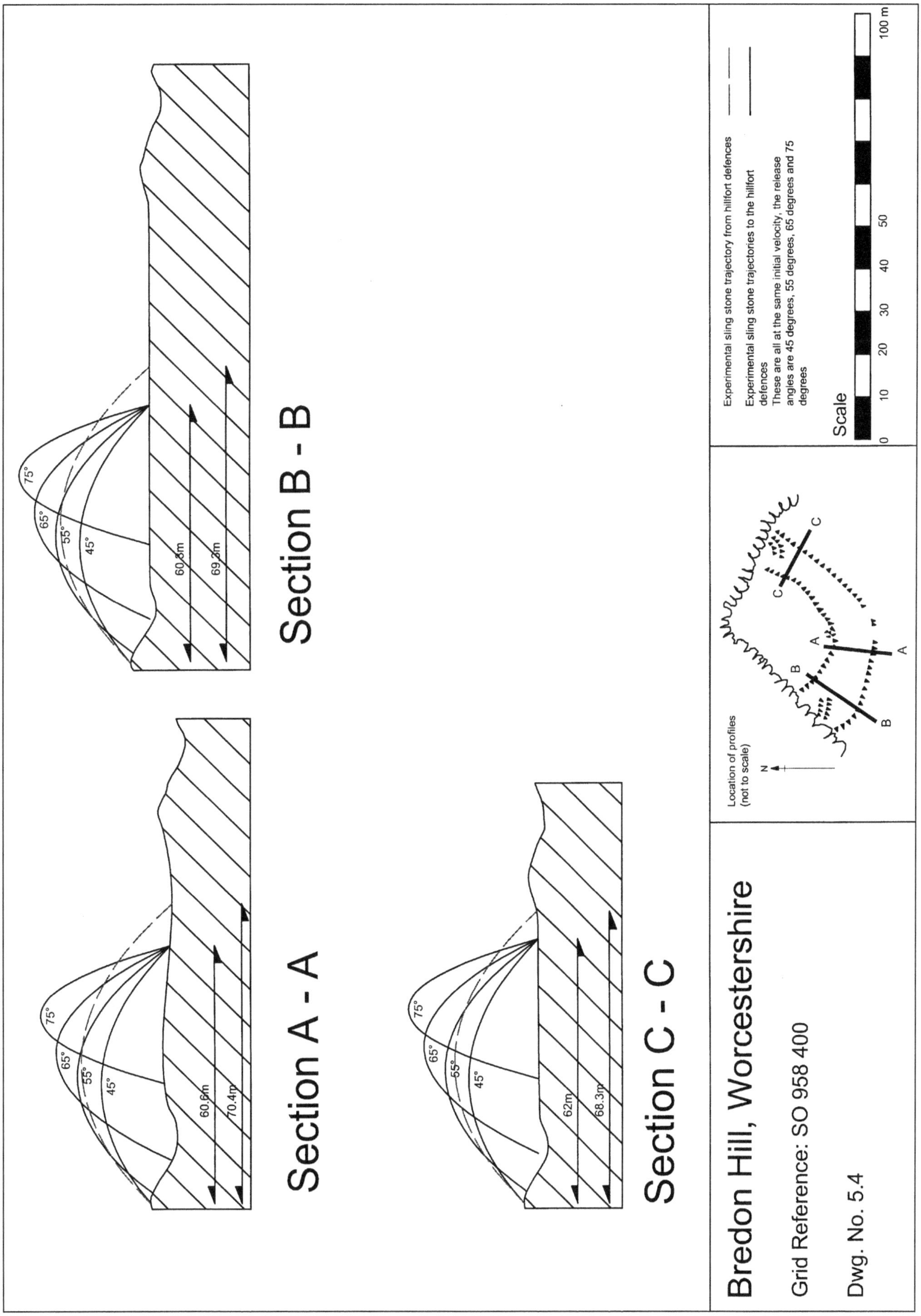
Section A - A
75°
65°
55°
45°
60.6m
70.4m
Section B - B
60.3m
69.3m
Section C - C
62m
68.3m
Experimental sling stone trajectory from hillfort defences
Experimental sling stone trajectories to the hillfort defences
These are all at the same initial velocity, the release angles are 45 degrees, 55 degrees, 65 degrees and 75 degrees
Scale
0
10
20
30
40
50
100 m
Location of profiles (not to scale)
N
A
B
C
Bredon Hill, Worcestershire
Grid Reference: SO 958 400
Dwg. No. 5.4

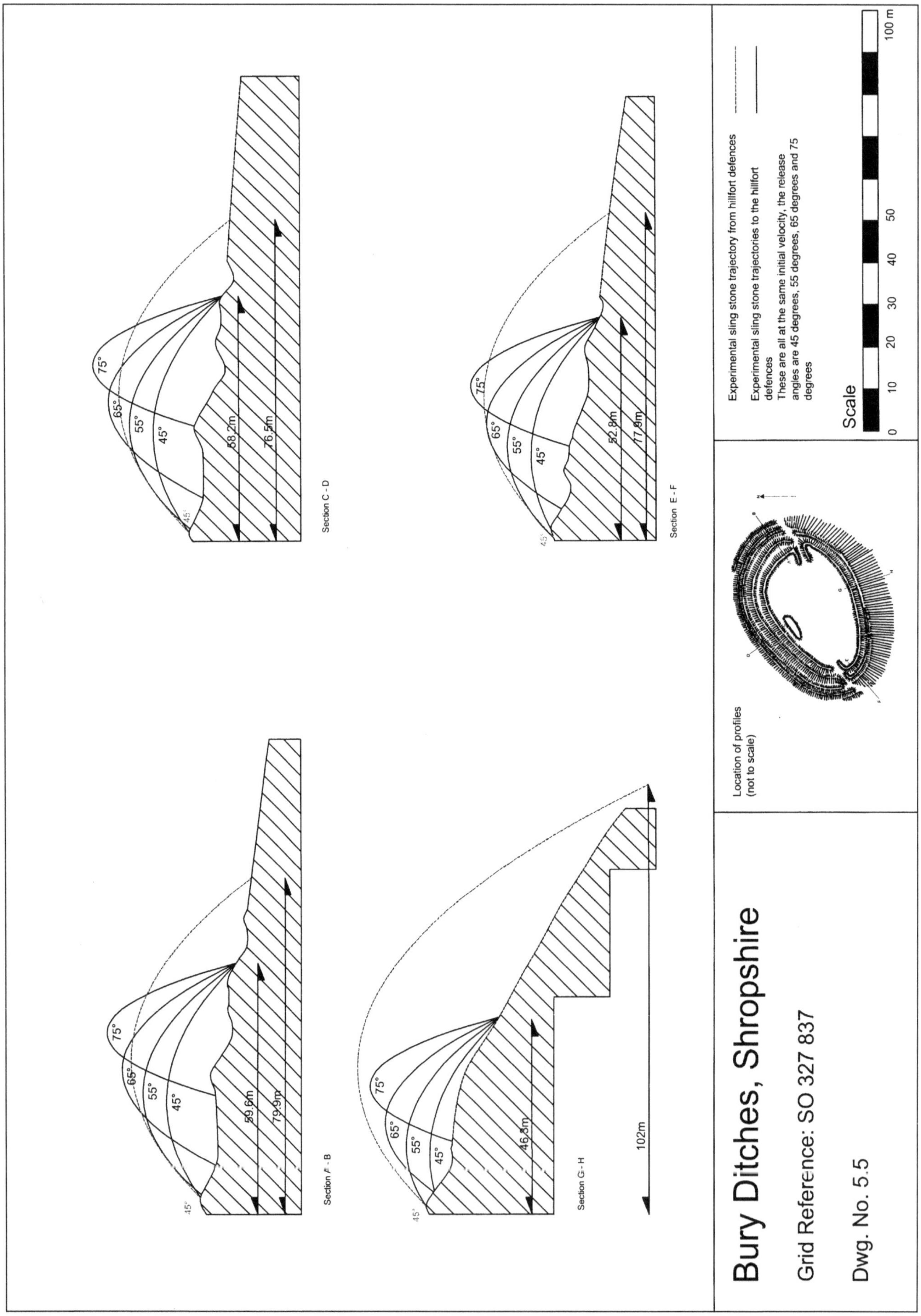

Bury Ditches, Shropshire
Grid Reference: SO 327 837
Dwg. No. 5.5
Location of profiles
(not to scale)
Experimental sling stone trajectory from hillfort defences
Experimental sling stone trajectories to the hillfort defences
These are all at the same initial velocity, the release angles are 45 degrees, 55 degrees, 65 degrees and 75 degrees
Scale
0 10 20 30 40 50 100 m
Section C - D
Section E - F
Section A - B
Section G - H
58.2m
76.5m
52.8m
77.9m
59.6m
79.9m
46.3m
102m
75°
65°
55°
45°

Site: Castle Ring, Oakhill		Grid Ref. SJ 373 012	
Reference: None			
Sling projectiles recovered from excavation: None			
Drawing No. 5.6			
Section	Would any assailant have to enter an area defined by external earthworks before being able to strike the innermost bank?		Distance factor in favour of the defenders
A - A	No		1.2
B - B	No		2.5
C - C	No		2.4
D- D	No		2.8

Additional notes:

Like Bodbury Ring, Castle Ring is an inland promontory fort. As there are no man-made defences on the eastern side, exact measurement of the internal area is difficult but would appear to be about 4ha. There has been no recorded excavation of Castle Ring, and the 'fort' is dated to the Iron Age on purely typological grounds. The defences consist of a ditch and bank, which survive to a maximum surveyed height of 3.38m, measured from the summit of the bank to the invert of the ditch. The ditch and bank cut across the hill's spur at its narrowest point (80m), effectively dividing the land in two, leaving a flat area to the south of the earthworks. The banks are arranged so that the western arm slightly overlaps the termination of the eastern one. This would have concentrated any assault on the entrance (and by extension the gate) into a narrow defile. The route to the entrance would mean that the assailants would present their right 'unshielded' side to the earthworks. Along the western side a terrace similar to that at Bodbury Ring has been constructed. The southern side is defined by a low bank, with no apparent ditch. The eastern side has no identifiable constructed features.

The entrance way faces almost due south, and is aligned with the quartzite tor on the Stiperstones know as the Devil's Chair. As there are Bronze Age cairns adjacent to the Devil's Chair, it is possible that this view was not obscured by trees or similar growth in the Middle Iron Age. The soils along the ridge are thin and overlie a mass of shattered rock fractured by frost during the last glaciation. Once any tree growth had been removed, it would be difficult for it to re-establish. The view to the Devil's Chair may have been a deciding factor relating to the location of the fort (Boast and Evans 1986, 196), as other apparently similar hills or spurs exist in the immediate vicinity.

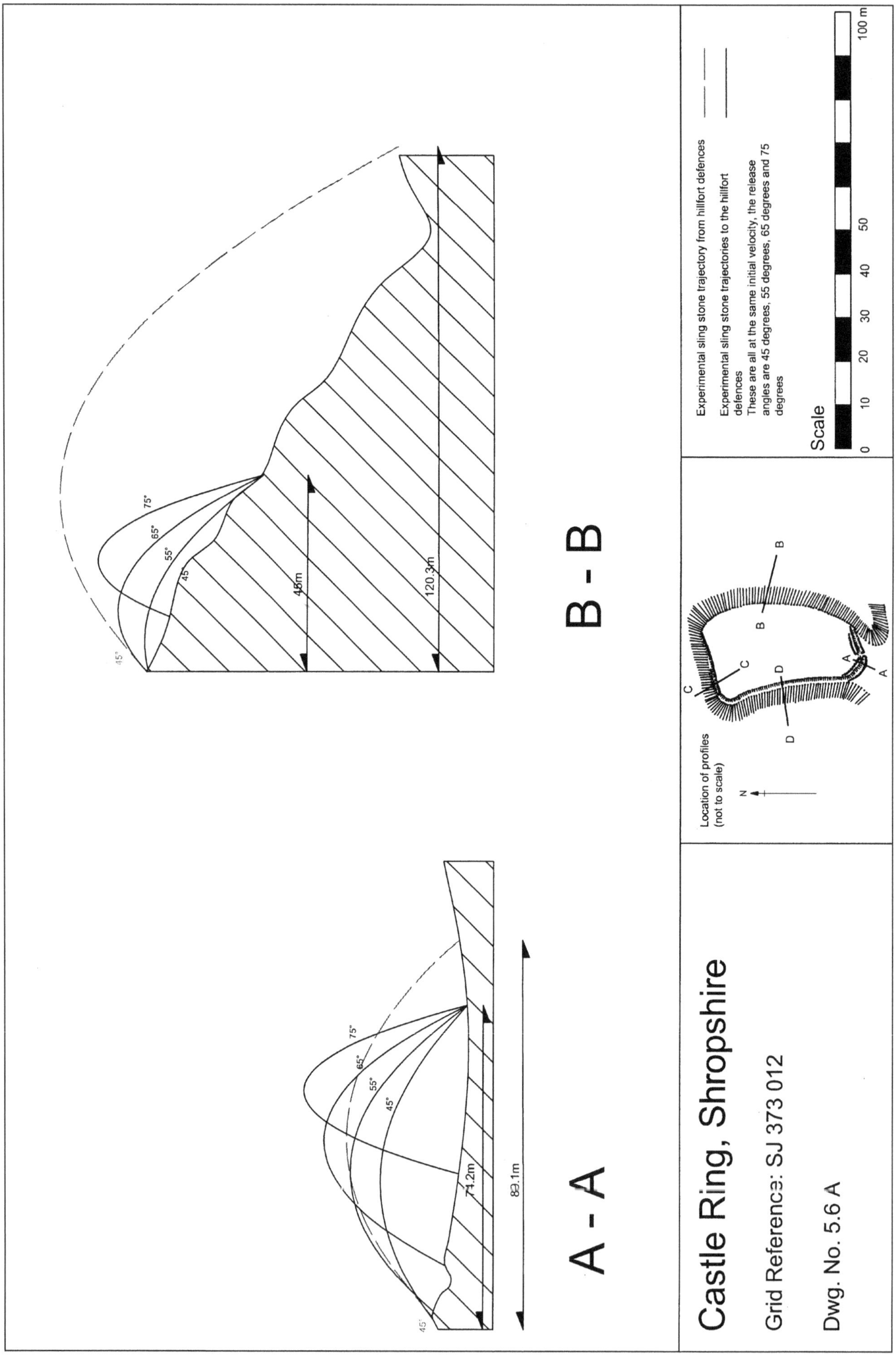
A - A
B - B
75°
65°
55°
45°
71.2m
89.1m
45m
120.3m
Location of profiles (not to scale)
N
A
B
C
D
Experimental sling stone trajectory from hillfort defences
Experimental sling stone trajectories to the hillfort defences
These are all at the same initial velocity, the release angles are 45 degrees, 55 degrees, 65 degrees and 75 degrees
Scale
0 10 20 30 40 50 100 m
Castle Ring, Shropshire
Grid Reference: SJ 373 012
Dwg. No. 5.6 A

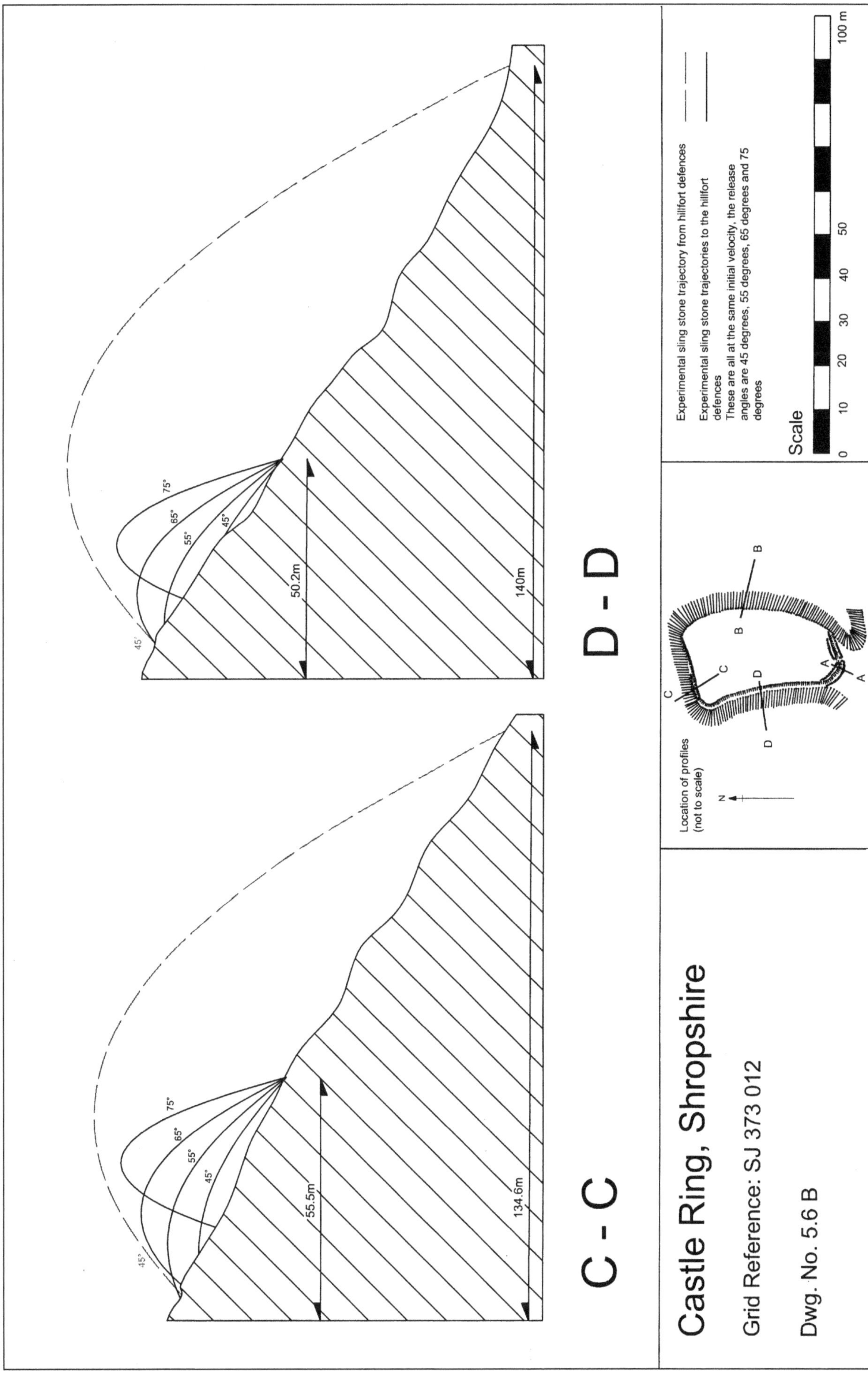

C - C
D - D
75°
65°
55°
45°
45°
55.5m
134.6m
50.2m
140m
Castle Ring, Shropshire
Grid Reference: SJ 373 012
Dwg. No. 5.6 B
Location of profiles (not to scale)
N
A
B
C
D
Experimental sling stone trajectory from hillfort defences
Experimental sling stone trajectories to the hillfort defences
These are all at the same initial velocity, the release angles are 45 degrees, 55 degrees, 65 degrees and 75 degrees
Scale
0
10
20
30
40
50
100 m

Site: Conway Mountain		Grid Ref. SH 760 778	
Reference: Griffiths and Hogg 1956			
Sling projectiles recovered from excavation: Yes			
Drawing No. 5.7			
Section	Would any assailant have to enter an area defined by external earthworks before being able to strike the innermost bank?		Distance factor in favour of the defenders
A - A	No		1.2
B - B	No		3.0
C - C	No		3.6

Additional notes:
The steep slope that surrounds Conway Mountain consists of broken boulders and larger rocks, which, today at least, make walking difficult. If these conditions applied in the Middle Iron Age, which the construction of the fort's wall from such frost-shattered debris suggests would have been the case, they would have served as a further deterrent to assault. The height of the walls has been taken to be 2.1m, to match those reconstructed at The Breiddin (Musson 1991, 33), as the excavators did not undertake any reconstruction or calculation of the volume of rock available.

Although the double walled western end seems to operate as a multi-walled defined area, the entrance to the outer section is from the hillside itself and there appears to have been no direct access from the inner camp. A slinger from the 'inner' defence would almost be able to strike an assailant as they came into range of the 'outer' walls.

Only three sections were taken. The northern side of the fort is protected by steep (nearly vertical) cliffs and no section was taken at this location.

Site: Crickley Hill		Grid Ref. SO 927 161	
Reference: Dixon 1994			
Sling projectiles recovered from excavation: Yes			
Drawing No. 5.8			
Section	Would any assailant have to enter an area defined by external earthworks before being able to strike the innermost bank?		Distance factor in favour of the defenders
A - A		No	1.2
B - B		No	2.0
C - C		No	2.0
D- D		No	5.1

Additional notes: As the linear defensive works on the north-western side were unfinished when the hillfort was apparently attached and abandoned, it is not possible to say if any outer earth works or walls were intended to be built. The distance factors on the natural slopes indicate that the site was chosen for its defendable nature when using the sling.

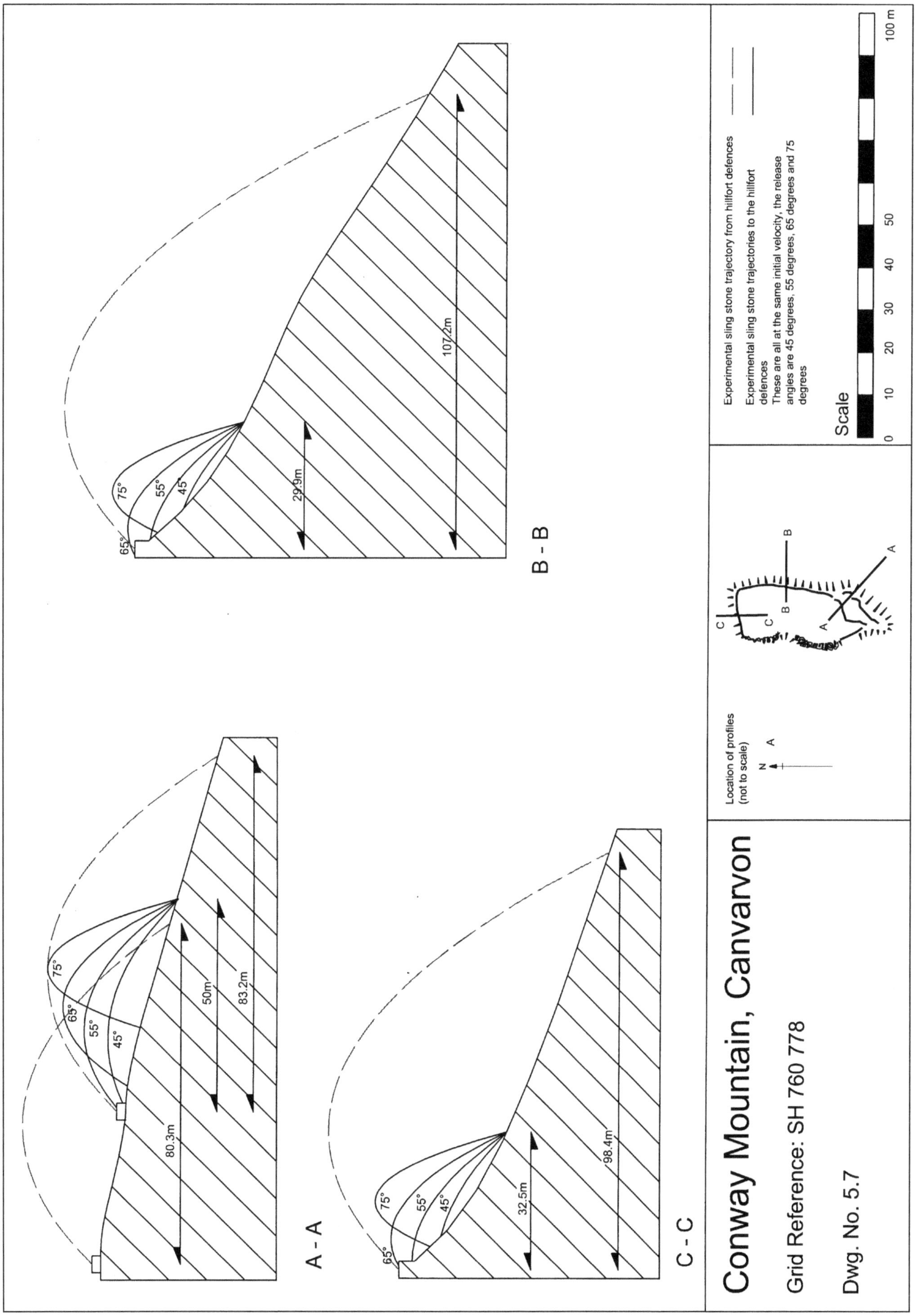
A - A
75°
65°
55°
45°
80.3m
50m
83.2m
B - B
75°
55°
45°
65°
29.9m
107.2m
C - C
75°
55°
45°
65°
32.5m
98.4m
Conway Mountain, Canvarvon
Grid Reference: SH 760 778
Dwg. No. 5.7
Location of profiles
(not to scale)
N
A
B
C
Experimental sling stone trajectory from hillfort defences
Experimental sling stone trajectories to the hillfort defences
These are all at the same initial velocity, the release angles are 45 degrees, 55 degrees, 65 degrees and 75 degrees
Scale
0 10 20 30 40 50 100 m

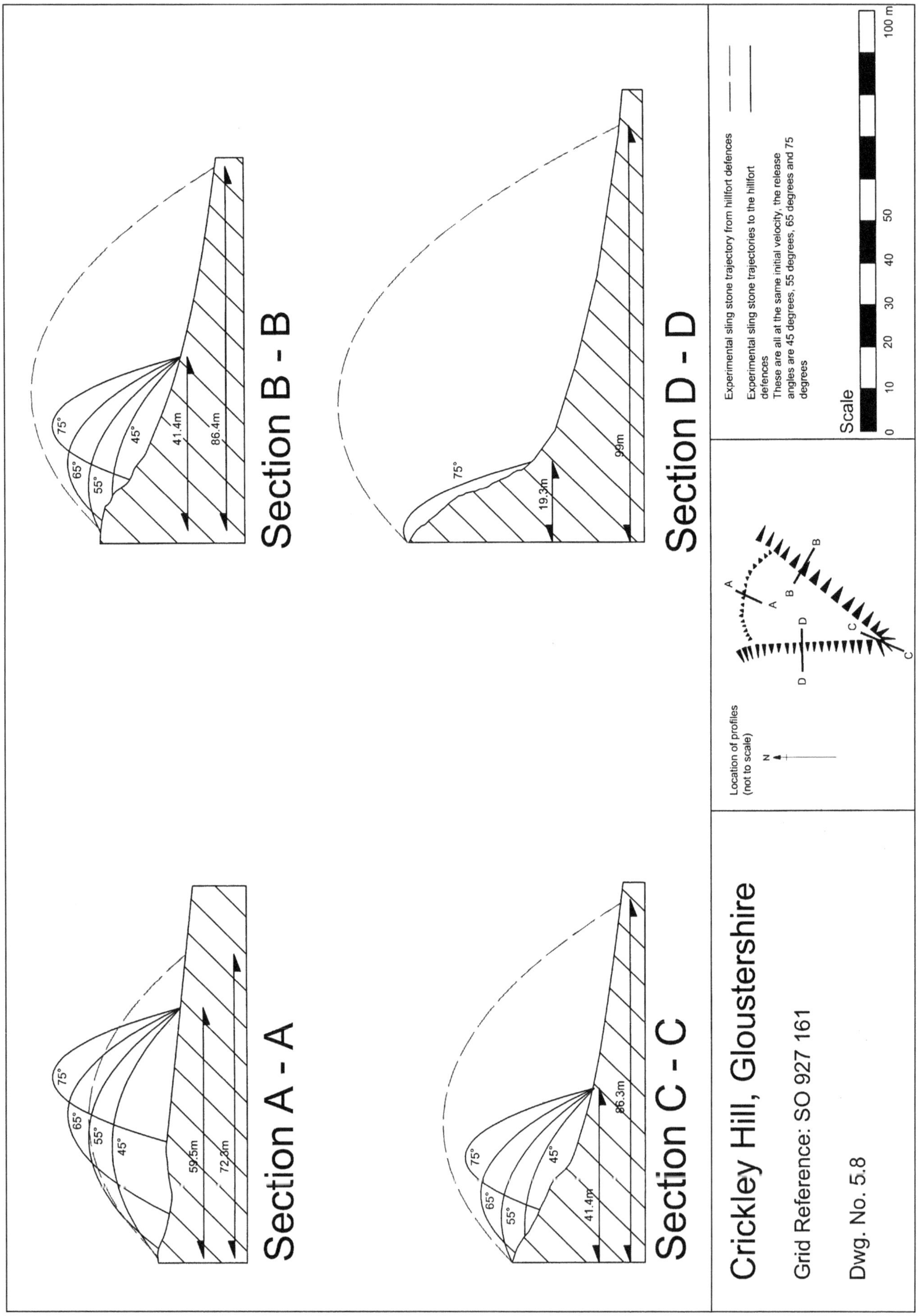
Section A - A
Section B - B
Section C - C
Section D - D
75°
65°
55°
45°
59.5m
72.3m
41.4m
86.4m
86.3m
19.3m
99m
Experimental sling stone trajectory from hillfort defences
Experimental sling stone trajectories to the hillfort defences
These are all at the same initial velocity, the release angles are 45 degrees, 55 degrees, 65 degrees and 75 degrees
Scale
0 10 20 30 40 50 100 m
Location of profiles (not to scale)
N
Crickley Hill, Gloustershire
Grid Reference: SO 927 161
Dwg. No. 5.8

<table>
<tr><td colspan="2">Site: Croft Ambrey</td><td colspan="2">Grid Ref. SO 445 668</td></tr>
<tr><td colspan="4">Reference: Stanford 1974</td></tr>
<tr><td colspan="4">Sling projectiles recovered from excavation: No</td></tr>
<tr><td colspan="4">Drawing No. 5.9</td></tr>
<tr><td>Section</td><td colspan="2">Would any assailant have to enter an area defined by external earthworks before being able to strike the innermost bank?</td><td>Distance factor in favour of the defenders</td></tr>
<tr><td>A - A</td><td>Yes</td><td></td><td>1.3</td></tr>
<tr><td>B - B</td><td></td><td>No</td><td>2.0</td></tr>
<tr><td>C - C</td><td></td><td>No</td><td>1.7</td></tr>
<tr><td colspan="4">Additional notes:
Outside the main defences an area was annexed by a smaller ditch and bank. The enclosing earthworks were not excavated, but nothing would prevent them from being part of the original design of at least the later fortification (Stanford 1974, 26). Stanford (ibid, 43-4) attempted to demonstrate that the defences were constructed to give clear advantage to the defenders while using spears. This model, though valid for the inner banks and ditches, does not take into consideration the outer annexed area.

Due to the overgrown nature of the fort only three sections were recorded. Section A- A runs though the main defences and across the area enclosed by the smaller outer ditch. This feature clearly defines an area, which an assailant would have to enter in order to strike the inner bank. Section C - C was taken over the complex western gateway. Here, the defenders would have been able to command the access road that leads to the fort. There is no clearly defined area where any assailant would have to enter before being able to strike the inner defences. However, the natural slope beyond the road is steep, at 1:3 (ibid, Figure 1).</td></tr>
</table>

<table>
<tr><td colspan="2">Site: Danebury</td><td colspan="2">Grid Ref. SU 323 376</td></tr>
<tr><td colspan="4">Reference: Cunliffe 1984a; Cunliffe and Poole 1991</td></tr>
<tr><td colspan="4">Sling projectiles recovered from excavation: Yes</td></tr>
<tr><td colspan="4">Drawing No.5.10</td></tr>
<tr><td>Section</td><td colspan="2">Would any assailant have to enter an area defined by external earthworks before being able to strike the innermost bank?</td><td>Distance factor in favour of the defenders</td></tr>
<tr><td>A - A</td><td>Yes</td><td></td><td>1.2</td></tr>
<tr><td>B - B</td><td>Yes</td><td></td><td>1.8</td></tr>
<tr><td>C - C</td><td>Yes</td><td></td><td>2.0</td></tr>
<tr><td>D- D</td><td>Yes</td><td></td><td>1.9</td></tr>
<tr><td colspan="4">Additional notes:
The location of the outer earthwork is of particular note. It does not maintain a constant distance from the main inner defences but its distance varies in relation to the slope of the hill. Simply put, the steeper the hill the closer the inner and outer bank are.

An assailant would have to enter the space defined by the outer earthwork before they could strike the inner bank, even at Section D - D where the inner and outer earthworks are only some 50m apart, as compared with 90m at Section C - C.</td></tr>
</table>

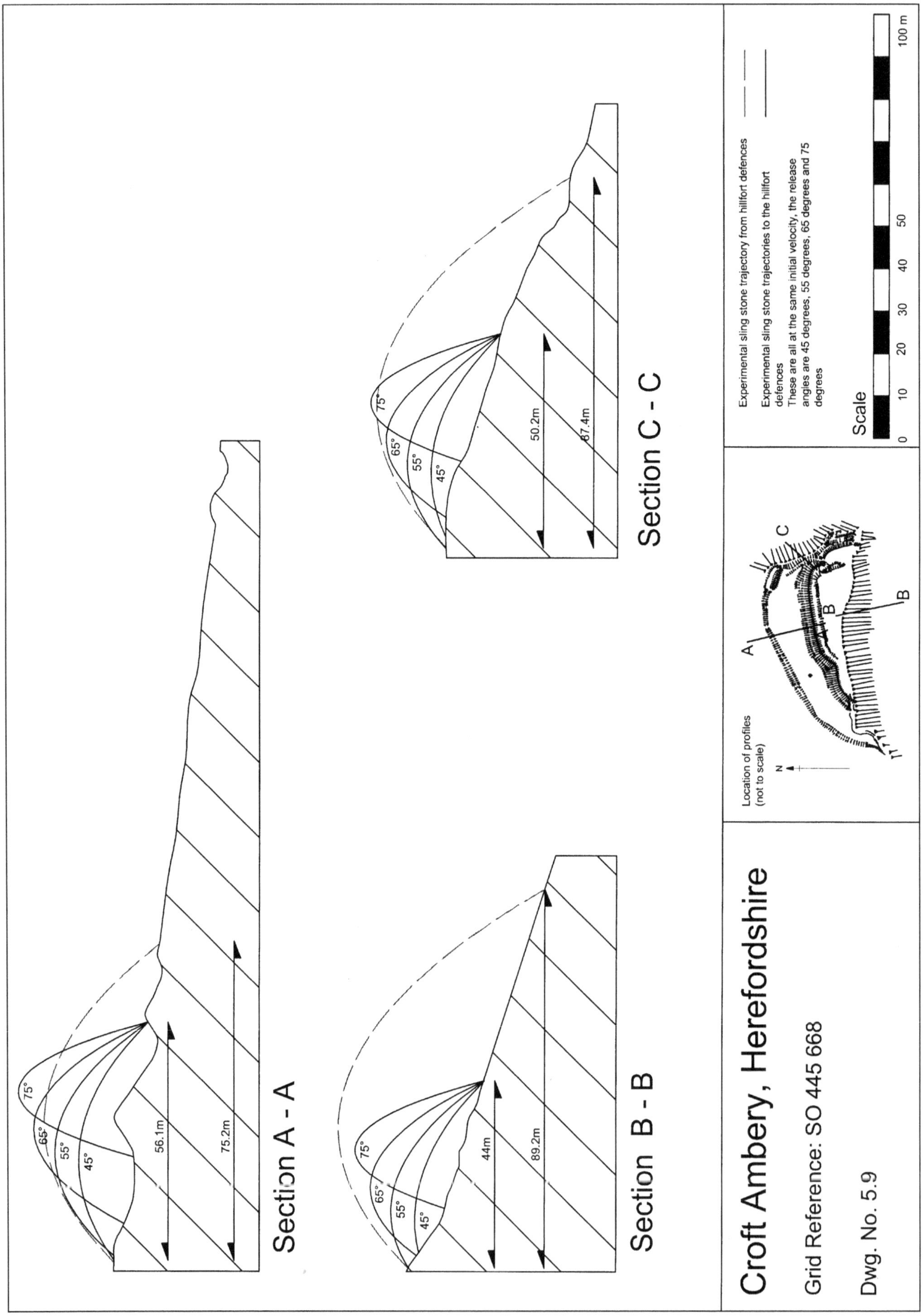
Section A - A
56.1m
75.2m
75°
65°
55°
45°
Section B - B
44m
89.2m
75°
65°
55°
45°
Section C - C
50.2m
87.4m
75°
65°
55°
45°
Experimental sling stone trajectory from hillfort defences
Experimental sling stone trajectories to the hillfort defences
These are all at the same initial velocity, the release angles are 45 degrees, 55 degrees, 65 degrees and 75 degrees
Scale
0 10 20 30 40 50 100 m
Location of profiles (not to scale)
N
A
B
C
Croft Ambery, Herefordshire
Grid Reference: SO 445 668
Dwg. No. 5.9

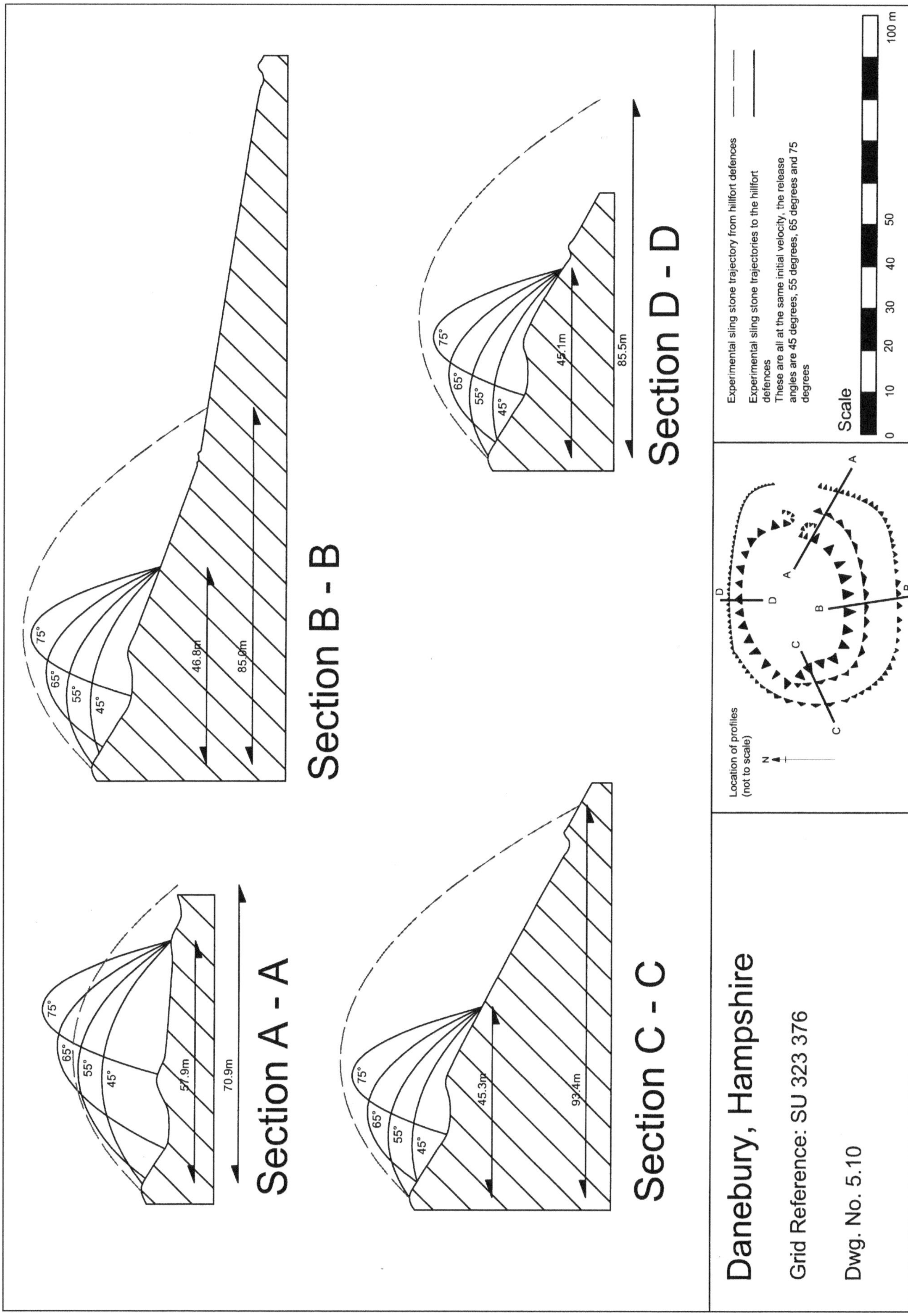
Section A - A
75°
65°
55°
45°
57.9m
70.9m
Section B - B
46.8m
85.0m
Section C - C
45.3m
93.4m
Section D - D
45.1m
85.5m
Experimental sling stone trajectory from hillfort defences
Experimental sling stone trajectories to the hillfort defences
These are all at the same initial velocity, the release angles are 45 degrees, 55 degrees, 65 degrees and 75 degrees
Scale
0 10 20 30 40 50 100 m
Location of profiles (not to scale)
N
Danebury, Hampshire
Grid Reference: SU 323 376
Dwg. No. 5.10

Site: Earls Hill		Grid Ref. SJ 409 049	
Reference: Forde-Johnston 1962			
Sling projectiles recovered from excavation: No			
Drawing No. 5.11			
Section	Would any assailant have to enter an area defined by external earthworks before being able to strike the innermost bank?		Distance factor in favour of the defenders
A - A	Yes		1.8
B - B		No	2.1
C - C		No	1.8
D- D		No	2.5

Additional notes:
Though no excavation has been undertaken at Earls Hill, the earthworks strongly suggest that there were at least two phases of construction. The first enclosed the summit of the ridge with a ditch and bank. Later, the entire ridge was enclosed with a terrace similar to those at Bodbury Ring and Castle Ring. The south-eastern side of the original enclosure was apparently allowed to silt up following the enclosure of the remaining section of the ridge. The eastern flank of the fort is defined by steep outcrops of rock creating sections of cliff. There are two outlying ditches that have no stratigraphic relationship to the other earthworks. Their nature certainly implies that they are connected to the construction of the fort, but during which phase is at present impossible to say.

Site: Eggardon		Grid Ref. SY 541 947	
Reference: Marsh 1901			
Sling projectiles recovered from excavation: Yes			
Drawing No. 5.12			
Section	Would any assailant have to enter an area defined by external earthworks before being able to strike the innermost bank?		Distance factor in favour of the defenders
A - A	Yes		1.4
B - B	Yes		1.8
C - C	Yes		1.2
D- D	Yes		1.4

Additional notes:
A prominent triple-banked and ditched multivallate hillfort, Eggardon encloses an area of approximately 8ha and lies on a long ridge, rising to 190m above sea level. The main entrance faces south, towards the clearly visible sea. The earthworks enclose a single round barrow. Though there is evidence of some collapse on the southern side, leading to a new bank and ditch being constructed, no excavation through the earthworks has been undertaken, so no relative dating is possible. Marsh's excavation concentrated on the central area and recovered pits (interpreted as dwellings), some pottery and some beach pebbles.

Though all the profiles indicated that an assailant would have to enter a space defined by a ditch and bank, the layout of these features on the relatively level north-western approach is particularly noteworthy. Here, two relatively small banks and their associated ditches have been constructed. An assailant would have to enter both areas (as defined by the ditches and banks) before striking the innermost bank. This certainly indicates that sling warfare may have been one of the primary factors dictating the layout.

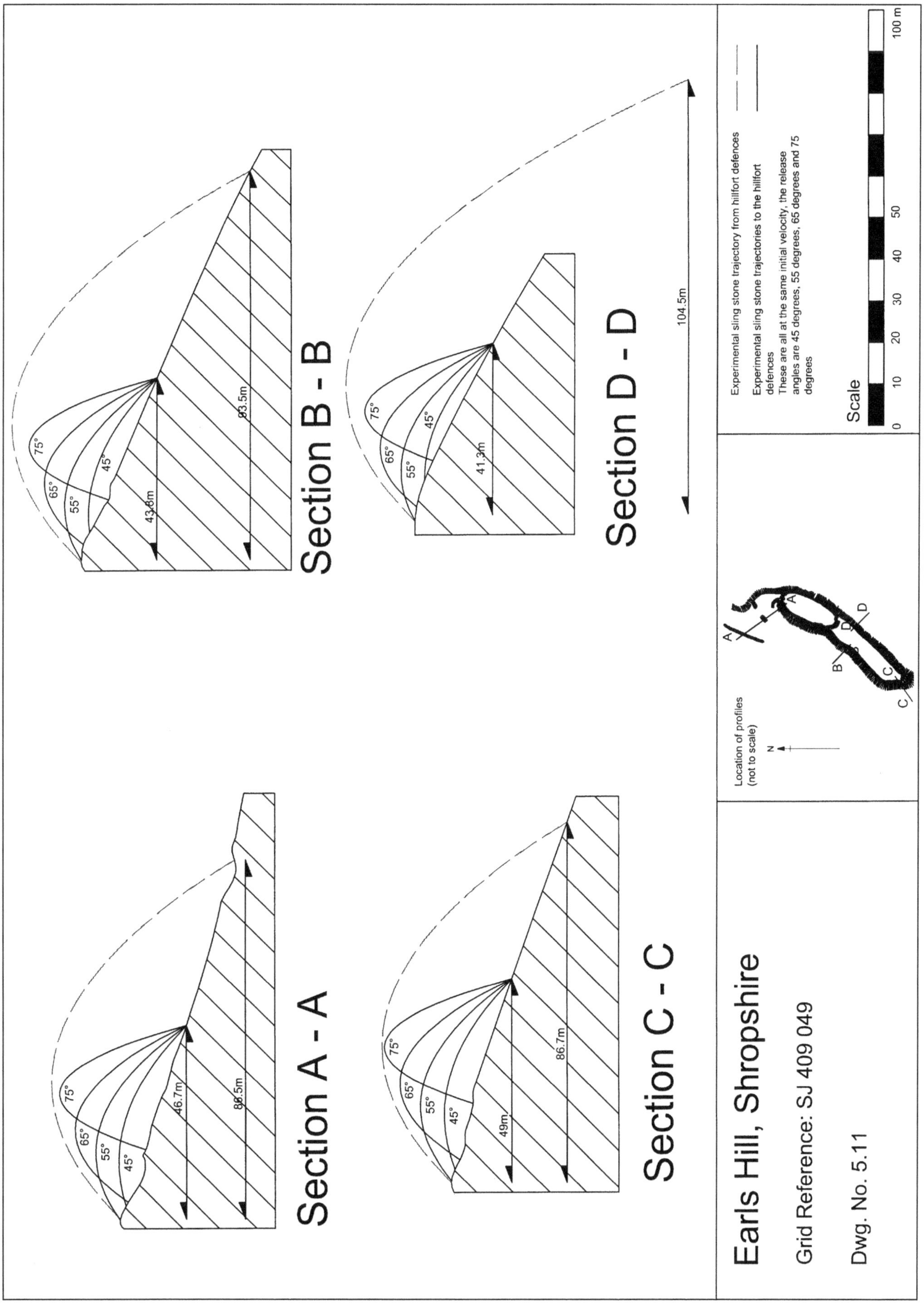
Section A - A
75°
65°
55°
45°
46.7m
86.5m
Section B - B
75°
65°
55°
45°
43.8m
93.5m
Section C - C
75°
65°
55°
45°
49m
86.7m
Section D - D
75°
65°
55°
45°
41.3m
104.5m
Earls Hill, Shropshire
Grid Reference: SJ 409 049
Dwg. No. 5.11
Location of profiles
(not to scale)
N
A
B
C
D
Experimental sling stone trajectory from hillfort defences
Experimental sling stone trajectories to the hillfort defences
These are all at the same initial velocity, the release angles are 45 degrees, 55 degrees, 65 degrees and 75 degrees
Scale
0
10
20
30
40
50
100 m

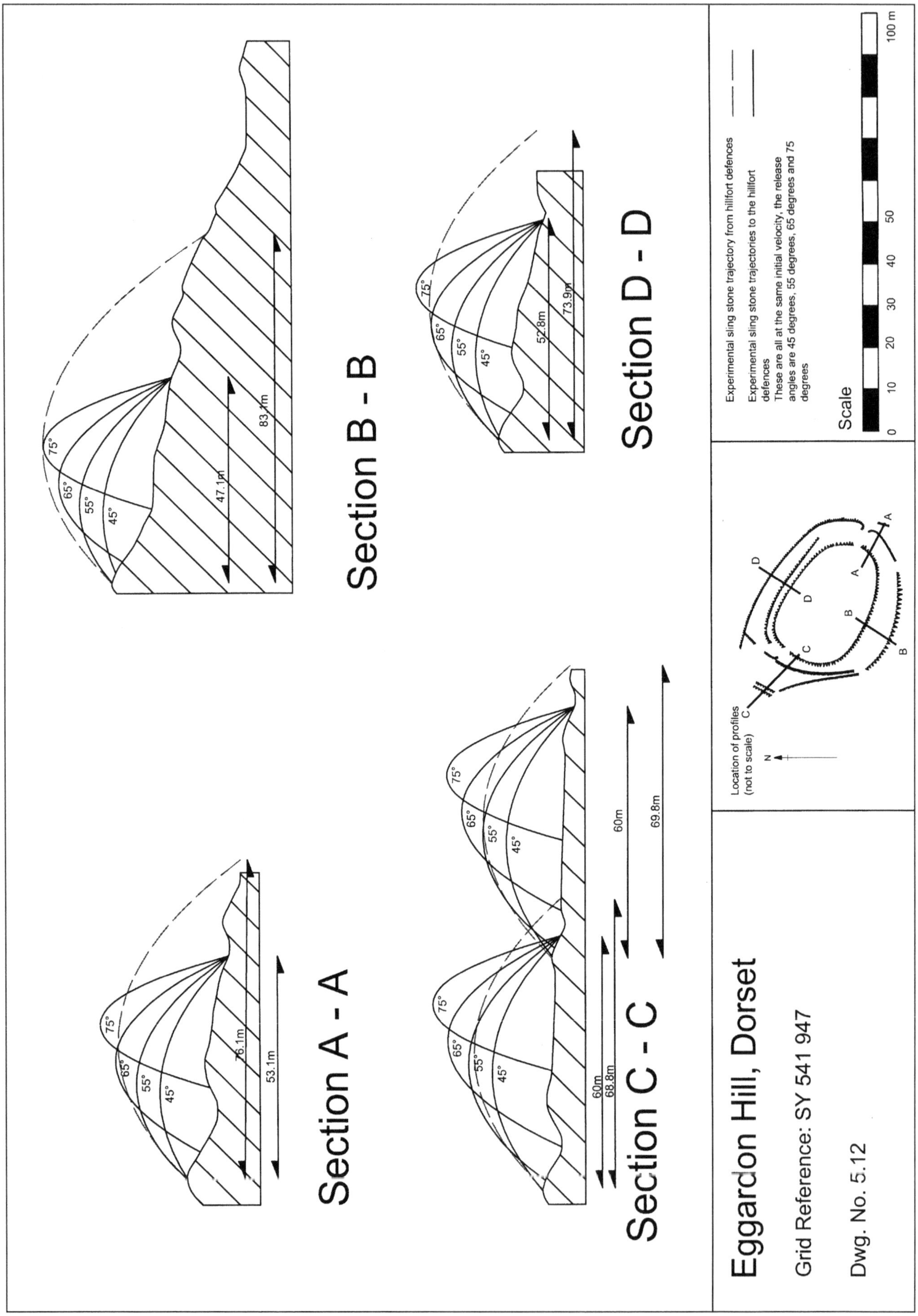
Section A - A
75°
65°
55°
45°
76.1m
53.1m
Section B - B
47.1m
83.1m
Section C - C
60m
68.8m
60m
69.8m
Section D - D
52.8m
73.9m
Experimental sling stone trajectory from hillfort defences
Experimental sling stone trajectories to the hillfort defences
These are all at the same initial velocity, the release angles are 45 degrees, 55 degrees, 65 degrees and 75 degrees
Scale
0 10 20 30 40 50 100 m
Location of profiles (not to scale)
N
Eggardon Hill, Dorset
Grid Reference: SY 541 947
Dwg. No. 5.12

<table>
<tr><td colspan="2">Site: Hambledon Hill</td><td colspan="2">Grid Ref. ST 845 126</td></tr>
<tr><td colspan="4">Reference: Cunnington 1895</td></tr>
<tr><td colspan="4">Sling projectiles recovered from excavation: Yes</td></tr>
<tr><td colspan="4">Drawing No. 5.13</td></tr>
<tr><td>Section</td><td colspan="2">Would any assailant have to enter an area defined by external earthworks before being able to strike the innermost bank?</td><td>Distance factor in favour of the defenders</td></tr>
<tr><td>A - A</td><td>Yes</td><td></td><td>1.4</td></tr>
<tr><td>B - B</td><td>Yes</td><td></td><td>2.1</td></tr>
<tr><td>C - C</td><td>Yes</td><td></td><td>1.1</td></tr>
<tr><td>D- D</td><td>Yes</td><td></td><td>1.2</td></tr>
<tr><td colspan="4">Additional notes:
The easier approach, also with the most elaborate entrance, faces towards Hod Hill on the south-eastern side of the fort. The fort has at least two identifiable phases of construction. However, as no systematic excavation has been undertaken, the absolute and relative dating of these remains unresolved. It is clear that the original earthworks enclosed about 3ha of the lower section of the spur of the hill. At some time, the entire hill was enclosed with three large ditches and banks. Surface examination shows that these were probably of glacis construction. The final hillfort enclosed approximately 10ha and, possibly significantly, included a Neolithic long mound.

The excavation carried out by Cunnington (1895) recovered evidence of storage pits, pottery and sling stones (attributed to the Romans in the original report; see the Gazetteer for further information).</td></tr>
</table>

<table>
<tr><td colspan="2">Site: Hod Hill</td><td colspan="2">Grid Ref. ST 857 126</td></tr>
<tr><td colspan="4">Reference: Richmond 1968</td></tr>
<tr><td colspan="4">Sling projectiles recovered from excavation: Yes</td></tr>
<tr><td colspan="4">Drawing No. 5.14</td></tr>
<tr><td>Section</td><td colspan="2">Would any assailant have to enter an area defined by external earthworks before being able to strike the innermost bank?</td><td>Distance factor in favour of the defenders</td></tr>
<tr><td>A - A</td><td>No</td><td></td><td>1.4</td></tr>
<tr><td>B - B</td><td>No</td><td></td><td>1.5</td></tr>
<tr><td>C - C</td><td>No</td><td></td><td>2.6</td></tr>
<tr><td>D- D</td><td>No</td><td></td><td>1.6</td></tr>
<tr><td colspan="4">Additional notes:
As the unfinished earthworks to the north-western and eastern sides of the fort are dated to the later Late Iron Age, they have not been included within the profiles.

The fields to the west and south of the fort had been ploughed, which could conceivably have removed any trace of a smaller outer earthwork. However, the northern approach is laid to pasture and is too steep to plough. Despite extensive investigation of this area, no earthworks (other than the unfinished ones) were identified. Therefore, the most parsimonious view is that the only earthworks in use during the Middle Iron Age are those in existence today.</td></tr>
</table>

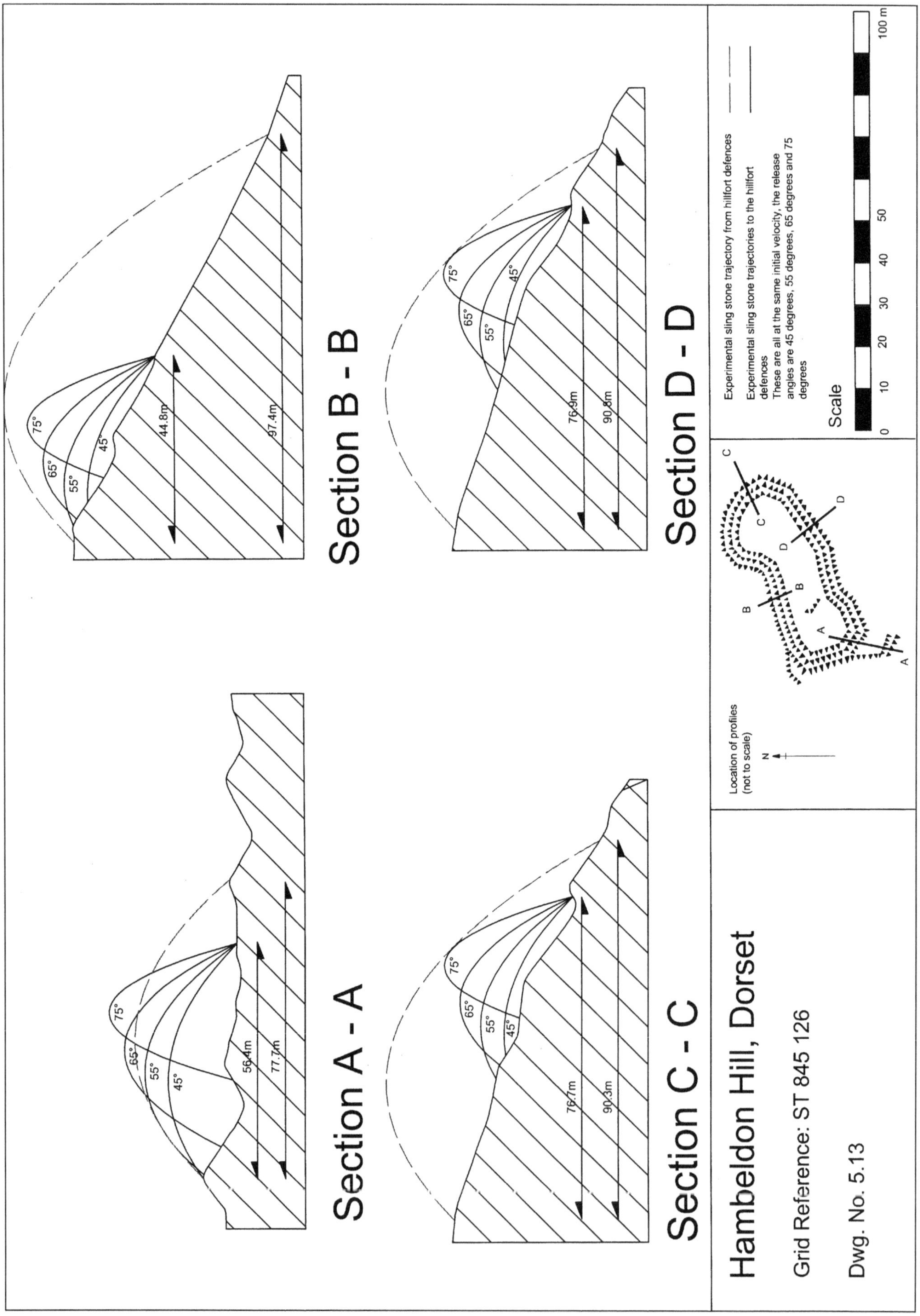
Section A - A
75°
65°
55°
45°
56.4m
77.7m
Section B - B
75°
65°
55°
45°
44.8m
97.4m
Section C - C
75°
65°
55°
45°
76.7m
90.3m
Section D - D
75°
65°
55°
45°
76.9m
90.8m
Hambeldon Hill, Dorset
Grid Reference: ST 845 126
Dwg. No. 5.13
Location of profiles (not to scale)
N
A
B
C
D
Experimental sling stone trajectory from hillfort defences
Experimental sling stone trajectories to the hillfort defences
These are all at the same initial velocity, the release angles are 45 degrees, 55 degrees, 65 degrees and 75 degrees
Scale
0
10
20
30
40
50
100 m

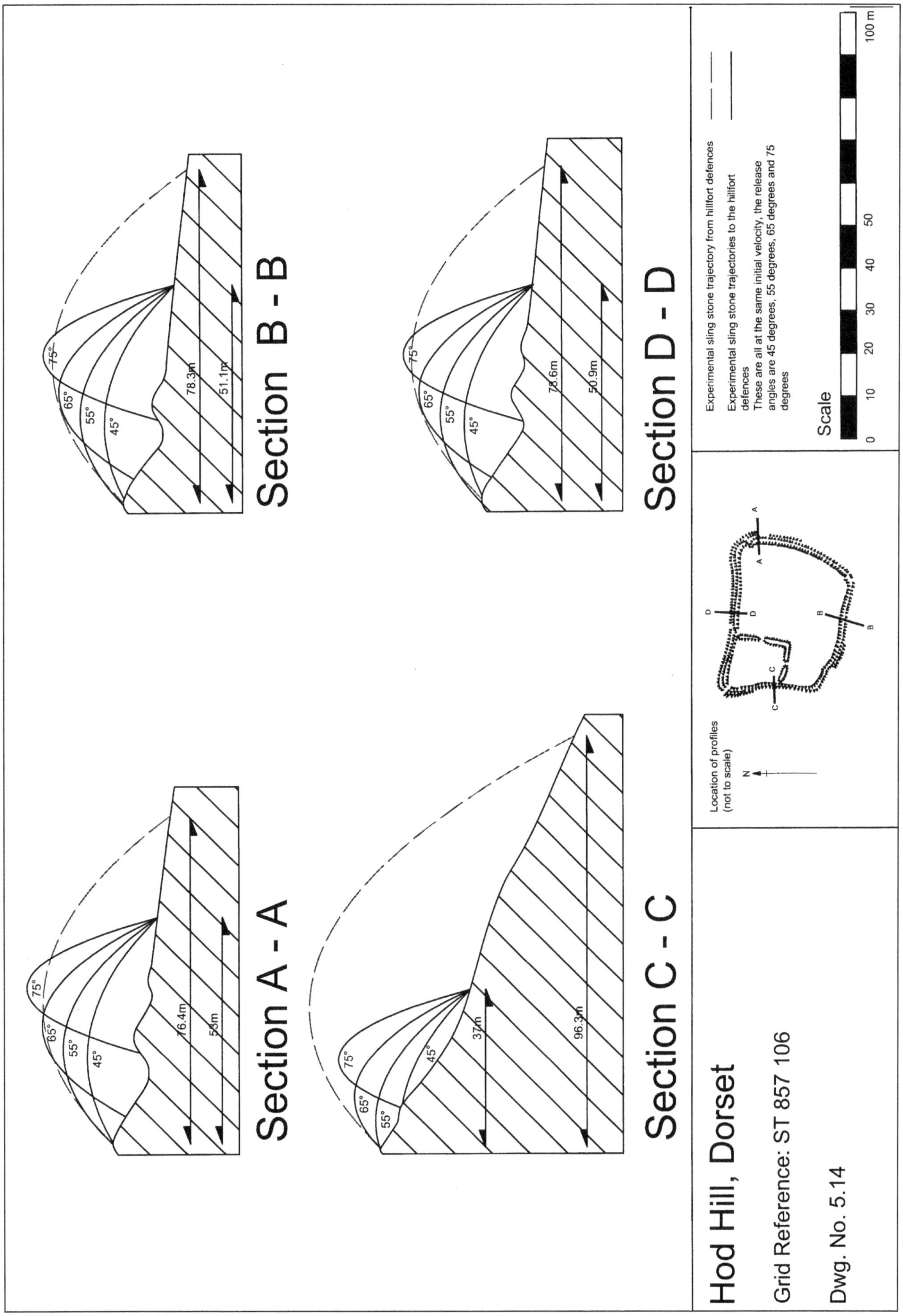
Section A - A
75°
65°
55°
45°
76.4m
53m
Section B - B
78.3m
51.1m
Section C - C
37m
96.3m
Section D - D
78.6m
50.9m
Experimental sling stone trajectory from hillfort defences
Experimental sling stone trajectories to the hillfort defences
These are all at the same initial velocity, the release angles are 45 degrees, 55 degrees, 65 degrees and 75 degrees
Scale
0 10 20 30 40 50 100 m
Location of profiles (not to scale)
N
Hod Hill, Dorset
Grid Reference: ST 857 106
Dwg. No. 5.14

Site: Ivinghoe Beacon		Grid Ref. SP 960 169	
Reference: Cotton and Frere 1968			
Sling projectiles recovered from excavation:			
Drawing Number 5.15			
Section	Would any assailant have to enter an area defined by external earthworks before being able to strike the innermost bank?		Distance factor in favour of the defenders
A - A		No	1.8
B - B		No	1.6
C - C		No	1.3
D- D		No	1.9
Additional notes: Ivinghoe Beacon appears to be protected on three flanks by steep slopes, which would have given a defender a significant range advantage over any assailant. However, the entrance appears to be vulnerable to attack using the sling. There is no evidence of any smaller outer earthwork on the spur of land that leads to the main entrance, despite the survival of a small barrow, the shallow ditch of which is still clearly visible, indicating that any slight feature associated with the fort should not only have survived but would also be visible.			

Site: Maiden Castle		Grid Ref. SY 669 885	
Reference: Wheeler 1943			
Sling projectiles recovered from excavation: Yes			
Drawing No. 5.16 (a and b)			
Section	Would any assailant have to enter an area defined by external earthworks before being able to strike the innermost bank?		Distance factor in favour of the defenders
A - A		Yes	1.3
B - B		Yes	1.4
C - C		Yes	1.7
D- D			
Additional notes: Profile A-A seems to indicate the interconnected banks at the western entrance had been designed to provide 'covering fire' from the banks behind, as any assailant would be out of range of the defending slingers, even if occupying one of the outer banks, while they were easily in range of the assailants. Such measures may represent a response to the more organised warfare of the Late Iron Age.			

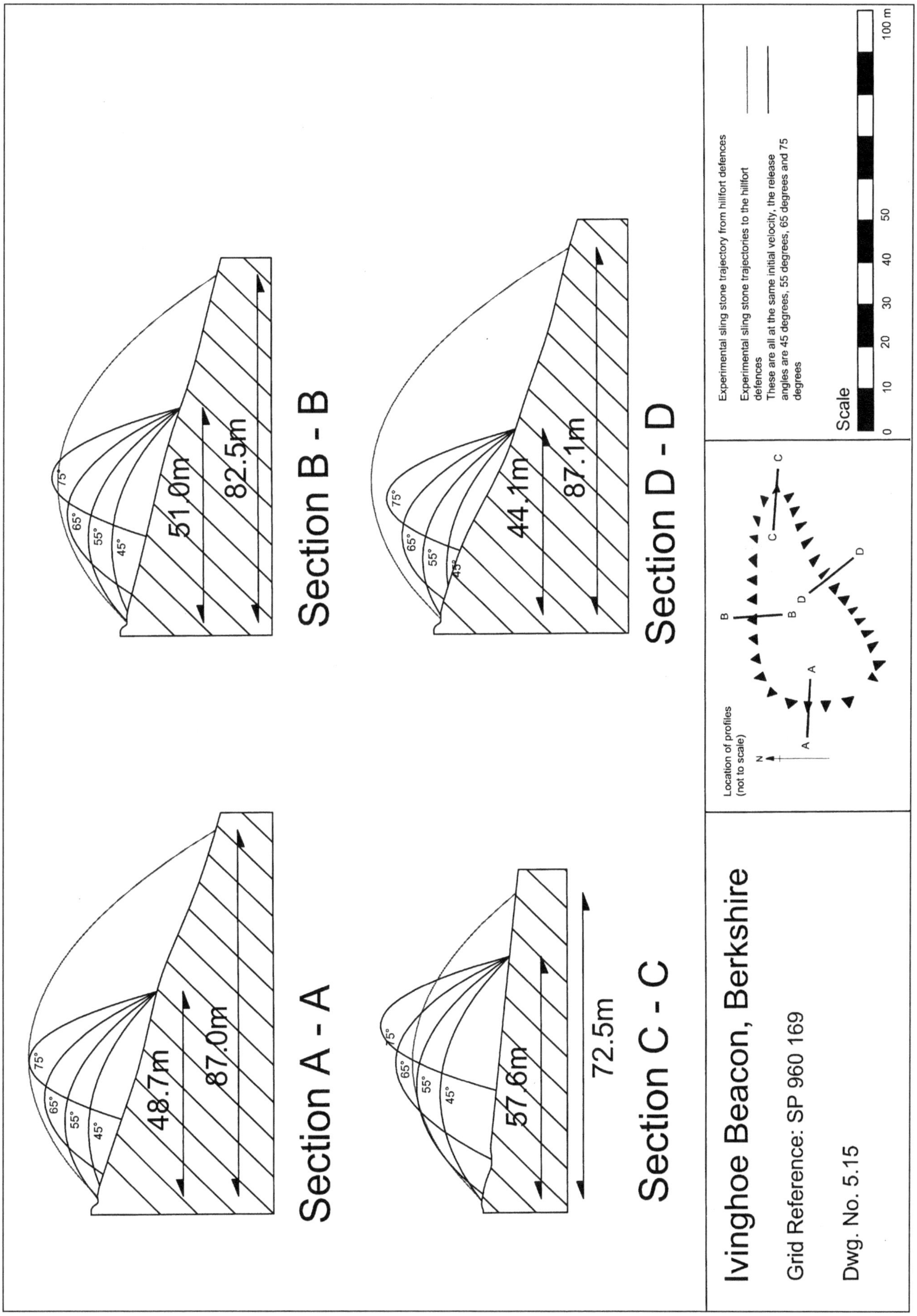
Section A - A
75°
65°
55°
45°
48.7m
87.0m
Section C - C
75°
65°
55°
45°
57.6m
72.5m
Section B - B
75°
65°
55°
45°
51.0m
82.5m
Section D - D
75°
65°
55°
45°
44.1m
87.1m
Ivinghoe Beacon, Berkshire
Grid Reference: SP 960 169
Dwg. No. 5.15
Location of profiles
(not to scale)
N
A
B
C
D
Experimental sling stone trajectory from hillfort defences
Experimental sling stone trajectories to the hillfort defences
These are all at the same initial velocity, the release angles are 45 degrees, 55 degrees, 65 degrees and 75 degrees
Scale
0
10
20
30
40
50
100 m

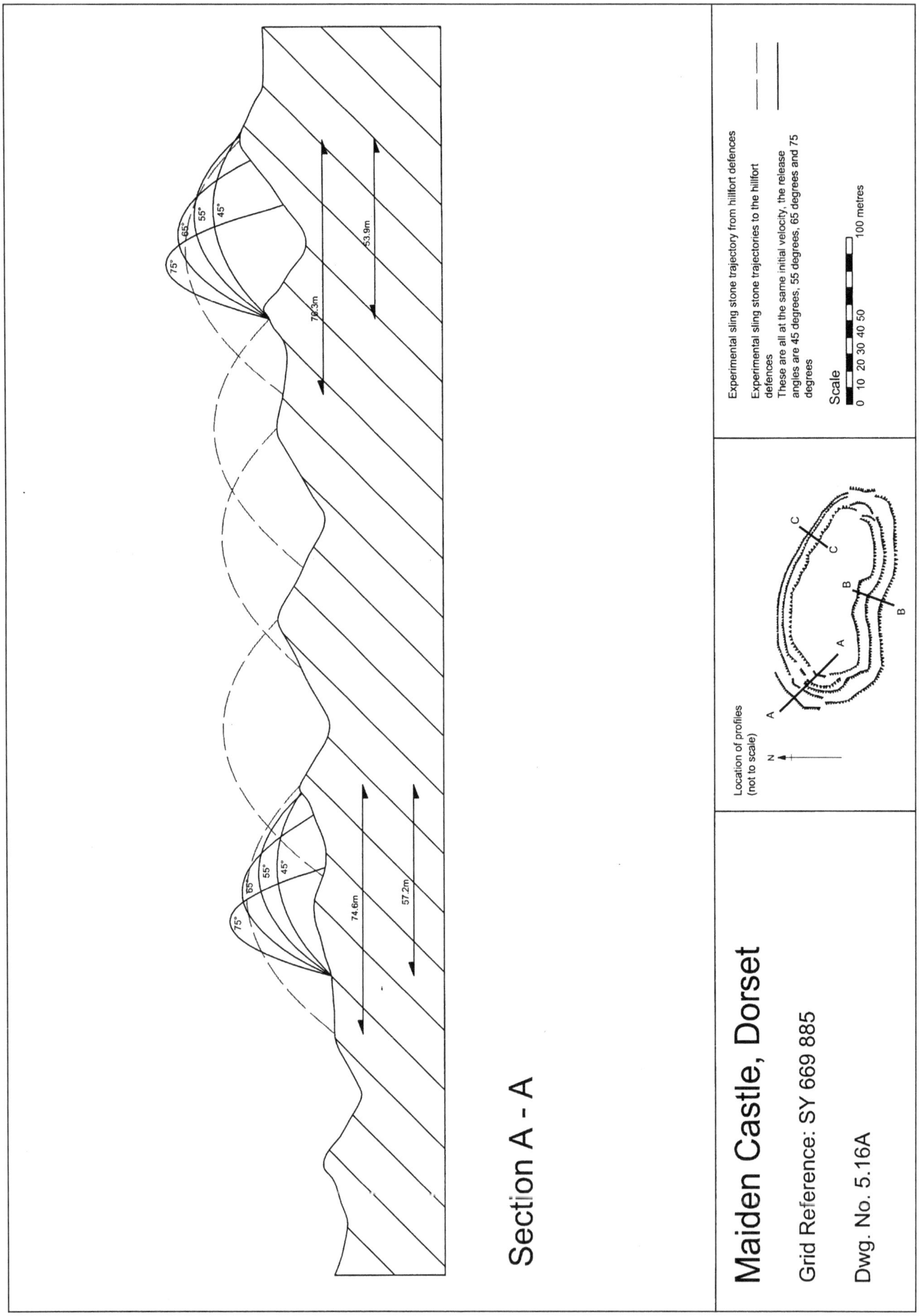
75°
65°
55°
45°
76.3m
53.9m
74.6m
57.2m
Section A - A
Experimental sling stone trajectory from hillfort defences
Experimental sling stone trajectories to the hillfort defences
These are all at the same initial velocity, the release angles are 45 degrees, 55 degrees, 65 degrees and 75 degrees
Scale
0 10 20 30 40 50 100 metres
Location of profiles (not to scale)
N
A
B
C
Maiden Castle, Dorset
Grid Reference: SY 669 885
Dwg. No. 5.16A

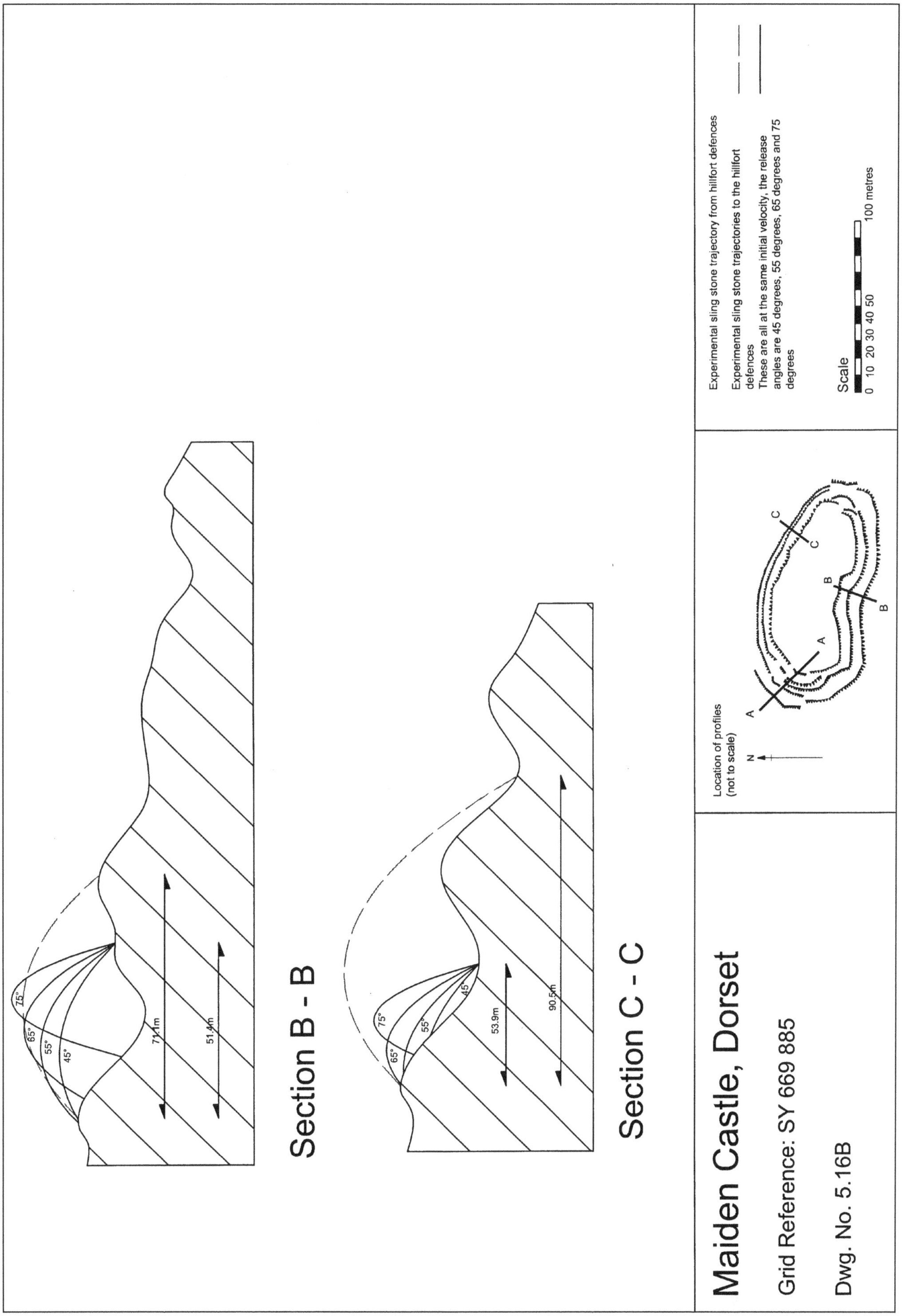
Maiden Castle, Dorset
Grid Reference: SY 669 885
Dwg. No. 5.16B
Section B - B
Section C - C
75°
65°
55°
45°
71.1m
51.4m
53.9m
90.5m
Location of profiles
(not to scale)
N
A
A
B
B
C
C
Experimental sling stone trajectory from hillfort defences
Experimental sling stone trajectories to the hillfort defences
These are all at the same initial velocity, the release angles are 45 degrees, 55 degrees, 65 degrees and 75 degrees
Scale
0 10 20 30 40 50
100 metres

Site: Nordy Bank		Grid Ref. SO 577 847	
Reference: None			
Sling projectiles recovered from excavation: No			
Drawing No. 5.17			
Section	Would any assailant have to enter an area defined by external earthworks before being able to strike the innermost bank?		Distance factor in favour of the defenders
A - A	No		1.5
B - B	No		1.6
C - C	No		1.8
D- D	No		1.6

Additional notes:

A univallate hillfort occupying a spur of Brown Clee Hill that rises some 100m from the land below, the earthworks enclose an area of about 2.8ha. No recorded excavation has taken place. The main entrance faces south, and is oriented in such a manner as to require anybody approaching it to present their right/unshielded side to the ramparts.

The earthworks that form the hillfort are well preserved, standing to a maximum surveyed height of 5.16m from the top of the bank to the invert of the ditch. Along the summit of the bank, larger stones appear to have been laid horizontally, creating a pavement. In areas where the section of the bank is exposed, it would appear that these irregular flags do not extend into the construction matrix of the bank.

The interior also seems well preserved, although an undated rectilinear structure has been constructed. Unfortunately, there has been widespread damage to the land adjacent to the fort. Here, there has been extensive small-scale bell-pit coal mining, which may have commenced as early as the thirteenth century but was certainly in severe decline by the middle of the nineteenth century (Chapman *et al.* 1995). This has left a landscape pockmarked with roughly circular depressions and spoil heaps often in excess of 2m in height. These are concentrated on the southern/eastern flanks but occur sporadically over the whole area. The workings have not only destroyed any archaeological features in the area but have distorted, significantly in places, the natural ground level. None of the linear features visible on the ground appear to have any direct relationship to the layout of the earthworks. This can clearly be seen from aerial photographs (eg. Watson and Musson 1999, Plate 30).

None of the profiles seem to give the defenders any significant advantage and the whole fort, as it stands, would appear to be vulnerable to attack using the sling.

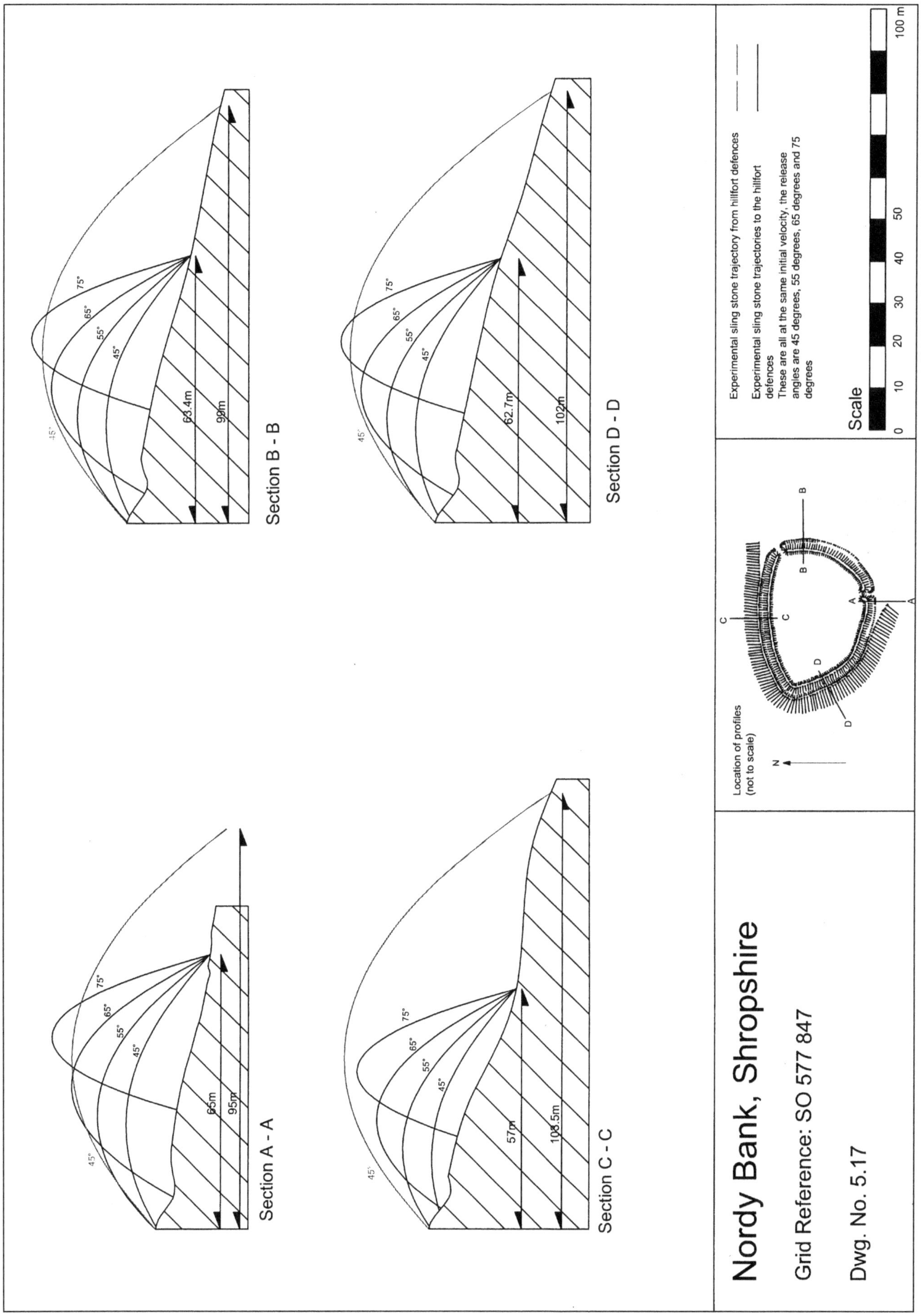

Section A - A
75°
65°
55°
45°
45°
65m
95m
Section B - B
63.4m
99m
Section C - C
57m
105.5m
Section D - D
62.7m
102m
Location of profiles
(not to scale)
N
A
B
C
D
Experimental sling stone trajectory from hillfort defences
Experimental sling stone trajectories to the hillfort defences
These are all at the same initial velocity, the release angles are 45 degrees, 55 degrees, 65 degrees and 75 degrees
Scale
0
10
20
30
40
50
100 m
Nordy Bank, Shropshire
Grid Reference: SO 577 847
Dwg. No. 5.17

Site: Old Oswestry		Grid Ref. SJ 269 310	
Reference: Varley 1948			
Sling projectiles recovered from excavation: No			
Drawing No. 5.18			
Section	Would any assailant have to enter an area defined by external earthworks before being able to strike the innermost bank?		Distance factor in favour of the defenders
A - A	Yes		1.4
B - B	Yes		2.0
C - C	Yes		1.6
D- D	Yes		1.7

Additional notes:
The only recorded excavations were carried out by Prof. W. J. Varley in 1939-40 (Varley 1948). No formal finds report has been published. Therefore, no information relating to potential finds of sling stones is available.

The western entrance presents something of a paradox. It is clearly elaborated beyond purely functional requirements. However, it is in comparison with the eastern entrance that it is (in military terms) the most interesting. Both the eastern and western approaches were presumably in use throughout the hillfort's history. The eastern entrance only goes through limited phases of reshaping, seemingly more as a result of the requirements of other ditch and bank building than as a purposeful construction.

The level of construction of the eastern entrance should presumably represent an acceptable level of defence. The approach to the eastern entrance is narrow and fairly steep. The banks even today tower over the approacher, giving a significant feeling of enclosure. This presumably would have been increased if the entrance walls were revetted in stone, as were those at the western entrance. There is a short extension of the south-western bank northwards that characteristically requires the right (unshielded side) of an approacher to be presented to the ramparts above.

Section A-A: the large bank constructed as part of the annexe prevents any view from the inner bank. It is from the top of this bank that an assailant would first come into range of the inner bank. It is not without significance that this feature may be a rebuilding on a much larger scale of a smaller bank, originally constructed during Varley's Period Three. The bank appears to negate any military advantage in constructing the outermost banks, allowing an assailant to advance to this point, unseen and unchallenged.

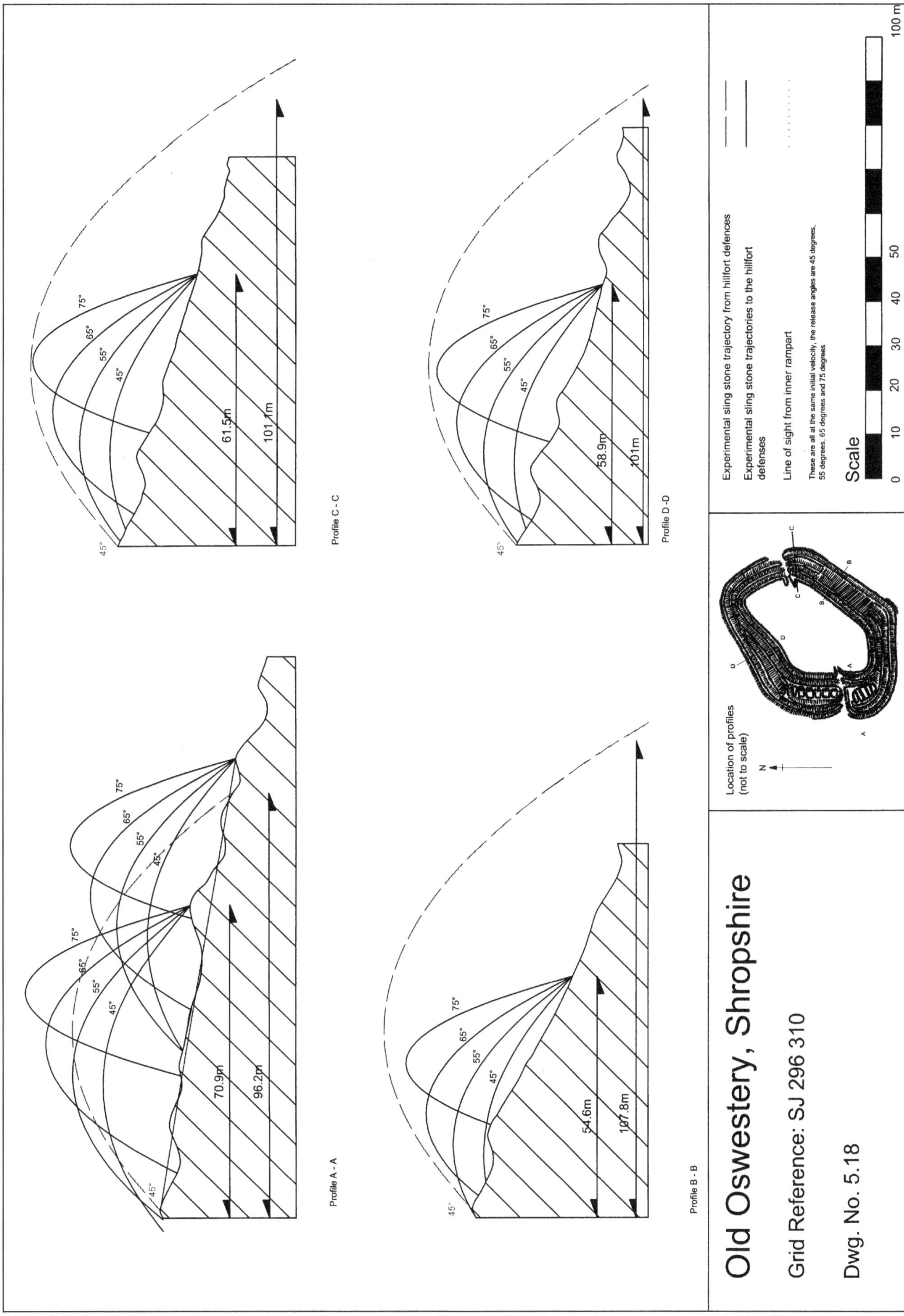
Old Oswestery, Shropshire
Grid Reference: SJ 296 310
Dwg. No. 5.18
Location of profiles
(not to scale)
N
A
B
C
D
Experimental sling stone trajectory from hillfort defences
Experimental sling stone trajectories to the hillfort defenses
Line of sight from inner rampart
These are all at the same initial velocity, the release angles are 45 degrees, 55 degrees, 65 degrees and 75 degrees
Scale
0
10
20
30
40
50
100 m
Profile A - A
70.9m
96.2m
Profile B - B
54.6m
107.8m
Profile C - C
61.5m
101.1m
Profile D -D
58.9m
101m
45°
55°
65°
75°

Site: Pilsdon Pen		ST 412 013 Grid Ref.	
Reference: Gelling 1977			
Sling projectiles recovered from excavation: Yes			
Drawing No. 5.19			
Section	Would any assailant have to enter an area defined by external earthworks before being able to strike the innermost bank?		Distance factor in favour of the defenders
A - A	No		1.8
B - B	No		1.4
C - C	No		1.5
D- D	No		1.8
Additional notes: The excavation showed Pilsdon Pen to date to the 1st Century BC, when the use of the sling appears to have been declining in the Wessex region (Sharples 1991a, 83). The hillfort shows little evidence of occupation except for the construction (during the Iron Age) of a square structure. This may be analogous with the square feature within Nordy Bank.			

Site: Uffington Castle		Grid Ref. SU 299 863	
Reference: Miles *et al.* 2003			
Sling projectiles recovered from excavation: Yes			
Drawing No 5.20			
Section	Would any assailant have to enter an area defined by external earthworks before being able to strike the innermost bank?		Distance factor in favour of the defenders
A - A		No	1.6
B - B		No	1.4
C - C		No	1.3
D- D		No	1.4
Additional notes: Despite the recovery of sling stones from Uffington Castle (Miles *et al.* 2003, 185), the hillfort seems to be vulnerable to assault using the sling, as there are no surviving outer earthworks. It is possible that a small outer ditch or bank may have been ploughed away during the nineteenth and early twentieth centuries, when the area surrounding the hillfort was more intensively farmed. However, geophysical survey outside the hillfort (*ibid*, 134) indicated that there was significant activity that is no longer visible on the surface. Most of the geophysically surveyed features appear to date to the Romano-British period, but the area surveyed was some 150m from the hillfort, so is likely to have missed any smaller ditch and bank. Further evidence that may indicate that initially, at least the area to the front of the hillfort was protected from grazing comes from the molluscan assemblage from the ditches. In the secondary fill of the ditches were species that prefer shaded conditions, suggesting tall herbaceous vegetation (*ibid,* 199). Although this may result from the ditch being too steep to allow sheep easy access to graze (*ibid*, 199), it could also result from wider exclusion of grazing animals. There is some potential evidence that an outer earthwork may have existed, but this is at best hypothetical at present.			

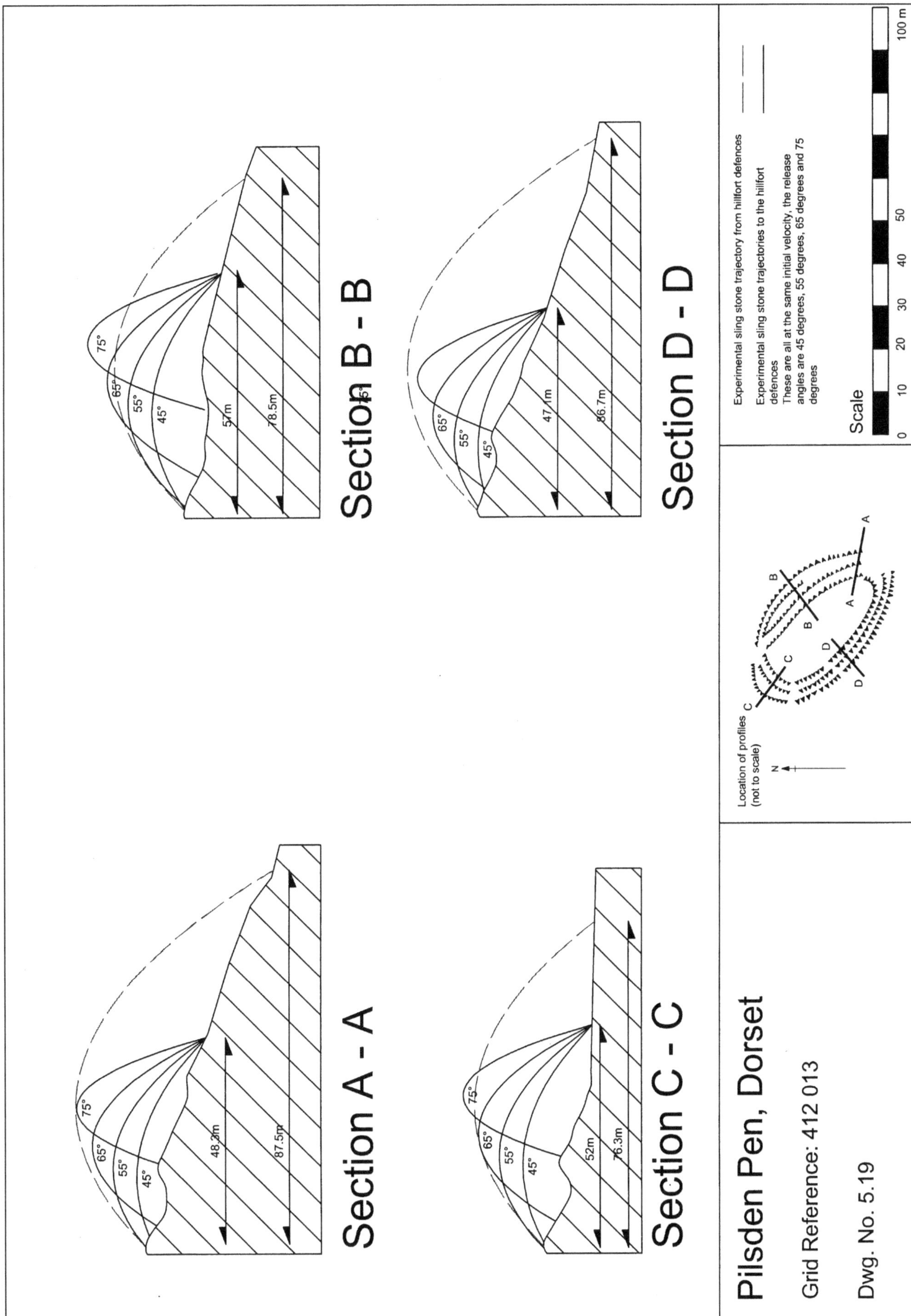
Section A - A
75°
65°
55°
45°
48.3m
87.5m
Section B - B
57m
78.5m
Section C - C
52m
76.3m
Section D - D
47.1m
86.7m
Pilsden Pen, Dorset
Grid Reference: 412 013
Dwg. No. 5.19
Location of profiles
(not to scale)
N
Experimental sling stone trajectory from hillfort defences
Experimental sling stone trajectories to the hillfort defences
These are all at the same initial velocity, the release angles are 45 degrees, 55 degrees, 65 degrees and 75 degrees
Scale
0 10 20 30 40 50 100 m

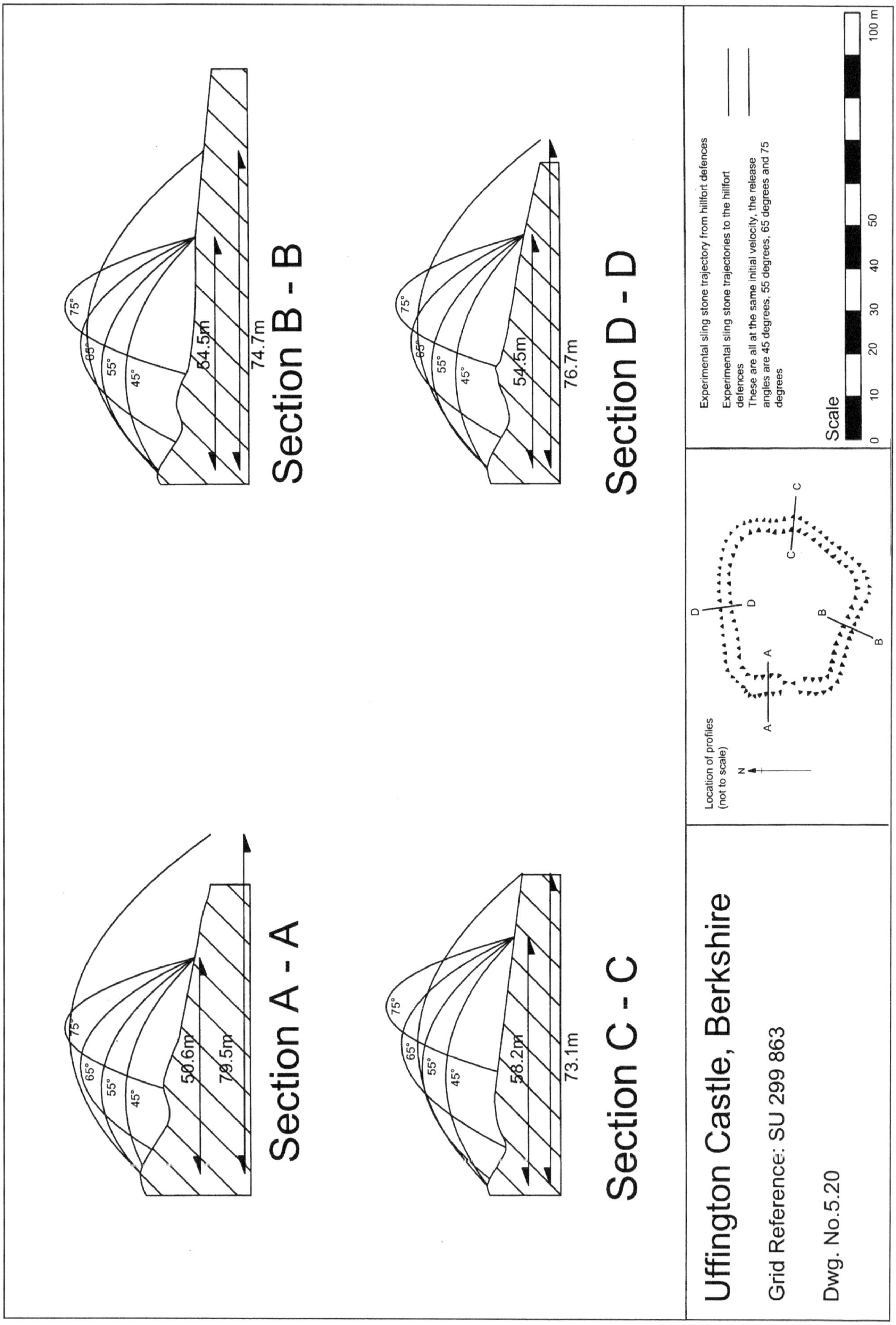
Section A - A
75°
65°
55°
45°
50.6m
79.5m
Section B - B
75°
65°
55°
45°
54.5m
74.7m
Section C - C
75°
65°
55°
45°
58.2m
73.1m
Section D - D
75°
65°
55°
45°
54.5m
76.7m
Uffington Castle, Berkshire
Grid Reference: SU 299 863
Dwg. No.5.20
Location of profiles
(not to scale)
N
A
B
C
D
Experimental sling stone trajectory from hillfort defences
Experimental sling stone trajectories to the hillfort defences
These are all at the same initial velocity, the release angles are 45 degrees, 55 degrees, 65 degrees and 75 degrees
Scale
0
10
20
30
40
50
100 m

Bibliography

All references to classical authors are from the Loeb Classical Library unless otherwise stated

Abrams, E. M. 1989: Architecture and energy: an evolutionary perspective. In Schiffer, M. B. (ed.) *Archaeological Method and Theory: Volume 1* (Tucson), 47-88.

Ahlberg, G. 1971: *Fighting on Land and Sea in Greek Geometric Art* (Lund).

Aitchison, N. B. 1994: *Armagh and the Royal Centres in Early Mediaeval Ireland: Monuments, Cosmology and the Past.* (Glasgow).

Arkush, E. and Stanish, C. Interpreting Conflict in the Ancient Andes. *Current Anthropology* 46 (1) 3-28

Alcock, L. 1968: Excavations at South Cadbury Castle. *Antiquaries Journal* 48, 6 -17.

Alcock, L. 1972a: Excavations at Cadbury-Camelot, 1966-70. *Antiquity* 46, 29-38.

Alcock, L. 1972b: *'By South Cadbury is that Camelot...' Excavations at Cadbury Castle 1966-1970* (London).

Alva, C. and Donnan, C. B. 1994: *Royal Tombs of Sipan* (Los Angeles)

Allen, D. F. 1967: Iron currency bars in Britain. *Proceedings of the Prehistoric Society* 33, 307-35.

Allen, T. G. and Robinson, M. A. 1993: *The Prehistoric Landscape and Iron Age Enclosed Settlement at Mingies Ditches Hardwick-with-Yelford, Oxon* (Oxford).

Anderson, J. K. 1991: Hoplite weapons and offensive arms. In Hanson, V. D. (ed.) *Hoplites: the Classical Greek Battle Experience* (London), 15-37.

Antonaccio, C. 1993: The archaeology of ancestors. In Dougherty, C. and Kurke, L. (eds.), *Cultural Poetics in Archaic Greece: Cult, Performance, Politics* (Cambridge), 46-70.

Ascham, R. 1985 [1545] *Toxophillus.* (Manchester).

Avery, M. 1976: Hillforts of the British Isles: a student's introduction. In Harding, D. W. (ed.) *Hill-forts: Later Prehistoric Earthworks in Britain* (London) 1-58.

Avery, M. 1986: 'Stoning and fire' at hillfort entrances of Southern Britain. *World Archaeology* 18 (2), 216-30.

Avery, M. 1993: *Hillfort Defences of Southern Britain* (BAR British Series 231: Oxford).

Avery, M., Sutton, J. E. G. and Banks, J. W. 1967: Rainsborough, Northants, England: excavations 1961-5. *Proceedings of the Prehistoric Society* 33, 207-306.

Bailey, C. J. 1967: An Early Iron Age/Romano - British site at Pins Knoll, Litton Cheney, Final Report. *Proceedings of the Dorset Natural History and Archaeological Society* 89, 147-59.

Bankes, G. 1980: *Moche Pottery from Peru* (London).

Bankes, G. 1996: *South America, Pre-Columbian VIII, 1 (ii) Arms and Armour: Isthmian* In Dictionary of Art Vol. 29 (London).

Barrett, J. 1999 The mythological landscapes of the British Iron Age. In Ashmore, W. and Knapp, A, B, (eds.) *Archaeologies of Landscape: Contemporary Perspectives* (Oxford), 253-65.

Barrett, J. C. and Bradley, R. J. 1980: Later Bronze Age settlement in south Wessex and Cranborne Chase. In Barrett, J. C. and Bradley, R. J. (eds.) *Settlement and Society in the British Later Bronze Age.* (BAR British Series. 83: Oxford), 181-208.

Barrett, J. C. and Needham, S. P. 1988: Production, circulation and exchange: problems in the interpretation of Bronze Age bronzework. In Barrett, J. C. and Kinnes, I. A. (eds.) *The Archaeology of Context in the Neolithic and Bronze Age: Recent Trends.* (Sheffield), 127-40.

Barrett, J., Woodward, A and Freeman, P. 2000: *Cadbury Castle, Somerset: the Later Prehistoric and Early Historic Archaeology* (London).

Baring-Gould, S., Burnard, R. and Anderson, I. K. 1900: Exploration of Moel Trigarn. *Archaelogia Cambrensis* 17, 189 -211.

Bauer, B. S. 1996: *The Development of the Inca State* (Austin)

Bauer, B. S. 2004 *Ancient Cuzco: Heartland of the Inca* (Austin).

Bauer, B. and Stanish, C. 2001: *Ritual and Pilgrimage in the Ancient Andes The islands of the Sun and the Moon* (Austin)

Bayne, N. 1957: Excavation at Lyneham Camp, Lyneham, Oxon. *Oxoniensia* 22, 1-10.

Beavitt, P.1997: Fighters and foragers, warfare and the spread of agriculture in Borneo. In Carman, J. (ed.) *Materila Harm archaeological studies of war and violence* (Glasgow).

Becker, C. J. 1982: Siedlungen der Bronzezeit und der Vorrömischen Eisenzeit in Dänemark *Offa* 39, 53-71.

Bell, M. 1977: Excavations at Bishopstone. *Sussex Archaeological Collections* 115, 1-299.

Berrin, K 1997: *The Spirit of Ancient Peru* (New York)

Bersu, G. 1940: Excavations at Little Woodbury, part I. *Proceedings of the Prehistoric Society* 6, 30-111.

Betanzos, J. de 1996 [1557]: *Narrative of the Incas.* (translated, Hamilton R. and Buchanan, D.) (Austin).

Biel, J. 1981: The late Hallstatt chieftain's grave at Hochdorf. *Antiquity* 55, 16-8.

Billman, B. R. 2001: Understanding the timing and tempo of the evolution of political centralization on the Central Andean coastline and beyond. In Hass, J. (ed.). *From Leaders to Rulers* (New York), 177-204.

Bishop, M. C. and Coulston, J. C. N. 1993: *Roman Military Equipment.* (London).

Black, L. 1994: Early Christain settlrment in the Briad and Upper Glenarm valleys. Unpublished BA thesis, The Queens University of Belfast

Blenkinsopp, J. 1995: *Sage, Priest, Prophet: Religious and Intellectual Leadership in Ancient Israel.* (Westminster).

Blundell, S. 1995: *Women in Ancient Greece* (Harvard).

Boardman. J, 1983: Symbol and story in Greek Geometric Art. In Moon, W. G. (ed.) *Ancient Greek Art and Iconography* (Madison). 15-36.

Boardman. J, 1998: *Early Greek Vase Painting* (London).

Boast, R. and Evans, C. 1986: The transformation of space: two examples from British prehistory. *Archaeological Review from Cambridge* 5, 193-205.

Bourget, S. 2001: Rituals of Sacrifice: Its Practice at Huaca de la Luna and Its Representation in Moche Iconography. In Pillisbury, J. (ed.) *Moche Art and Archaeology in Ancient Peru* (Washington). 87-101

Bowden, M. and McOmish, D.1987: The required barrier. *Scottish Archaeological Review* 4, 76-84.

Bowden, M. and McOmish, D.1989: Little boxes: more about hillforts. *Scottish Archaeological Review* 6, 12-6.

Bradbury J. 1985: *The Mediaeval Archer* (Woodbridge).

Bradley, R. 1971: Economic change in the growth of early hillforts. In Jesson, M. and Hill, D. (eds.) *The Iron Age and Its Hill-forts* (Southampton). 71-84.

Bradley, R. 1984: *The Social Foundations of Prehistoric Britain: Themes and Variations in the Archaeology of Power* (London).

Bradley, R. 1991: The pattern of change in British prehistory. In Earle, T. (ed.) *Chiefdoms: Power, Economy and Ideology* (Cambridge), 44-70.

Brailsford, J. W. 1962: *Hod Hill, Vol. 1: Antiquities from Hod Hill in the Durden Collection* (London).

BBC 1996: *Dances with Llamas.* (Video)

BBC 1998: *Wildlife on One, Andalusia.* (Video)

Bitnell, W. 1989: The Collfryn hillslope enclosure, Llansantffraid Deuddwr, Powys: excavations 1980-1982. *Proceedings of the Prehistoric Society* 55, 89-134.

Brown, G., Field, D. and McOmish, D. 1994: East Chisenbury midden complex. In Fitzpatrick, A. P. and Morris, E. L. (eds.) *The Iron Age in Wessex: recent Work* (Salisbury), 46-9.

Brown, L. 1984: Objects of stone. In Cunliffe, B. *Danebury: an Iron Age Hillfort in Hampshire: Volume 2: the Excavations 1969-1978: the Finds.* (CBA Research Report 52: London), 407-25.

Brown, M. A. 1982: Swords and sequence in the British Bronze Age. *Archaeologia* 107, 1-42.

Browne, J. 1940: Projectile points. *American Antiquity* 5, 209-13.

Brundage, B. C. 1963: *Empire of the Inca* (Norman).

Brundage, B. C. 1967: *Lords of Cuzco* (Norman).

Buchholz, H-G. 1965: Die Schleuder als Waffe im agaischen Kulturkreis. *Anadolu Arastirmalari* II, 133-59.

Bulleid, A. and Gray, H. St. G. 1917: *The Glastonbury Lake Village Vol.2* (Glastonbury).

Burger, R. L. 1983: Pojoc and Waman Wain: Two Early Horizon Villages in the Chavin Heartland, *Nawpa Pacha* 20, 3-40

Burger, R. L. 1992: *Chavin and the Origins of Andean Civilization* (London).

Burgess, C. 1980: *The Age of Stonehenge* (London).

Burgess, C. B., Coombs. D. and Davies, D. G. 1972: The Broadward Complex and barbed spearheads. In Burgess, C. and Lynch, F. (eds.) *Prehistoric Man in Wales and the West* (Bath), 211-83.

Burket, W. 1992: *The Orientalizing Revolution. Near Eastern Influence on Greek Culture in the Early Archaic Age* (trans. Pinder, M.) (London).

Burn. P, 1995: From chiefdom to state organization in Celtic Europe. In Arnold, B. and Gibson, B. (eds.) *Celtic Chiefdom, Celtic State* (Cambridge), 13-25.

Butler, W. B. 1975: The atl-atl: The physics of function and performance, *Plains Anthropolgist* 20(68), 105-10

Byrne, F. J. 1987: *Irish Kings and High Kings* (London).

Cahlander, A. 1980: *Sling Braiding of the Andes* (Weavers Journal Monograph 4: Boulder, Colorado).

Carman, J. (ed.) 1997: *Material Harm: Archaeological Studies of War and Violence* (Glasgow).

Carman, J. and Harding, A. (eds.) 1999: *Ancient Warfare: Ancient Perspectives* (Stroud).

Carmichael, P. H. 1995: Nasca Burial Patterns: Social Structure and Mortuary Ideology. In Dillehay, T. D. (ed.) *Tombs for the Living: Andean Mortuary Practices* (Washington)

Carneiro, R. L. 1981: The chiefdom as precursor of the state. In Jones, G. D. and Kautz, R. R. (eds.) *The Transition to Statehood in the New World.* (Cambridge), 37-79.

Carneiro, R. L. 1990: Chiefdom-level warfare as exemplified in Fiji and the Cauca Valley. In Haas, J. (ed) *The Anthropology of War* (Cambridge), 190-211.

Carpenter, T. H.1998: *Art and Myth in Ancient Greece*: a Handbook (London).

Carroll, R. P. 1995: War in the Hebrew Bible. In Rich, J. and Shipley, G. (eds.) *War and Society in the Greek World* (London), 25-44.

Chagnon, N. 1990: Reproductive and somatic conflicts of interest in the genesis of violence and warfare among tribesmen. In Haas, J. (ed.) *The Anthropology of War* (Cambridge)

Chapman, C. H. and Chapman, E. F. 1983: *Indians and Archaeology of Missouri* (Columbia).

Chapman, N., Evans, R., Poyner, D. and Powell. S, 1995: Forest of Wyre and Clee Hill Coalfields. In Pearce, A. (ed) *Mining in Shropshire* (Telford), 14-26.

Charlton, D. B. and Mitcheson, M. M. 1983: Yardhope. A shrine to Cocidius? *Britannia* 14, 143-53.

Childe, V. G. 1951: The Significance of the sling for Greek prehistory. *Studies Presented to David Moore Robinson* (St. Louis), 1-5.

Cieza de León, P. 1959 [1553]: *The Incas* (translation, de Onis, H.) (Norman).

Clark, D. 1969: *Plane and Geodetic Surveying for Engineers Vol. 1 Plane Surveying, Metric Edition (SI Units)* (Sixth Edition) (Revised by Jackson, J. E.) (London).

Clark, J. G. D. 1963: Neolithic bows from Somerset, England and the prehistory of archery in North West Europe. *Proceedings of the Prehistoric Society* 29, 50-98.

Clay, R. C. C. 1924: An Early Iron Age Site on Fifield Bavant Down. *Wiltshire Archaeological Magazine* 42, 457-96.

Clay, R. C. C 1925: An inhabited site of La Tène I date on Swallowcliffe Down. *Wiltshire Archaeological Magazine* 43, 59-93.

Clutton-Brock, J. 1987: *A Natural History of Domesticated Mammals.* (Cambridge).

Cobo, B. 1990 [1653]: *Inca Religion and Customs* (translation Hamilton, R.) (Austin)

Coldstream, J. N.1976: Hero cults in the age of Homer. *Journal of the Hellenic Society* 96, 8-17.

Coldstream, J. N.1977: *Geometric Greece* (London).

Coldstream, J. N. 1991: The Geometric style: birth of the picture. In Rasmussen, T. and Spivey, N. (eds.) *Looking at Greek Vases* (Cambridge), 37-56.

Coles, J. M. 1962: European Bronze Age shields. *Proceedings of the Prehistoric Society* 28, 156-190.

Coles, J. M. 1973: *Archaeology by Experiment* (London).

Coles, J. M. 1979: *Experimental Archaeology* (London).

Coles, J. M., Heal, S. V. and Orme, B. 1978: The use and character of wood in prehistoric Britain and Ireland *Proceedings of the Prehistoric Society* 44, 1-45.

Collingwood, R. G. 1930: *The Archaeology of Roman Britain* (London).

Collins, A. E. P. 1953: Excavations on Blewburton Hill, 1948 and 1949. *The Berkshire Archaeological Journal* 53, 21-64.

Collins, A. E. P. 1959: Excavations on Blewburton Hill, 1953, *The Berkshire Archaeological Journal* Vol. 59, 252-73.

Collis, J. R. 1968: Excavations at Owslebury Hants: an interim report. *Antiquaries Journal* 48, 18-31.

Collis, J. R. 1970: Excavations at Owslebury, Hants: a second interim report. *Antiquaries Journal* 50, 246-61.

Collis, J. R. 1973: Burials with weapons in Iron Age Britain. *Germania* 51, 121-33.

Collis, J. R. 1984a *The European Iron Age* (London).

Collis, J. R. 1984b: *Oppida: Earliest Towns North of the Alps* (Sheffield).

Collis, J. R.1996: Hill-Forts, enclosures and boundaries. In Champion, T. C. and Collis, J. R. (eds.) *The Iron Age in Britain and Ireland: Recent Trends* (Sheffield). 87-94.

Collis, J. R. 1997: Celtic myths. *Antiquity* 71, 195-201

Colquhoun, I. and Burgess, C. B. 1988: *The Swords of Britain* (Prähistorische Bronzefunde 4 (5): Munich).

Condit, T. 1995: Hillfort discoveries near Killaloe, Co. Clare. *Archaeology Ireland* 9 (1), 34-7.

Connolly, P. 1991: *Greece and Rome at War* (London).

Conrad, G. W. and Demarest, A. A. 1984: *Religion and Empire: the Dynamics of Aztec and Inca Expansionism* (Cambridge).

Cook, N. D. 1981: *Demographic Collapse Indian Peru, 1520- 1620* (Cambridge)

Coombs, D. 1975: Bronze Age weapon hoards in Britain. *Archaeologia Atlantica,* 1 (1), 49-81.

Costin, C. L., Earle, T. K., Owen, B., and Russell, G. S. 1989: Impact of Inka conquest on local Technology in the Upper Mantaro Valley, Peru. In van der Leeuw, S. E. and Torrance, R. (eds.) *What's New? : A Closer Look at the Process of Innovation* (London) 107-139

Cotter, C. 1994: Atlantic fortifications - The Dùns of the Arran Islands. *Archaeology Ireland* 8 (1), 24-8.

Cotterell, B. and Kamminga, J. 1990: *Mechanics of Pre-Industrial Technology: an Introduction to the Mechanics of Ancient and Traditional Material Culture* (Cambridge).

Cotton, M. A. and Frere, S. S. 1968: Ivinghoe Beacon: excavations 1963-5. *Records of Buckinghamshire* 18, 187-260.

Cowen, J. D. 1967: The Hallstatt sword of bronze: on the continent and in Britain. *Proceedings of the Prehistoric Society* 33, 377-454.

Craig, R., Knusel, C. J. and Carr, G. Fragmentation, mutilation and dismemberment: an interpretation of human remains on Iron Ages sites. In Parker Pearson, M. and Thorpe, I. J. N. (eds.) *Warfare, Violence and Slavery in Prehistory* (BAR International Series 1374: Oxford), 165-80

Creamer. W, 2001: The Origins of centralization: changing features of local and regional control during the Rio Grande Classic Period, A.D. 1325-1540. In Haas, J. (ed.) *From Leaders to Rulers* (New York), 37-58.

Credland, A. G. 1985: Sir Ralph Payne-Galloway and the Turkish Bow. *Journal of the Society of Archer-Antiquaries* 28, 37-48.

Cross, F. M. 1980: Newly found inscriptions in old Canaanite and early Phoenician scripts. *Bulletin of the American Schools of Oriental Research* 238, 1-21.

Cross, F. M. 1992: An inscribed arrowhead of the Eleventh Century BCE in the Bible Land Museum in Jerusalem. *Eretz Israel* (the Biran Vol.), 21-6.

Cunliffe, B. 1968: Excavations at Eldon's Seat, Encombe, Dorset. *Proceedings of the Prehistoric Society* 34, 191-237.

Cunliffe, B. 1971: Some Aspects of hill-forts and their cultural environments. In Jesson, M. and Hill, D. *The Iron Age and its Hill-Forts* (Southampton), 53 - 70.

Cunliffe, B. 1984a: *Danebury: an Iron Age Hillfort in Hampshire. Vol. 1: the Excavations 1969-1978: the Site* (CBA Research Report 52 (1): London).

Cunliffe, B. 1984b: *Danebury: an Iron Age Hillfort in Hampshire. Vol. 2: the Excavations 1969-1978: the Finds* (CBA Research Report 52 (2): London).

Cunliffe, B. 1984c: Iron Age Wessex: continuity and change. In Cunliffe, B. and Miles, D. (eds.) *Aspects of the Iron Age in Central Southern Britain* (Oxford University Committee for Archaeology Monograph 2: Oxford). 12-45.

Cunliffe, B. 1984d: Relations between Britain and Gaul in the first century BC and early first century AD. In Macready, S. and Thompson, F. H. (eds.) *Cross-Channel Trade between Gaul and Britain in the Pre-Roman Iron Age.* (Society of Antiquaries Occasional Paper 4: London), 3-23.

Cunliffe, B. 1986: *Danebury: the Anatomy of a Hillfort.* (London).

Cunliffe, B. 1987: *Hengistbury Head Dorset, Volume 1: the Prehistoric and Roman Settlement, 3500 BC-AD 500* (Oxford University Committee for Archaeology Monograph 13: Oxford).

Cunliffe, B. 1991: *Iron Age Communities in Britain (3rd Edition)* (London).

Cunliffe, B. 1992: Pits, preconceptions and propitiation in the British Iron Age. *Oxford Journal of Archaeology,* 11 (1), 69-82.

Cunliffe, B. 1994: After hillforts. *Oxford Journal of Archaeology,* 13 (1), 71-84.

Cunliffe, B. 1995: *Danebury: an Iron Age Hillfort in Hampshire, Vol. 6: a Hillfort Community in Perspective* (CBA Research Report 102: London).

Cunliffe, B. and Poole, C. 1991: *Danebury: an Iron Age Hillfort in Hampshire, Vol. 5: the Excavations 1979-88: the Finds* (CBA Research Report 73: London).

Cunnington, E. 1895: Excavations on Hambledon Hill, *Proceedings of the Dorset Natural History and Antiquarian Field Club* 16, 156-8.

Cunnington, M. E. 1923a: *The Early Iron Age Inhabited Site at All Cannings Cross* (Devizes).

Cunnington, M. E. 1923b: Pits in Battlesbury Camp. *Wiltshire Archaeological Magazine* 42, 368-73.

Cunnington, M. E.1933: Excavations at Yarnbury Castle Camp, 1932. *Wiltshire Archaeological Magazine* 46, 198-213.

Cunnington, M. E. and Cunnington, B. H. 1913: Casterley Camp excavations. *Wiltshire Archaeological Magazine* 38, 53-105.

Cunnington, M. E. and Cunnington, B. H. 1917: Lidbury Camp. *Wiltshire Archaeological Magazine* 40, 12-36.

Curwen, E and Curwen, E. C. 1927: Excavations in the Caburn, near Lewes. *Sussex Archaeological Collection* 68, 1-56.

Curwen, E. C. 1929: Excavations in the Trundle, Goodwood, 1928. *Sussex Archaeological Collection* 70, 33-85.

Curwen, E. C. 1933: Excavations on Thundersbarrow Hill. *Antiquaries Journal* 13, 109-33.

Curwen, E. C. and Williamson, R. P. R. 1931: The date of Cissbury Camp. *Antiquaries Journal* 11, 14-36.

D'Altroy, T. N. 2001: The archaeological context. In D'Altroy, T. N. and Hastorf, C. A. (eds.) *Empire and Domestic Economy* (New York), 65-96.

D'Altroy, T. N. 2003: *the Incas* (Oxford).

D'Altroy, T. N. and Earle, T. K. 1985: Staple finance, wealth finance and storage in the Inka political economy. *Current Anthropology* 26, 187-97.

Dark, K. R. 1995: *Theoretical Archaeology* (London).

Davidson. H. R. E. 1988: *Myths and Symbols in Pagan Europe: Early Scandinavian and Celtic Religions* (Syracuse).

Davidson. H. R. E, and Fisher, P. 1999: *Saxo Grammaticus, The History of the Danes Books I-IX* (London).

Davies, O. 1939: Excavations at Lissachiggel. *County Louth Archaeological Journal* 9, 209-43.

Davies, S. M. 1981: Excavations at Old Down Farm, Andover, part II: prehistoric and Roman. *Proceedings of the Hampshire Field Club and Archaeological Society* 37, 81-163.

Dawkins, R. M. 1929: *The Sanctuary of Artemis Orthia at Sparta*. The Journal of Hellenic Studies Supplement 5: London.

De Gamboa, P. S. 1999 [1572]: *History of the Incas* (trans. Markham, C.) (New York).

DeMarrais, E., Castillo, L. J. and Earle, T. 1996: Ideology, materialization and power strategies. *Current Anthropology* 37, 15-31.

Dent, J. 1982: Cemeteries and settlement patterns of the Iron Age in the Yorkshire Wolds. *Proceedings of the Prehistoric Society* 48, 437-57.

Dent, J. 1983: Weapons, wounds and war in the Iron Age. *Archaeological Journal* 140, 120-8.

Dent, J. 1985: Three cart burials from Wetwang, Yorkshire. *Antiquity* 52, 85-92.

de Polignac, F. 1995: *Cults, Territory, and the Origins of the Greek City-State* (translation, Lloyd, J) (Chicago).

Dixon, P. 1994: *Crickley Hill: The Hillfort Defences* (Nottingham).

Donlan, W. and Thompson, J. 1976: The charge at Marathon: Heroditus 6.112. *Classical Journal* 71, 339-43.

Donnan, C. B. 1976: *Moche Art and Iconography* (Los Angles).

Downes, J. 1997: The shrine at South Cadbury Castle: belief enshrined? In Gwilt, A. and Haselgrove, C. (eds.) *Reconstructing Iron Age Societies* (Oxbow Monograph 71: Oxford), 145-52.

Drennan, R. D. 1991: Pre-Hispanic chiefdom trajectories in Mesoamerica, Central America and Northern South America. In Earle, T. (ed.) *Chiefdoms: Power, Economy and Ideology* (Cambridge), 263-87.

Drusini, A. G. 1991: Anthropological study of Nazca trophy heads. *Homo*. 41 (3), 251-65.

Duday, H. 1998: Le charnier gaulois de Ribemont-sur-Ancre (Somme). *Etudes et Documents Fouilles 4, Les Celtes: rites funeraires en Gaule du nord entree VIe et le 1er siecle avant Jesus-Christ.* Namur, (113-9)

Dugas, C. 1928: *Delos: Fascicule 10: les Vases de l' Heraion* (Paris).

Dwyer, E. B. 1971: The Early Inca occupation of the valley of Cuzco, Peru. Unpublished Ph.D dissertation. Department of Anthropology, University of California, Berkely

Earle, T. K. 1978: *Economic and Social Organisation of a Complex Chiefdom: The Halelea District, Kaua'I Hawaii* (Ann Arbor).

Earle, T. K. 1990: Style Iconography as Legitimation in Complex Chiefdoms. In Conkey, M. and Hastorf, C. A. *The Uses of Style in Archeaology* (Cambridge). 73-81

Earle, T. K. 1991a: The Evolution of chiefdoms. In Earle, T. K. (ed.) *Chiefdoms: Power, Economy, and Ideology* (Cambridge), 1-15.

Earle, T. K. 1997: *How Chiefs Come to Power: the Political Economy in Prehistory* (Standford).

Earle, T. K. 2001: Institutionalization of chiefdoms: why landscapes are built. In Haas, J. (ed.) *From Leaders to Rulers* (New York), 105-24.

Earle, T. K. 2002: *Bronze Age Economics: the Beginnings of Political Economies* (Boulder).

Earle, T., D'Altroy, T., Hastorf, C. A., Scott, C., Costin, C., Russell, G. and Sandefur, E. 1987: *The Impact of the Inka Conquest on the Wanka Domestic Economy* (Los Angeles).

Earle, T., D'Altroy, T., LeBlanc, C., Hastorf, C. A. and LeVine, T. Y. 1980: Changing settlement patterns in the upper Mantaro Valley, Peru. *Journal of New World Archaeology* 4, 1-49.

Edlin, H. L. 1970: *Collins Guide to Tree Planting and Cultivation* (Newton Abbott).

Edmonds, M. and Thomas, J. 1987: The archers: an everyday story of country folk. In Brown, A. G. and

Edmonds, M.R. (eds.) *Lithic Analysis and Later British Prehistory: some Problems and Approaches* (BAR British Series 162: Oxford), 187-99.
Edwards, N. 1990: *The Archaeology of Early Mediaeval Ireland* (London).
Ehrenberg, M. R. 1977: *Bronze Age Spearheads from Berkshire, Buckinghamshire and Oxfordshire* (BAR British Series 34: Oxford).
Ehrenreich, R. M.1986: Blacksmithing technology in Iron Age Wessex. *Oxford Journal of Archaeology* 5, 165-84.
Ehrenreich, R. M.1991: Metalworking in Iron Age Britain: hierarchy or heterarchy? In Ehrenreich, R. M. (ed.) *Metals in Society: Theory beyond Analysis* (Philadelphia), 69-80.
El Inca, G. de La Vega 1987 [1609]: *Royal Commentaries of the Incas and General History of Peru* (trans. by Livermore, H. V.) (Austin).
Enriquez De Guzman, Don A. 1862 [1543]: *Libro de la Vida y Costumbres de don Alonso Enriquez De Guzman* (Trans. Markham, C.R.) (Cambridge).
Eogan, G. 1984: *Excavations at Knowth 1* (Dublin).
Ethelberg, P. 1993: The chieftains' farms of the Over Jerstal group. *Journal of Danish Archaeology* 11, 111-35.
Evans, A, 1928: *The Palace of Minos at Knossos II* (London).
Evans, C. 2003: *Power and Island Communities: Excavations at the Wardy Hill Ringwork, Coveney, Ely* (East Anglia Archaeology 103: Cambridge).
Evans, J. G. 1975: *The Environment of Early Man in the British Isles* (London)
Evans, S. S. 1997: The *Lords of Battle: Image and Reality of the 'Comitatus' in Dark-Age Britain.* (Woodbridge).

Fagan, B. M. 1991: *Ancient North America, The Archaeology of a Continent* London
Fagles, R. 1990: *The Iliad* (London).
Farrar, R. A. H. 1955: An Early Iron Age fort on Shipton Hill, Shipton Gorge. *Proceedings of the Dorset Natural History and Archaeological Society* 77, 135-6.
Fasham, P. J. 1985: *The Prehistoric Settlement at Winnall Down, Winchester: Excavations of MARC3 Site R17 in 1976 and 1977* (Hampshire Field Club Monograph 2: Winchester).
Feest, C. 1980: *The Art of War* (London).
Ferguson, R. B. (ed.) 1984: *Warfare, Culture and Environment* (London).
Ferguson, R. B. 1990: Explaining war. In Haas, J. (ed.) *The Anthropology of War* (Cambridge), 26-55.
Field, N. and Parker Pearson, M. 2003: *Fiskerton: an Iron Age Timber Causeway with Iron Age and Roman Votive Offerings: the 1981 Excavations* (Oxford).
Finley, M. I. 1979: *The World of Odysseus* (London).
Fisher, N. 2000: Hybris, revenge and stasis in the Greek city-states. In van Wees, H. (ed.) *War and Violence in Ancient Greece* (London), 83-123.
Fisher, N. and van Wees, H. (eds.)1998: *Archaic Greece: New Approaches and New Evidence* (London).
Fitzpatrick, A. P. 1997: Everyday life in Iron Age Wessex. In Gwilt, A. and Haselgrove, C. (eds.) *Reconstructing Iron Age Societies* (Oxbow Monograph 71: Oxford), 73-86.
Flannery, K. V. 1972: The cultural evolution of civilization. *Annual Review of Ecology and Systematics* 3, 399-426.
Forde, D. C., Griffith, W. E., Hogg, A. H. A. and Houlder, C. H. 1963: Excavations at Pen Dinas, Aberystwyth. *Archaeologia Cambrensis* 112, 125-53.
Forde-Johnston, J. 1962: Earl's Hill, Pontesbury, and related hillforts in England and Wales. *Archaeological Journal* 119, 66 -91.
Forde-Johnston, J. 1976: *Hillforts of the Iron Age in England and Wales: a Survey of the Surface Evidence* (Liverpool).
Forrest, W. G. 1980: *A History of Sparta* (London)
Fox, C. 1946: *A Find of the Early Iron Age from Llyn Cerrig Bach, Anglesey: interim Report* (Cardiff).
Frere, S. S., Hassall, M. W. C. and Tomlin, R. S. O. 1977: Roman Britain in 1976. *Britannia* 8, 356-425.
Fried, M. H. 1967: *The Evolution of Political Society: an Essay in Political Anthropology* (New York).

Gantz, J. 1981: *Early Irish Myths and Sagas* (London).
Gantz, T. 1993: *Early Greek Myth* (Baltimore).
Garcia, L. P. 1972: *The Balearic Islands*, (Translation, Brown, M.) (London).
Gardner, W. 1910: Pen-y-Corddyn, near Abergele. *Archaeologia Cambrensis* 10, 79-156.
Gardner, W. and Savory, H. N. 1964: *Dinorban, a Hillfort Occupied in Early Iron Age and Roman Times* (Cardiff).
Gelling, P.S. 1977: Excavations at Pilsdon Pen, Dorset, 1964-71. *Proceedings of the Prehistoric Society* 43, 263-86.
Gent, H. 1983: Centralized storage in later prehistoric Britain. *Proceedings of the Prehistoric Society* 49, 243-67.
Gibson, A. M. 2002: *Prehistoric Pottery in Britain and Ireland* (Stroud).
Gibson, T. 1990: Raiding, trading and tribal autonomy in insular Southeast Asia. In Hass, J. (ed.) *The Anthropology of War* (Cambridge), 125-45.
Gifford, D. and Hoggarth, P. 1976: *Carnival and Coca Leaf Some Traditions of the Quechua Ayllu* (Edinburgh).
Glob P. V. 1969: *The Bog People: Iron-Age Man Preserved* (trans. Bruce-Mitford, R. L. S) (London).
Grant, A. 1984: Animal husbandry in Wessex and the Thames Valley. In Cunliffe, B. and Miles, D. (eds.) *Aspects of the Iron Age in Central Southern Britain* (Oxford University Committee for Archaeology Monograph 2: Oxford), 102-19.
Grant, M. 1995: *The Ancient Historians* (London).
Green, H. S. 1980: *The Flint Arrowheads of the British Isles* (BAR British Series 75: Oxford).
Green, M. 1989: *Symbol and Image in Celtic Religious Art* (London).
Greenhaugh, P. A. L. 1973: *Early Greek Warfare: Horsemen and Chariots in the Homeric and Archaic Ages* (Cambridge).
Greenland, H. 1994: *The Traditional Archers Handbook: a Practical Guide* (Bristol).

Greenwell, W. and Brewis, W. P. 1909: The origin, evolution and classification of bronze spearheads in Great Britain and Ireland. *Archaeologia* 61, 439-72.

Greep, S. J. 1987: Lead sling-shot from Windridge Farm, St. Albans and the use of the sling by the Roman Army in Britain. *Britannia* 18, 183-200.

Gregory, L. [1904] 1970: *Gods and Fighting Men* Gerrards Cross

Gregory, T. 1991: *Excavations in Thetford, 1980-82, Fison Way* (East Anglia Archaeology 53: Dereham).

Gresham, C. A. 1939: Spettisbury Rings, Dorset. *Archaeological Journal* 96, 114 -31.

Grice, E. 2006: A chilling tale of ritual murder. *The Daily Telegraph* Saturday January 7 2006

Griffith, P. 1995: *The Viking Art of War* (London).

Griffiths, W. B. 1989: The sling and its place in the Roman Imperial Army. In van Driel-Murray, C. (ed.) *Roman Military Equipment: the Sources of Evidence* (BAR International Series 476: Oxford), 255-79.

Griffiths, W. B. and Carrick, P. 1994: Reconstructing Roman slings. *The Arbeia Journal* 3, 1-11.

Griffiths, W. E. and Hogg, A. H. A. 1956: The hill-fort on Conway Mountain, Caernarvonshire. *Archaeologia Cambrensis* 105, 49-80.

Grogan, E. and Condit, T. 1994: New hillfort date gives clue to late Bronze Age. *Archaeology Ireland* 8 (2), 7.

Gush, G. 1975: *Renaissance Armies 1480-1650* (Cambridge).

Haas, J. 1990: Warfare and the evolution of tribal polities in the prehistoric Southwest. In Haas, J. (ed.) *The Anthropology of War* (Cambridge), 171-89.

Hall, E. 1995: Asia unmanned: images of victory in classical Athens. In Rich, J. and Shipley, G. (eds.) *War and Society in the Greek World* (London), 108-133.

Hambleton, E. 1999: *Animal Husbandry Regimes in Iron Age Britain: a Comparative Study of Faunal Assemblages from British Iron Age Sites* (BAR British Series 282: Oxford).

Hamilton, J. R. C. 1968: Iron Age forts and epic literature. *Antiquity* 42, 103-8.

Hamilton, S. and Gregory, K. 2000: Updating the Sussex Iron Age. *Sussex Archaeological Collections* 138, 57-74.

Hamm, J. 1989: *Bows and Arrows of the Native Americans* (New York).

Hanson, V. D. 1991: *Hoplites: the Classical Greek Battle Experience* (London).

Hanson, V. D. 1998: *Warfare and Agriculture in Classical Greece* (2nd Edition) (Berkeley and Los Angeles).

Hanson, V. D. 2000: Hoplite battle as Ancient Greek warfare: when, where and why? In van Wees, H. (ed.) *War and Violence in Ancient Greece* (London), 201-32.

Harbison, P. 1969: *The Daggers and Halberds of the Early Bronze Age in Ireland.* (Prähistorische Bronzefunde 6 (I) Munich).

Harbison, P. 1971: Wooden and stone *chevaux-de-frise* in Central and Western Europe. *Proceedings of the Prehistoric Society* 37, 195-225.

Harbison, P. 1976: *Bracers and V-perforated buttons in the Beaker and Food Vessel Cultures of Ireland* (Appendix B: a tentative list of British bracers) Archaeologia Atlantica Research Report 1: Hamburg.

Harding, A. F. 1984: *The Mycenaens and Europe* (London).

Harding, A. F. 1999: Swords, shields and scholars: Bronze Age warfare, past and present. In Harding, A. F. (ed.) *Experiment and Design, Archaeological Studies in Honour of John Coles* (Oxford), 87-93.

Harding, D. W. 1972: *The Iron Age in the Upper Thames Basin.* (Oxford).

Harding, D. W. 1974: *The Iron Age in Lowland Britain* (London).

Harding, D. W. 1980: *Celts in Conflict, Hillfort Studies 1927-1977* (University of Edinburgh, Department of Archaeology Occasional Paper 3: Edinburgh).

Harding, D. W., Blake, I. M. and Reynolds, P. J. 1993: *An Iron Age Settlement in Dorset: Excavation and Reconstruction* (University of Edinburgh Department of Archaeology Monograph 1: Edinburgh).

Hardy, R. 1995: *Longbow: a Social and Military History* (3rd Edition) (Frome).

Harner, M. J. 1972: *The Jívaro: People of the Sacred Waterfalls* (Berkeley).

Harris, M. 1984: A cultural materialism: theory of band and village warfare: The Yanomamö test. In Ferguson, B. (ed.) *Warfare, Culture and Environment* (Orlando), 111-40.

Hartridge, R. 1977: Excavations at the prehistoric and Romano-British site on Slonk Hill, Shoreham Sussex. *Sussex Archaeological Collections* 116, 69-141.

Haselgrove, C. 1996: Iron Age coinage: recent work. In Champion, T. C. and Collis, J. R. (eds.) *The Iron Age in Britain and Ireland: Recent Trends* (Sheffield), 67-85.

Hastorf, C. A. 1993: *Agriculture and the Onset of Political Inequality before the Inka* (Cambridge).

Hastorf, C. A. 2001 (a): The Natural Environment. In *Empire and Domestic Economy*. In D'Altroy, T. N. and Hastorf, C. A. (eds) New York. 55-64

Hastorf, C. A. 2001 (b): Agricultural Production and Consumption. In *Empire and Domestic Economy*. In D'Altroy, T. N. and Hastorf, C. A. (eds) New York. 155-78

Hastorf, C.A. 2001 (c): The Xauxa Andean life. In D'Altroy, T. N. and Hastorf, C. A. *Empire and Domestic Economy* (New York), 315-24.

Hastorf, C. A. and D'Altroy, T. D. 2001: The domestic economy, households and Imperial transformation. In D'Altroy, T.N. and Hastorf, C.A. (eds.) *Empire and Domestic Economy* (New York), 3-25.

Hastorf, C. A., Earle, T., Wright, H., LeCount, L., Russell, G. and Sandefur, E. 1989: Settlement archaeology in the Jauja region of Peru. *Andean Past* 2, 81-130.

Hastorf, C. A., and Johannessen, S. 1989: Prehispanic tree cropping in the Andes: The development of fuel and family. Paper presented at the 29th Annual Meeting of the Institue of Andean Studies (Berkeley).

Haubold, J. 2000: *Homer's People: Epic Poetry and Social Formation* (Cambridge).

Hawkes, C. F. C. 1931: Hill forts. *Antiquity* 5, 60-97.

Hawkes, C. F. C. 1939: The Excavations at Quarley Hill, 1938. *Proceedings of the Hampshire Field Club* 14, 136-94.

Hawkes, C. F. C. 1959: The ABC of the British Iron Age. *Antiquity* 33, 170-81.

Hawkes, C. F. C. 1971: Fence, wall, dump, from Troy to Hod. In Jesson, M. and Hill, D (eds.) *The Iron Age and its Hill-Forts* (Southampton), 5-18.

Hawkes, C. F. C., Myres, J. N. L. and Stevens, C. G. 1930: *St. Catherine's Hill, Winchester* (Winchester).

Haynes, C. V. Jnr. 1966: Elephant-hunting in North America. *Scientific American* 214, 104-12.

Heaney. M, 1994: *Over Nine Waves: a Book of Irish Legends* (London).

Hedges, J. W. 1984: *Tomb of the Eagles: a Window on Stone Age Tribal Britain* (London).

Helgeland, G. (ed.) 1975: *Complete Guide to Bowhunting* (Eaglewood Cliffs).

Helms, M. 1988: *Ulysses' Sail. An ethnographic odyssey of power, knowledge and geographical distance.* New Jersey

Hemming, J. 1983: *The Conquest of the Incas* (London)

Hencken, H. 1950: Lagore Crannog: an Irish Royal Residence of the 7th to 10th Centuries AD. *Proceedings of the Royal Irish Academy* 53C, 1-247.

Hencken, T. C. 1938: The Excavation of the Iron Age Camp on Bredon Hill, Gloucestershire, 1935-1937. *Archaeological Journal* 95, 1-111.

Hey, G., Bayliss, A. and Boyle, A. 1991: Iron Age inhumation burials at Yarton, Oxfordshire. *Antiquity* 73, 551-62

Heyerdahl, T Sandweiss, D. H., and Narvaez, A.1995: *Pyramids of Tucume The quest for Peru's forgotten City* (London)

Hill, H. n.d: *My Method of Shooting, a Bow and Arrow* (Wilsonville).

Hill, J. D. 1989: Rethinking the Iron Age. *Scottish Archaeological Review* 6,16-24.

Hill, J. D. 1995a: *Ritual and Rubbish in the Iron Age of Wessex: a Study on the Formation of a Specific Archaeological Record* (BAR, British Series 242: Oxford).

Hill, J. D. 1995b: How should we understand Iron Age societies and hillforts? A contextual study from Southern Britain. In Hill, J. D. and Cumberpatch, C. G. (eds.) *Different Iron Ages: Studies on the Iron Age in Temperate Europe* (BAR International Series 602: Oxford), 45-66.

Hill, J. D. 1996: Hill-forts and the Iron Age of Wessex. In Champion, T. C. and Collis, J. R. (eds.) *The Iron Age in Britain and Ireland: Recent Trends* (Sheffield), 95-116.

Hingley, R. 1984 Towards social analysis in archaeology: Celtic society in the Iron Age of the Upper Thames Valley. In Cunliffe, B. and Miles, D. (eds.) *Aspects of the Iron Age in Central Southern Britain* (Oxford University Committee for Archaeology Monograph 2: Oxford), 72-88.

Hodder, I. R. 1982a: *The Present Past: an Introduction to Anthropology for Archaeologists* (London).

Hodder, I. R. 1982b: *Symbols in Action* (Cambridge).

Hodder, I. R. 1986: *Reading the Past: Current Approaches to Interpretation in Archaeology* (Cambridge).

Hodder, I. R. 1990: *The Domestication of Europe* (Cambridge).

Hodges, H. 1989: *Artefacts, An Introduction to Early Materials and Technology* (London).

Hodges, R. 1982: *Dark Age Economics, the Origins of Towns and Trade AD 600-1000* (London).

Hoerner, S. F. 1965: *Fluid-Dynamic Drag: Practical Information on Aerodynamic Drag and Hydrodynamic Resistance* (London).

Hogg, A. H. A. 1973: Excavations at Harding's Down West Fort, Gower. *Archaeologia Cambrensis* 122, 55-68.

Holladay, J. S. Jnr. 1995: The Kingdoms of Israel and Judah: political and economic centralization in the Iron Age IIA-B (ca.1000-750 BCE). In Levy, T. E. (ed.) *The Archaeology of Society in the Holy Land* (Leicester), 368-98.

Hooley, R. W. 1921: Note on a hoard of iron currency-bars found on Worthy Down, Winchester. *The Antiquaries Journal* 1, 320-7.

Hooper, B. Anatomical Considerations. In Cunliffe, B. and Poole, C. 1991: *Danebury: an Iron Age Hillfort in Hampshire, Vol. 5: the Excavations 1979-88: the Finds* (CBA Research Report 73: London). 425-31

Howard, C. D. 1974: The atlatl: function and performance. *American Antiquity* 39 (1), 102-4.

Hoyle, 1966: *Peru* (London).

Hubrecht, A. V. M. 1964: The use of the sling in the Balearic Isles. *Bulletin Antieke Beschavung* 39, 92-3.

Hughes, H. and Gardner 1906: Exploration of Pen-Y-Gaer, above Llanbedr-Y-Cenin. *Archaeologia Cambrensis* 6, 241-67.

Hull, V. 1938: 'Aided Meidbe': The Violent Death of Medb. *Speculum* 13, 52-61.

Hunt, P. 1998: *Slaves, Warfare, and Ideology in the Greek Historians* (Cambridge).

Hurwit, J. M. 1985: *The Art and Culture of Early Greece, 1110-480 BC* (Ithaca and London).

Isbell, W. H. 1997: *Mummies and Mortuary Monuments*: a Postprocessual Prehistory of Central Andean Social Organization. (Austin).

Jackson, A. H. 1995: War and raids for booty in the world of Odysseus. In Rich, J and Shipley, G. (eds.) *War and Society in the Greek World* (London), 64-76.

Jackson, K. H. 1964: *The Oldest Irish Tradition: a Window on the Iron Age* (Cambridge).

Jacobsen, T. 1987: *The Harps that Once... Sumerian Poetry in Translation* (New Haven).

James, N. D. G. 1989: *The Forester's Companion* (Oxford).

James, S. 1998: Celts, politics and motivation in archaeology. *Antiquity* 72, 200-9.

James, S. 1999: *The Atlantic Celts: Ancient People or Modern Invention?* (London).

Jelicic, L. and O'Connell, M. 1992: History of vegetation and land use from 3200 B.P to the present in the north-west Burren, a karstic region of western Ireland. *Vegetation History and Archaeobotany* 1, 119-40.

Jennings, J. D. 1978: *Ancient South Americans* (San Francisco)

Jensen: J, 1982: *The Prehistory of Denmark* (London).

Jewish Publication Society 1978: *Tanakh: a new Translation of the Holy Scriptures according to the Traditional Hebrew Text* (New York).

Jobey, G. 1978: Burnswark Hill. *Transactions of the Dumfrieshire and Galloway Natural History and Archaeological Society* 53, 57-108.

Jones, M. 1984: Regional patterns in crop production. In Cunliffe, B. and Miles, D. (eds.) *Aspects of the Iron Age in Central Southern Britain* (Oxford University Committee for Archaeology Monograph 2: Oxford), 120-25.

Jones, M. 1996: Plant exploitation. In Champion, T. C. and Collis, J. R. (eds.) *The Iron Age in Britain and Ireland: Recent Trends* (Sheffield), 29-40.

Jope, E. M. 1961: Daggers of the Early Iron Age in Britain. *Proceedings of the Prehistoric Society* 27, 307-43.

Julien, C. J. 1993: Finding a Fit: Archaeology and Ethnohistory of the Incas. In Malpas, M. A. (ed.) *Provincial Inca Archaeological and Ethnohistorical Assessment of the Impact of the Inca State* (Iowa City), 177-233

Karageorghis, V. 1976: *Kition: Mycenaean and Phoenician Discoveries in Cyprus* (London).

Kaul, F. 2003: The Hjortspring find. In Crumlin-Pedersen, O. and Trakadas, A. (eds.) *Hjortspring, a Pre-Roman Iron Age Warship in Context* (Roskilde), 141-86.

Keatinge, R. W. 1988: *Peruvian Prehistory* (Cambridge)

Keeley, L. H. 1996: *War before Civilisation: the Myth of the Peaceful Savage* (Oxford).

Kelleher J.V. 1971: The Táin and the annals. *Ériu* 22, 107-27.

Kelly, F. 1988: *A Guide to Early Irish Law* (Dublin).

Kelly, F. 1998: *Early Irish Farming, a Study Based Mainly on the Law-Texts of the 7th and 8th centuries AD* (Dublin).

Kelly, P, 1992: The Táin as literature. In Mallory, J. P. (ed.) *Aspects of the Táin* (Belfast), 69-102.

Kern, P. B. 1999: *Ancient Siege Warfare* (London).

Kenyon, K. 1969: *Archaeology in the Holy Land* (London).

Kibble, T. W. B. 1985: *Classical Mechanics* (3rd Edition) (London).

King, D. G. 1967: Bury Wood Camp. *Wiltshire Archaeological Magazine* 62, 1-15.

Kinsella, 1969: *The Tain* (Oxford)

Kirch. P, 1991: Chiefship and competitive involution: the Marquesas Islands of eastern Polynesia. In Earle, T. K. (ed.) *Chiefdoms: Power, Economy, and Ideology* (Cambridge), 119-45.

Knight, D. 1984: *Late Bronze Age and Iron Age Settlement in the Nene and Great Ouse Basins* (BAR British Series 130: Oxford).

Knight, R. W., Browne, C. and Grinsell, L. V. 1972: Prehistoric skeletons from Tormarton. *Transactions of the Bristol and Gloucester Archaeological Society* 91, 14-7.

Kolata, A. L. 1983: The South Andes. In Jennings, J. D. (ed.) *Ancient South Americans* (San Francisco), 241-85.

Kolata, A. L. 2004: The Flow of Cosmic Power, Religion, Ritula and the People of Tiwanaku. In Yong-Sanchez, M. *Tiwanaku Ancestors of the Inca* Lincoln

Kolb, M. J. and Snead, J. E. 1997: It's a small world after all: comparative analyses of community organization in archaeology. *American Antiquity* 62 (4), 609-28.

Korfmann, M. 1973: The sling as a weapon. *Scientific American* 229 (4), 34-42.

Kroeber, T. 1987: *Ishi in Two Worlds: a Biography of the Last Wild Indian in North America* (London).

Kroeber, A. L. and Collier, D. 1998: *The Archaeology and Pottery of the Nazca, Peru Alfred L. Kroeber's 1926 Expedition* (Walnut Creek).

Kubler, G. 1962: *Art and Architecture of Ancient America* (London)

Laing, L. 1975: *The Archaeology of Late Celtic Britain and Ireland c. 400-1200 A.D.* (London).

Lake, F. and Wright, H. 1974: *A Bibliography of Archery* (Manchester).

Lambrick, G. 1984: Pitfalls and possibilities in Iron Age pottery studies - experiences in the Upper Thames Valley. In Cunliffe, B. and Miles, D. (eds.) *Aspects of the Iron Age in Central Southern Britain* (Oxford University Committee for Archaeology Monograph 2: Oxford), 162-77.

Lang, J. 1987: The technology of Celtic iron swords. In Scott, B. G. and Cleere, H. (eds.) *The Crafts of the Blacksmith* (Belfast), 61-72.

Langlet, H. 1927: *Riff* (Stockholm).

Lanning, E. P. 1967: *Peru before the Incas* (Eaglewood Cliffs)

Laughlin, J. C. H. 2000: *Archaeology and the Bible* (London).

Lawson, A. 2000: *Potterne 1982-5: Animal Husbandry in Later Prehistoric Wiltshire* (Wessex Archaeology Report 17: Salisbury).

Leach, E. R. 1976: *Culture and Communication: the Logic by which Symbols are Connected* (Cambridge).

LeBlanc, S. A. 2003 (a): *Constant Battles: the Myth of the Peaceful, Noble Savage.* (New York).

LeBlanc, S. A. 2003 (b) Warfare in the American Southwest and Mesoamerica. In Brown, M. K. and Staton, T. W. (eds.) *Ancient Mesomerican Warfare* (Walnut Creek)

Lechtman, H. 1979: Issues in Andean metallurgy. In Benson, E. (ed.) *Precolumbian Metallurgy of South America* (Washibgton D.C.). 1 - 40

Lendon, J. E. 2000: Homeric vengeance and the outbreak of Greek wars. In van Wees, H. (ed.) *War and Violence in Ancient Greece* (London), 1-30.

Leonard, J. N. 1973: *The First Farmers* (New York).

Lepper, F. and Frere, S. 1988: *Trajan's Column* (Gloucester).

Levitan, B. 1982: *Excavations at West Hill Uley, 1979: the Sieving and Sampling Programme* (Bristol).

Liddell, D. M. 1933: Excavations at Meon Hill *Proceedings of the Hampshire Field Club and Archaeological Society* 12, 127-62.

Lindblom, K. G. 1927: *Die Schleduder in Afrika und Anderwarts* (Stockholm).

Lindblom, K. G. 1940: *The Sling Especially in Africa (Additional Notes to a Previous Paper)* (Stockholm).

Linton, R. 1913: *The Tanala, a Hill Tribe of Madagascar* (Chicago).

Littauer, M. A. and Crouwell, J. H. 1983: Chariots in Late Bronze Age Greece. *Antiquity* 57, 87-192.

Lock, G. and Gosden, C. 1997: Hillforts of the Ridgeway Project: Excavations at Segsbury Camp, 1996. *South Midlands Archaeology* 27, 69-77.

Lucas, A. T. 1989: *Cattle in Ancient Ireland* (Kilkenny).

Luckenbil, D. D. 1926: *Ancient Records of Assyria and Babylonia II* (Chicago).

Lynn, C. J. 1992: The Iron Age mound in Navan Fort: A physical realization of Celtic religious beliefs? *Emania* 10, 33-57.

McElrath, D. L., Emerson, T. E. and Fortier A. C. 2000: Social Evolution or Social Response? A Fresh Look at the "Good Gray Cultures" After Four Decades of Midwest Research. In McElrath, D. L., Emerson, T. E. and Fortier A. C. (eds.) *Late Woodland Societies Tradition and Transformation across the Midcontinent* (Lincoln).

McEwan, C., Borrero, L. A. and Prieto, A. 1997: *Patagonia* (London).

McIntyre, L. 1975: *The Incredible Incas and their Timeless Land* (Washington D.C.).

Mac Niocaill, G. 1972: *Ireland before the Vikings* (Dublin).

McOmish, D. 1996: East Chisenbury: ritual and rubbish at the British Bronze Age-Iron Age transition. *Antiquity* 70, 68-76.

Machiavelli, N. 2005 [1521]: *Art of War* (translation Lynch, C.) (Chicago).

Madsen, O. 1997: Hedegård: a rich village and cemetery complex of the Early Iron Age on the Skjern River: interim report. *Journal of Danish Archaeology* 13, 57-93.

Mallory, J. P. 1981: The sword of the Ulster Cycle. In Scott, B. G. (ed.) *Studies on Early Ireland: Essays in Honour of MV Duignan* (Belfast), 99-114.

Mallory, J. P. 1986: Silver in the Ulster Cycle of Tales. In Evans, D. E., Griffith, J. G., and Jope, E. M. (eds.) *Proceedings of the Seventh International Congress of Celtic Studies, Oxford 1983* (Oxford), 31-78.

Mallory, J. P. 1991: Excavations at Haughey's Fort, 1989-1990. *Emania* 8, 10-26.

Mallory, J. P. 1992: Táin Bo Cuailnge: an outline of the plot. In Mallory J.P. (ed.) A*spects of the Táin* (Belfast), 9-28.

Mallory, J. P. and McNeill, T. 1991: *The Archaeology of Ulster: from Colonization to Plantation* (Belfast).

Maltby, M. 1985: Animal husbandry. In Fasham, P. J. *The Prehistoric Settlement at Winnall Down, Winchester: Excavations of MARC3 Site R17 in 1976 and 1977* (Hampshire Field Club Monograph 2: Winchester), 97-112.

Maltby, M. 1995: Animal Bone. In Wainwright, G. J. and Davies. S. M. *Balksbury Camp, Hampshire: Excavations 1973 and 1981* (English Heritage Archaeological Report 4: London), 83-7.

Malinowski, B. 1964: Anthropological analysis of war. In Bramson, L. and Goethals, G. W. (eds.) *War: Studies from Psychology, Sociology, Anthropology* (New York), 245-68.

Marinatos, N. 2000: *The Goddess and the Warrior: the Naked Goddess and the Mistress of Animals in Early Greek Religion* (London).

Marsden, E. W. 1969: *Greek and Roman Artillery (Volume 1)* (Oxford).

Marsh, H. C. 1901: Report of the excavation on Eggardun, 1900. *Proceedings of the Society of Antiquaries* 18, 258-62.

Mason, J. A. 1969: *The Ancient Civilizations of Peru* (London).

Mayer, E. 1985: Production zones. In Masuda, S., Shimada, I. and Morris, C. (eds.) *Andean Ecology and Civilisation, an Interdisciplinary Perspective on Andean Ecological Complementarily* (Tokyo), 45-84.

Mead, M. 1940: Warfare is only an invention - not a biological necessity. *Asia* 40, 402-5.

Means, P. A. 1919: Distribution and use of slings in pre-Columbian America, with descriptive catalogue of ancient Peruvian slings in the US National Museum. *Proceedings of the United States National Museum (Washington D.C.)* 55, 317-49.

Meek, C. K. 1925: *The Northern Tribes of Nigeria* (Oxford).

Megaw, J. V. S. and Megaw, M. R. 1996: Ancient Celts and modern ethnicity. *Antiquity* 70, 175-81.

Megaw, J. V. S. and Megaw, M. R. 1998: 'The mechanism of (Celtic) dreams?' a partial response to our critics. *Antiquity* 72, 432-5.

Megaw, J. V. S. and Simpson, D. D. A. 1979: *Introduction to British Prehistory* (Leicester).

Meggars, B. J. 1972: *Prehistoric America* (Chicago).

Menzel, D. 1964: Style and Time in the Middle Horizon *Nawpa Pacha* 2, 1-105

Mercer, R. J. 1970: Metal arrowheads in the European Bronze and Early Iron Age. *Proceedings of the Prehistoric Society* 36, 171-213.

Mercer, R. J. 1989: The earliest defences in Western Europe. *Fortress* 2, 2-11.

Mercer, R. J. 1999: The origins of warfare in the British Isles. In Carman, J. and Harding, A. (eds.) *Ancient Warfare* (Stroud), 143-56.

Miles, D., Palmer, S., Lock, G., Gosden C. and Cromarty, A. M. 2003: *Uffington White Horse and its Landscape: Investigations at White Horse Hill Uffington, 1989-95 and Tower Hill Ashbury, 1993-4* (Oxford).

Milik. J, T, 1956: An unpublished arrow-head with Phoenician inscription of the 11th - 10th Century B.C. *Bulletin of the American Schools of Oriental Research* 143, 3-6.

Miller, J. M. 1991: Is it possible to write a history of Israel without relying on the Hebrew Bible? In Edelman, D. V. (ed.) *The Fabric of History: Text, Artifact and Israel's Past* (Sheffield), 93-102.

Miller, R., McEwan, E. and Bergman, C. 1986: Experimental Approaches to Ancient Near Eastern Archery. *World Archaeology* 18 (2), 178-95.

Milne, A. A. 2000: [1928]: *The House at Pooh Corner* (London).

Milner, N. P. 1993: *Vegetius: Epitome of Military Science* (Liverpool).

Mitchell, T. C. 1986: Another Phoenician inscribed arrowhead. In Tubb, J. N. (ed.) *Palestine in the Bronze and Iron Ages: Papers in Honor of Olga Tufnell* (London), 150-3.

Moliniè-Fioravanti, A. 1986: The Andean community today. In Murra, J. V., Wachtel, N. and Revel, J. (eds.) *Anthropological History of Andean Polities,* (Cambridge).

Monckton, C. A. W. 1921: *Some Experiences of a New Guinea Resident Magistrate* (London).

Monk, M. 1995: A tale of two ringforts: Lisleagh I and II. *Journal of the Cork Historical and Archaeological Society* 100, 105-16.

Moore. T, 2003: Rectangular houses in the British Iron Age? – 'Squaring the circle'. In Humphrey, J. (ed.) *Re-searching the Iron Age* (Leicester), 47-58.

Morris, E. L. 1996: Artefact production and exchange in the British Iron Age. In Champion, T. C. and Collis, J. R. (eds.) *The Iron Age in Britain and Ireland: Recent Trends* (Sheffield), 41-65.

Morris, I. M. 1988: Tomb cult and the 'Greek renaissance': the past in the present in the 8th century BC. *Antiquity* 62, 750-61.

Morris, C. and Thompson, D. E. 1985: *Huánuco Pampa An Inca City and its Hinterland* (London).

Morris, C. and von Hagen, A. 1993: *The Inka Empire and its Andean Orgins* (Washington D. C.).

Moseley, M. E. 1992: *The Incas and Their Ancestors: The Archaeology of Peru* (London).

Munn, M. H. 1993: *The Defense of Attica, The Dema Wall and the Boiotian War of 378-375 BC* (Berkeley).

Murra, J. 1985: 1. "El Archipiélago Vertical" Revisited, 2. The Limits and Limitations of the "Vertical Archipelago" in the Andes. In Masuda, S., Shimada, I. and Morris, C. (eds.) *Andean Ecology and Civilization,* (Tokyo)

Murray, O. 1993: *Early Greece* (London).

Murray, T. and Walker, M. J. 1988: Like WHAT? A practical question of analogical inference and archaeological meaningfulness. *Journal of Anthropological Archaeology* 7, 248-87.

Musson, C. 1991: *The Breiddin Hillfort, a Later Prehistoric Settlement in the Welsh Marches* (CBA Research Report 76: London).

Musson, C. and Watson, M. 1993: *Shropshire from the Air* (Telford).

Mytum, H. 1992: *The Origins of Early Christian Ireland* (London).

Needham, S. P. and Burgess, C. 1980: The later Bronze Age in the Lower Thames Valley: the metalwork evidence. In Barrett, J. C. and Bradley, R. J. *Settlement and Society in the British Late Bronze Age* (BAR British Series 83: Oxford), 437-70.

Ni Bhrolchain, M. 1994: Re Toin Mna: in pursuit of troublesome women. In Mallory, J. P. and Stockman, G. *Ulidia: Proceedings of the First International Conference on the Ulster Cycle of Tales* (Belfast), 115-22.

Nicholls, R. V. 1958: Old Smyrna: the Iron Age fortifications and associated remains on the city perimeter. *Annual of the British School at Athens* 53-4, 35-137.

Noetling, F. 1911: Notes on hunting sticks (lughrana), spears (perenna) and baskets (tughbrana) of the Tasmain Aborigines. *Papers and Proceedings of the Royal Society of Tasmania, for the Year 1911,* 64-8.

Nowicki, K. 1994: Some remarks on pre- and proto-palatial peak sanctuaries in Crete. *Aegean Archaeology* 1, 31-48.

Ó Corráin, D. 1972: *Ireland Before the Normans* (The Gill History of Ireland 2: Dublin).

Ó Corráin, D. and Maguire, F. 1981: *Gaelic Personal Names* (Dublin).

O'Curry, E. 1873*: On the Manners and Customs of the Ancient Irish* (Dublin).

O'hUiginn, R. 1992: *The Background and Development of the Táin.* In Mallory, J. P. (ed.) Aspects of the Tain (Belfast), 29-68.

O'Neil, B. H. St J. 1934: Excavations at Titterstone Clee Camp, Shropshire 1932. *Antiquaries Journal* 14, 13-32.

O'Neill, B. H. 1937: Excavations at Breiddin Hill Camp, Montgomeryshire, 1933-35. *Archaeologia Cambrensis* 92, 86-128.

O'Rahilly, C. 1976: *Táin* Bó Cúalnge *Recension I* (Dublin).

O'Rahilly, T. F. 1984: *Early Irish History and Mythology* (Dublin).

O'Riain, P. 1994: The Táin: a clue to its origins. In Mallory, J. P. and Stockman, G. (eds.) *Ulidia, Proceedings of the First International Conference on the Ulster Cycle of Tales* (Belfast), 31-8.

Ó Ríordáin, S. P. 1960: *Tara: the Monuments on the Hill, 3rd Edition* (Dundalk).

Oakely, S. P. 1995: *The Hill-Forts of the Samnites* (British School at Rome Archaeological Monograph 10: London).

Ober, J. 1987: Early artillery towers: Messenia, Boiotia, Attica, Megarid. *American Journal of Archaeology* 91, 569-604.

Ochsenschlager, E. L. 1993: Village weavers: ethnoarchaeology at al-Hiba. *Bulletin of Sumerian Agriculture* 7, 43-62.

Oldfield, F., Higgitt, S. R., Richardson, N. and Yates, G. 1986: Pollen, charcoal, rhizopod and radiometric analyses. In Stead, I. M., Bouke, J. B. and Brothwell, D. (eds.) *Lindow Man: The Body in the Bog* (London), 82-85

Olmsted, G. 1992: The earliest narrative version of the Táin*:* seventh-century poetic references to Táin Bó Cuailnge. *Emania* 10, 5-17.

Olsen, S. L. 2003: The bone and antler artefacts: their manufacture and use. In Field, N. and Parker Pearson, M. (eds.) *Fiskerton: an Iron Age Timber Causeway with Iron Age and Roman Votive Offerings.* (Oxford), 92-110.

Orme, B. 1981: *Anthropology for Archaeologists: an Introduction* (London).

Osborne, R. 1996: *Greece in the Making 1200-479 BC* (London).

Osgood, R, 1998: *Warfare in the Late Bronze Age of North Europe* (BAR International Series 694: Oxford).

Osgood, R., Monks, S. and Toms, J. 2000: *Bronze Age Warfare* (Stroud).

Otterbien, K. F. 1970: *The Evolution of War: a cross-cultural Study* (New Haven).

Owen, B. 2001: The Economy of Metal and Shell Wealth Goods. In (D'Altroy, T. N. and Hastorf, C. A. (eds.) *Empire and Domestic Economy* (New York).

Paravicino, J. O. and Romero, M. C. 2002: Religious Ideology and Military Organization in the Iconography of a D-Shaped Ceremonial Precinct at Conochopata. In Silverman, H and Isbell, W. H. (eds.) *Andean Archaeology II Art, Landscape, and Society* (New York).

Pare, A. 1969 [1569]: *Ten Books of Surgery* (Athens G.A.).

Parry, A. (ed.) 1971: *The Making of Homeric Verse: the collected Papers of Milman Parry* (Oxford).

Payne-Galloway, R. 1981 [1903]: *The Crossbow* (London).

Peake, H. J. E. and Padel, J. 1934: Exploration of three round barrows on Wooley Down, Berkshire. *Transactions of the Newbury District Field Club* 7, 30-48

Peddie, J. 1997: *The Roman War Machine* (Stroud).

Perles, C. 2001: *The Early Neolithic in Greece: the first Farming Communities in Europe* (Cambridge).

Piggott, S. 1931: Ladle Hill - an unfinished hill-fort. *Antiquity* 5, 474-85.

Piggott, S. 1950: Swords and scabbards of the British Early Iron Age. *Proceedings of the Prehistoric Society* 16, 1-28.

Platt, T. 1986: Mirrors and maize. In Murra, J. V., Wachtel, N. and Revel, J. (eds.) *Anthropological History of Andean Politics* (Cambridge), 228-259

Pleiner, R. 1993: *The Celtic Sword* (Oxford).

Poma, Huamán. 1978 [1603] *Letter to a King: a Picture-History of the Inca Civilisation* (trans. and ed. Dilke, C.) (London).

Poole, C. 1995: Pits and propitiation. In Cunliffe, B. *Danebury: an Iron Age Hillfort in Hampshire Vol. 6: a Hillfort Community in Perspective* (CBA Research Report 102: London), 249-75.

Poole, C. 2000: Clay slingshot. In Barrett, J., Woodward, A and Freeman, P. (eds) *Cadbury Castle, Somerset: the Later Prehistoric and Early Historic Archaeology* (London).

Pope, S. T. 1923: *A Study of Bows and Arrows* (Berkeley and Los Angeles).

Popham, M., Touloupa, E. and Sacket, L. H. 1982: The hero of Lefkandi. *Antiquity* 56, 169-74.

Powell. W, 1884: *Wanderings in a Wild Country* (London).

Pozorski, S. and Pozorski, T. 1987: *Early Settlement and Subsistance in the Casma Valley, Peru* (Iowa City)

Pritchett, W. K. 1971-1985:*The Greek State at War* (5 vols.) (Berkeley and Los Angeles).

Pryor. F, 1984: *Excavation at Fengate Peterborough, England: the Fourth Report* (Northamptonshire Archaeological Society Monograph 2/ Royal Ontario Museum Archaeology Monograph 7).

Raaflaub, K. A. 1998: A historian's headache: how to read 'Homeric Society'? In Fisher, N. and van Wees, H. *Archaic Greece: new Approaches and new Evidence* (London), 169-94.

Raftery, B. 1969: Freestone Hill, Co. Kilkenny: an Iron Age hillfort and Bronze Age cairn. *Proceedings of the Royal Irish Academy* 68(C), 1-108.

Raftery, B. 1972: Irish hill-forts. In Thomas, C. (ed.) *The Iron Age in the Irish Sea Province* (CBA Research Report 9: London), 37-58.

Raftery, B. 1994: *Pagan Celtic Ireland - The Enigma of the Irish Iron Age* (London).

Raftery, B. 1998: Observations on the Iron Age in Munster. *Emania* 17, 21-4.

Rahtz, P. and ApSimon, A. M. 1962: Excavations at Shearplace Hill, Sydling St. Nicholas, Dorset. *Proceedings of the Prehistoric Society* 28, 289-328.

Rainey, A. F. 1973: The World of Sinuhe. *Israel Oriental Studies* 2, 369-408.

Randsborg, K. 1995: *Hjortspring: Warfare and Sacrifice in early Europe* (Aarhus).

Rankin, D. 1996: *Celts and the Classical World* (London).

Rausing, G. 1967: *The Bow: Some Notes on its Origin and Development* (Stockholm).

Rawlings, L.2000: Alternative agonies: hoplite martial and combat experiences beyond the phalanx. In van Wees, H. (ed.) *War and Violence in Ancient Greece* (London). 223-260

Raymond, A. 1986: Experiments in the function and performance of the weighted atlatl. *World Archaeology* 18 (2), 153-77.

Redfield, J. M. 1994: *Nature and Culture in the Iliad: the Tragedy of Hector* (Durham and London).

Reinhard, J. 1992: Sacred Peaks of the Andes. *National Geographic* 189(6), 62-81

Renfrew, C. 1974: Beyond a subsistence economy, the evolution of social organisation in prehistoric Europe. In Moore, C. B. (ed.) *Reconstructing Complex Societies* (Atlanta), 69-96.

Richardson, K. M. 1940: Excavations at Poundbury, Dorchester, Dorset, 1939. *Antiquaries Journal* 20, 429-48.

Richardson, K. M. 1951: The excavation of an Iron Age village on Boscombe Down West. *Wiltshire Archaeological Magazine* 54, 123-68.

Richmond, I. A. 1968: *Hod Hill, Vol.2: Excavations Carried Out between 1951 and 1958* (London).

Ritchie, W. F. and Ritchie, J. N. G. 1985: *Celtic Warriors* (Princes Risborough).

Rivet, A. L. F. 1971: Hillforts in Action. In Jesson, M. and Hill, D. (eds.) *The Iron Age and its Hill-Forts* (Southampton), 189-202.

Robinson, D. M. 1941: *Excavations at Olynthus, Part X, Metal and Minor Miscellaneous Finds: an Original Contribution to Greek Life* (London).

Robinson, M. 1981: The Iron Age to Early Saxon environment of the Upper Thames Terraces. In Jones, M. and Dimbelby, G. (eds.) *The Environment of Man: the Iron Age to Anglo-Saxon Periods* (BAR British Series 87:Oxford), 251-86.

Rockhill, W. 1895: *Notes on the Ethnology of Tibet* (Washington).

Roe, F. E. S. 1966: The battle axe series in Britain. *Proceedings of the Prehistoric Society* 32, 199-245.

Rolle, R. 1989: *The World of the Scythians* (London).
Rosenberg, G. 1937: *Hjortspringfundet* (Copenhagen).
Ross, A. 1958: The human head in insular Pagan Celtic Religion, *Proceedings of the Society of Antiquaries of Scotland* 91, 10-43.
Ross, A. 1967: *Pagan Celtic Britain* (London).
Rountree, H. C. 1989: *The Powhatan Indians of Virginia: their Traditional Culture* (Norman).
Rowe, J. H. 1946: Inca Culture at the time of the Sapnish Conquest. In Steward, J. (ed.) *Handbook of the South American Indians* 2, 183-330 (Washington).
Rowe, J. H. 1962: *Chavin Art: An Inquiry into its Form and Meaning* (New York)
Rynne, E. 1972: Celtic stone idols in Ireland. In Thomas, C. (ed.) *The Iron Age in the Irish Sea Province* (CBA Research Report 9: London), 79-93.

Sadler, A. J. and Thorning, D. W. S. 1996: *Understanding Mechanics* (Second Edition) (Oxford).
Salomon, F. 1986: *Native Lords of Quito in the Age of the Incas: the Political Economy of North Andean Chiefdoms* (Cambridge).
Saunders, P. R. 1997: The excavations of an Iron Age settlement site at Stockton. *Wiltshire Archaeological Magazine* 90, 13-25.
Savory, H. N. 1976: Welsh hillforts: a reappraisal of recent research. In Harding, D. W. (ed.) *Hillforts: Later Prehistoric Earthworks in Britain and Ireland* (London), 263-76.
Sandefur, E. C. 2001: Animal Husbandry and Meat Consumption. In D'Altroy, T. N. and Hastorf, C. A. (eds.) *Empire and Domestic Economy* (New York).
Schwimmer, E. 1966: *The World of the Maori* (Wellington).
Seaford, R. A. S. 1994: *Reciprocity and Ritual*: Homer and Tragedy in the developing City-State (Oxford).
Sellwood, L. 1984: Tribal boundaries viewed from the perspective of numismatic evidence. In Cunliffe, B. and Miles, D. (eds.) *Aspects of the Iron Age in Central Southern Britain* (Oxford University Committee for Archaeology Monograph 2: Oxford), 191-204.
Service, E. R. 1962: *Primitive Social Organisation: an Evolutionary Perspective* (New York).
Sessle, E. 1994: Misogyny and Medb: approaching Medb with feminist criticism. In Mallory, J. P. and Stockman, G. (eds.) *Ulidia, Proceedings of the First International Conference on the Ulster Cycle of Tales* (Belfast), 135-8.
Schobinger, J. 1991: Sacrifices of the High Andes. *Natural History* 100(4), 63-69
Sharples, N. 1991a: Warfare in the Iron Age of Wessex. *Scottish Archaeological Review* 8, 79-89.
Sharples, N. 1991b: *Maiden Castle: Excavation and Field Survey, 1985-86* (English Heritage Archaeological Report 19: London).
Shimada, I. 1999: The late prehispanic coastal states. In Minelli, L. L. (ed.) *The Inca World: the Development of Pre-Columbian Peru AD 1000-1534* (Norman), 49-110.
Sinos, R. H. 1998: Divine selection: Epiphany and politics in Archaic Greece. In Dougherty, C. and Kurke, L. (eds.) *Cultural Poetics in Archaic Greece: Cult, Performance, Politics* (Oxford), 73-91.
Sipes, R. G. 1973: War, sports and aggression: an empirical test of two rival theories. *American Anthropologist* 75, 64-86.
Smith, C. 1979: *Fisherwick: The Reconstruction of an Iron Age Landscape* (BAR British Series 61: Oxford).
Smith, K. 1977: Winklebury Camp, Basingstoke, Hampshire. *Proceedings of the Prehistoric Society* 43, 31-129.
Snodgrass, A. M. 1964: Early *Greek Armour and Weapons, from the End of the Bronze Age to 600BC.* (Edinburgh).
Snodgrass, A. M. 1980: Towards an interpretation of Geometric Figure scenes. *Athenische Mitteilungen* 95, 90-8.
Snodgrass, A. M. 2000: *The Dark Age of Greece: an Archaeological Survey of the Eleventh to the Eighth Centuries BC* (Edinburgh).
Stamp, D. 1979: *Field Archery* (London).
Stanford, S. C. 1974: *Croft Ambrey* (Hereford).
Stanford, S. C. 1981: *Midsummer Hill: an Iron Age Hillfort on the Malverns* (Hereford).
Stansfield, M. 1990: *The Herdsman Book* (London).
Stead, I. M. 1965: *The La Tène Cultures of Eastern Yorkshire* (York).
Stead, I. M. 1968: An Iron Age hill-fort at Grimthorpe, Yorkshire, England. *Proceedings of the Prehistoric Society* 34, 148-90.
Stead, I. M. 1986: Summary and conclusions. In Stead, I., Bourke, J. and Brothwell, D. (eds.) *Lindow Man: the Body in the Bog* (London), 177-80.
Stead, I. M. 1988: Chalk figurines of the Parisi. *Antiquaries Journal* 68, 9-29.
Stead, I. M. 2003: The Iron Age military items. In Field, N. and Parker Pearson, M. *Fiskerton: an Iron Age Timber Causeway with Iron Age and Roman Votive Offerings* (Oxford), 49-62.
Stead, I. M. 2006: *British Iron Age Swords and Scabbards*. London
Stead, I. M., Bourke, J. B. and Brothwell, D. 1986: *Lindow Man: The Body in the Bog* (London).
Stead, I. M., Hartwell, A. P., Lang, J. R. S., La Niece, S. C. and Meeks, N. D. 1981: An Iron Age sword and scabbard from Isleham. *Proceedings of the Cambridge Antiquarian Society* 70, 61-74.
Steponaitis, V. 1997: Contrasting patterns of Mississippian development. In Earle, T. K. (ed.) *Chiefdoms: Power, Economy, and Ideology* (Cambridge), 193-228.
Stevens, F. 1934: The Highfield pit dwellings, Fisherton, Salisbury. *Wiltshire Archaeological Magazine* 46, 579-624.
Stokes, W. 1901: The Destruction of Da Derga's Hostel *Revue Celtique* 22 (9), 165, 282, 390.
Stone, J. F. S. 1936: An enclosure on Boscombe Down East. *Wiltshire Archaeological Magazine* 47, 466-89.
Stone, G. C. 1961: *A Glossary of the Construction, Decoration and Use of Arms and Armour in all Countries and in all Times* (New York).
Stopford, J. 1987: Danebury an alternative view. *Scottish Archaeological Review* 4, 70-5.

Stout, G. 1997: *Atlas of the Irish Landscape* (Cork).

Strong, D. 1968: *The Early Etruscans* (London).

Strömberg, A. 1993: *Male or Female? A methodological study of grave-gifts as sex-indicators in Iron Age Athens* (Jonered).

Strutt, J. 1969 [1801]: *The Sports and Pastimes of the Peoples of England* (Bath).

Tabor, R. 1994: *Traditional Woodland Crafts* (London).

Tandy, D. W. 1997 *Warriors into Traders: the Power of the Market in Early Greece* (Berkeley).

Tarzia, W. 1987: No trespassing: border defence in the Táin Bó Cuailnge. *Emania* 3, 28-33.

Thomas, J. 1999: *Understanding the Neolithic* (London).

Thompson, F. H. 1979: Three Surrey hillforts: excavations at Anstiebury, Holmbury and Hascombe 1972-77. *Antiquaries Journal* 59, 245-318.

Thompson, T. L. 1992: *Early History of the Israelite People, from the written and archaeological Sources* (Leiden).

Time Team 2002: *Excavations at Gear Field, Cornwall* (Channel Four TV).

Todd. M, 1992: *The Early Germans* (London).

Toms, S. 1912: Excavations at the Beltout Valley Entrenchments. *Sussex Archaeological Collection* 55, 41-55.

Topic, J. and Topic, T. 1987: The archaeological investigation of Andean militarism: some cautionary observations. In Haas, J., Pozorski, S. and Pozorski, T. (eds.) *The Origins and Development of the Andean State* (Cambridge), 47-55.

Turney-High, H. H. 1971: *Primitive War: its Practice and Concepts (2nd Edition)* (Columbia).

Uhle, M. 1991 [1903] *Pachacamac: A Reprint of the 1903 Edition* Philadelphia

Upton, G. J. G. and Cook, I. T. 1998: *Introducing Statistics* (Oxford).

Urton, G. 1990:*The History of a Myth: Pacariqtambo and the Origin of the Inkas* (Austin).

Van der Standen, W *Though Nature to Eternity: the bog bodies of northwest Europe* Amsterdam

Vansina, J. 1985: *Oral Tradition as History* (Madison).

van Wees, H. 1992: *Status Warriors: War, Violence and Society in Homer and History* (Amsterdam).

van Wees, H. 1994: The Homeric way of war, *Greece and Rome* 41 (1), 1-18

van Wees, H. 1998: Greeks bearing arms: the state, the leisure class and the display of weapons in Archaic Greece. In Fisher, N. and van Wees, H. (eds.) *Archaic Greece: New Approaches and New Evidence* (London), 333-78.

van Wees, H. 2000: The development of the hoplite phalanx: iconography and reality in the seventh century. In van Wees, H. (ed.) *War and Violence in Ancient Greece* (London), 125-66.

Varley, W. J. 1950a: The Hill-forts of the Welsh Marches. *Archaeological Journal* 105, 41-66.

Varley, W. J. 1948: The Hill-forts of the Welsh Marches. *Archaeological Journal* 105, 41-66.

Vayda, A. 1969: Expansion and warfare among swidden agriculturalists. In Vayda, A. (ed.) *Environment and Cultural Behaviour: Ecological Studies in Cultural Anthropology* (Garden City), 202-20.

Vega, A. de. 1965 [1582]: La descripción que se hizo en la provincia de Xauxa por la instrucción de su majestad que a la dicha provincia se invió de molde. In Espada, M. J. de la (ed.) *Relaciones Geograficas de Indias: Peru I* (Madrid), 166-75.

Verano, J. W. 1986: A mass burial of mutilated individuals at Pacatnamu. In Donnan, C. B. and Cock, G. A. (eds.) *The Pacatnamu Papers, Volume I* (Los Angeles), 117-38.

Verano, J. W. 1995: Where do they rest? The treatment of human offerings and trophies in Ancient Peru. In Dillehay, T. D. (ed.) *Tombs for the Living: Andean Mortuary Practices* (Washington D.C.), 189-227.

Vernant, J.-P. 1990: *Myth and Society in Ancient Greece* (trans. Lloyd, J.) (London).

Vutiropulos, N. 1991: The sling in the Aegean Bronze Age. *Antiquity* 65, 279-86.

Wachtel, N. 1977: *The Vision of the Vanquished: the Spanish Conquest of Peru through Indian Eyes*, 1530-1570 (Hassocks).

Waddell, J. 1983: Rathcroghan – a royal site in Connacht. *Journal of Irish Archaeology* 1, 21-46.

Wailes, B. 1976: Dún Ailinne: an interim report. In Harding, D. W. (ed.) *Hillforts: Later Prehistoric Earthworks in Britain and Ireland* (London), 319-57.

Wailes, B. 1990: Dun Ailinne: a summary excavation report. *Emania* 7, 10-21.

Wainwright, G. J. 1970: An Iron Age promontory fort at Budbury, Bradford-on- Avon, Wiltshire. *Wiltshire Archaeological Magazine* 65, 108-66.

Wainwright, G. J. 1971: The excavation of a fortified settlement at Walesland Rath, Pembrokeshire. *Britannia* 2, 48-108.

Wainwright, G. J. 1979: *Gussage All Saints: an Iron Age Settlement in Dorset* (Department of the Environment Archaeological Report 10: London).

Wainwright, G. J. and Davies, S. M. 1995: *Balksbury Camp, Hampshire: Excavations 1973 and 1981* (English Heritage Archaeological Report 4: London).

Wainwright, G. J. and Spratling, M. 1973: The Iron Age settlement of Gussage All Saints. Part 1: the settlement. *Antiquity* 47, 109-30.

Wait, G. A. 1985: *Ritual and Religion in Iron Age Britain* (BAR British Series 149: Oxford).

Warner, R. B. 1998: Is there an Iron Age in Munster? *Emania* 17, 25-9.

Warren, P. 1989: *The Aegean Civilzations* (Oxford)

Wason, P. K. 1994: *The Archaeology of Rank* (Cambridge).

Waterman, D. M. 1997: *Excavations at Navan Fort 1961-71* (Belfast).

Watson, M. and Musson, C. 1993: *Shropshire from the Air: Man and the Landscape* (Telford).

Watson, P. J. 1979: The idea of ethnoarchaeology: notes and comments. In Kramer, C. (ed.) *Ethno-archaeology: Implications for Ethnography for Archaeology* (New York), 277-87.

Weaver, M. P. 1981 *The Aztecs, Maya and Their Predecessors, Archaeology of Mesoamerica* (New York)

Webb, M. C. 1987: Broader perspectives on Andean state origins. In Haas, J. Pozorski, S. and Pozorski, T. (eds.) *The Origins and Development of the Andean State* (Cambridge), 161-7.

Weir, D. A. 1994: The environment of Emain Macha. In Mallory, J. P. and Stockman, G. (eds.) *Ulidia: Proceedings of the First International Conference on the Ulster Cycle of Tales* (Belfast), 171-9.

Wells, C. 1964: *Bones, Bodies and Disease: Evidence of Disease and Abnormality in Early Man* (London).

Wells, P. S. 1984: *Farms, Villages and Cities: Commerce and Urban Origins in Late Prehistoric Europe* (London).

West, S. 1990: *West Stow, Suffolk: the Prehistoric and Romano-British Occupations* (East Anglian Archaeology 48: Bury St. Edmunds).

Wheeler, E. L. 1991: The General as hoplite. In Hanson, V. D. (ed.) *Hoplites: the Classical Greek Battle Experience* (London), 121-70.

Wheeler, R. E. M. 1943: *Maiden Castle, Dorset* (London).

Whimster, R. 1981: *Burial Practices in Iron Age Britain: a Discussion and Gazetteer of the Evidence* c. 700 BC-AD 43 (BAR British Series 90: Oxford).

Whitehead, N. 1990: The snake warriors – sons of the tiger's teeth: a descriptive analysis of Carib warfare ca. 1500-1820. In Haas, J. (ed.) *The Anthropology of War* (Cambridge), 146-70.

Whitley, M. 1943: Excavations at Chalbury Camp, Dorset, 1939. *Antiquaries Journal* 23, 98-121.

Whitney, A. J. M. 1988: Early states and hero cults: a reappraisal. *Journal of the Hellenic Society* 108,173-82.

Whittle, A. W. R. 1985: *Neolithic Europe: a Survey* (Cambridge).

Willey, G. R. 1953: *Prehistoric Settlement Patterns in the Viru Valley, Peru* (Washington)

Williams, A. 1951: Excavations at Allard's Quarry, Marnhull, Dorset. *Proceedings of the Dorset Natural History and Archaeological Society* 72, 20-75.

Wilson, D. J. 1987: Reoconstructing patterns of early warfare in the lower Santa Valley: new data on the role of conflict in the orgins of complex north-coast society. In Hass, J., Pozorski, S. and Pozorski, T. (eds.) *The Origins and developmnent of the Andean State* (Cambridge).

Wood, P. 1959: The Early Iron Age camp called Grimsbury Castle, near Hermitage, Berks. *The Berkshire Archaeological Journal* 59, 74-82.

Woolgar, C. M. 1999: *The Great Household in Late Medieval England* (London).

Wright, G. E. 1962: *Biblical Archaeology* (London).

Wright, R. P. and Jackson, K. H. 1968: A late inscription from Wroxeter. *Antiquaries Journal* 48, 296-300.

Wyatt, N, 1998: *Religious Texts From Ugarit* (Sheffield).

Wylie, A. M. 1985: The reaction against analogy. In Schiffer, B. (ed.) *Advances in Archaeological Method and Theory Volume 8* (New York), 63-111.

Yalden, D. 1999: *The History of British Mammals* (London).

Yardin, Y. 1963: *The Art of War in Biblical Lands in the Light of Archaeological Discovery* (London).

Yücel, Ü. 1997: Archery in the period of Sultan Mahmud II. *Journal of the Society of Archer-Antiquaries* 40, 68-80.

Zaidman, L. B. 1992: *Religion in the Ancient Greek City* (Cambridge).

Zamora, M. 1988: *Language, Authority and Indigenous History in the Comentarios Reales de Los Incas* (Cambridge).

Zirra, V. 1993: The cemetery of Çiumeşti and the chieftain's tomb. In Kruta, V., Frey, O. H., Raftery, B. and Szabó, M. (eds.) *The Celts* (London), 382-3.

www.ingramcontent.com/pod-product-compliance
Lightning Source LLC
LaVergne TN
LVHW071213210926
848482LV00015B/794